PROGRESS

PROGRESS

A History of Humanity's Worst Idea

SAMUEL MILLER MCDONALD

WILLIAM
COLLINS

William Collins
An imprint of HarperCollins*Publishers*
1 London Bridge Street
London SE1 9GF

WilliamCollinsBooks.com

HarperCollins*Publishers*
Macken House
39/40 Mayor Street Upper
Dublin 1
D01 C9W8, Ireland

First published in Great Britain in 2025 by William Collins

1

Copyright © Samuel Miller McDonald 2025

Samuel Miller McDonald asserts the moral right to
be identified as the author of this work in accordance
with the Copyright, Designs and Patents Act 1988

A catalogue record for this book is available from the British Library

ISBN 978-0-00-846247-5 (hardback)
ISBN 978-0-00-846248-2 (trade paperback)

All rights reserved. No part of this publication may be reproduced, stored in a retrieval system, or transmitted, in any form or by any means, electronic, mechanical, photocopying, recording or otherwise, without the prior written permission of the publishers.

Without limiting the exclusive rights of any author, contributor or the publisher of this publication, any unauthorised use of this publication to train generative artificial intelligence (AI) technologies is expressly prohibited. HarperCollins also exercise their rights under Article 4(3) of the Digital Single Market Directive 2019/790 and expressly reserve this publication from the text and data mining exception.

This book is sold subject to the condition that it shall not, by way of trade or otherwise, be lent, re-sold, hired out or otherwise circulated without the publisher's prior consent in any form of binding or cover other than that in which it is published and without a similar condition including this condition being imposed on the subsequent purchaser.

Typeset in Sabon MT Std by
Palimpsest Book Production Ltd, Falkirk, Stirlingshire

Printed and Bound in the UK using 100% Renewable Electricity
at CPI Group (UK) Ltd

This book contains FSC™ certified paper and other controlled sources
to ensure responsible forest management.

For more information visit: www.harpercollins.co.uk/green

Contents

What is Progress? — 1

BEFORE PROGRESS

Savagery and Civilisation of the Ohio River Valley — 32
The Vital Science of Human Ecology — 40
Animism and a Theory Of Everything — 44
Joy and the Meaning of Nature — 48
The Science of Hierarchy — 52

I HEAVEN

Context — 59
Myths of Mesopotamia — 67
The Ends of Zoroaster — 72
The Dominion of Abraham — 74
The Age of Athens — 78
Peace of Rome — 81
Kingdoms of Christ — 85
Vikings of Valhalla — 88
Empires of Islam — 93
Coda — 101

II NATION

Context — 109
Wealth of Europe — 112

Histories of Progress	122
Philosophies of Progress	134
The Forging of Liberalism	139
A Nation of Destiny	142
The Civilisation of Density	160
The Rise of Evolution	168
The Fall of Revolution	180
Coda	191

III SYSTEM

Context	195
The Future of Fascism	201
The Ambivalence of Modernism	209
The Destiny of Nazism	216
Stability of Keynesianism	226
Cults of Stalinism-Maoism	235
The Theocracy of Neoliberalism	244
Heavens of Tomorrow	253
Frontier of Ghosts	276
Coda	288

AFTER PROGRESS

Economies of Reciprocity	292
Cities and Farms of Wilderness	297
Equilibria of Energy	309
War of Life	313
Societies of Tomorrow	321
Meaning Beyond Progress	327
Acknowledgements	337
Endnotes	339
Index	409

What is Progress?

Let the philosophical observer commence a journey from the savages of the Rocky Mountains eastwardly towards our seacoast. These he would observe in the earliest stage of association, living under no law but that of nature . . . He would next find those on our frontiers in the pastoral state, raising domestic animals . . . and so in his progress he would meet the gradual shades of improving man until he would reach his, as yet, most improved state in our seaport towns. This, in fact, is equivalent to a survey, in time, of the progress of man from the infancy of creation to the present day. And where this progress will stop no one can say. Barbarism has, in the meantime, been receding before the steady step of amelioration, and will in time, I trust, disappear from the earth.[1]

This passage is from a letter Thomas Jefferson wrote in 1824, two years before his death and forty-one years after the end of the American Revolutionary War. Jefferson was the United States' third president and one of the country's most important Founders. This short text illuminates both the spiritual foundation of the country and the idea at the heart of this book. So let's deconstruct it.

If you were to travel from the Rocky Mountains in the west to the Atlantic coastline in the east, Jefferson suggests, the land you would pass through and the buildings dotted along your road would appear as they had at earlier points in human history, as if you were travelling along not just miles but centuries. Your journey would reflect the passage of time, the progress made by European settlers since they reached the east coast of North America. In other words, by 'savages ... living under no law but that of nature,' Jefferson means that at the Rockies there would be ancient wilderness housing violent fur-clad people without society who foraged for food and shelter, and dangerous beasts, representing life in humanity's earliest years. By the time you reached what are now the Midwestern states, you would find early agricultural societies, flocks of sheep, herds of cattle, and rows of corn and wheat surrounding simple towns and villages. Finally, reaching the end of your mirror-image quest of American westward frontier expansion, arriving in Washington, DC, New York City or Boston, you would find the as-yet 'most improved state' of human beings and their societies: laws, dense cities, bustling trade and sophisticated technology. There, you could rest assured that such developments would continue into a bright future. What Jefferson is sketching out is a grand narrative in a specific tradition that can be best captured in one word: progress.

Pick up any crime novel and you are likely to find a narrative formula. The details may change from story to story, but the general structure stays the same: a crime is committed, a detective begins the process of finding and piecing together clues, and the story culminates with the crime solved and the criminal brought to justice. Like crime novels, narratives of progress follow their own formula. This excerpt from Jefferson's letter offers an ideal distillation of that formula. Though the details have changed through time, from culture to culture, the formula's essential elements have remained remarkably consistent over not just centuries, but millennia.

The formula starts in the dark and wild beginning of humanity and moves forward and upward into a superior, more refined present, through changes that compound over time, culminating in some still vague, ever-future paradise. The story always parcels its characters into a binary, splitting those deemed civilised from the savage, the heathen from the blessed, the wild from the domesticated, the developed from the undeveloped. There is almost always some kind of frontier space, physical or metaphorical, into which the blessed must enter. The salvation awaiting in the future is set aside for the chosen, but only if they remain obedient to this quest, or, rather, to those leading it.

This narrative formula has served as the intellectual foundation on which Western civilisation itself has grown and spread. The American sociologist Robert Nisbet, the last author to publish a broad historical account of the idea of progress, wrote of the concept in 1980: 'No single idea has been more important in Western civilisation for nearly three thousand years.'[2]

The narrative formula of progress has been important for even longer than that, across many geographies and cultures. It has been important to how countless people over the last five thousand years have understood their place in the cosmos, the timeline reaching back to the beginning of all things and forward to the end of all things. It has been important to how armies are motivated, slaves and peasants are placated, gods invented and emperors unleashed. The formula has been foundational to those who have made major scientific discoveries or peeked beyond the planet's atmosphere, but also to those who have waged world wars and enslaved masses. Tracing the lineage of this narrative, we can not only see the evolution of an idea, but also understand more clearly the process that created a certain kind of society that we call civilisation, an anomaly that was sparked first in one place, and has since burned across time, peoples, and far stretches of the earth. Though two hundred years old now, Jefferson's letter appears in the latter part of this history. His worldview was grown out of a lineage that stretched back nearly five millennia, to the world's earliest civilisation in Mesopotamia. But that tradition did not end with Jefferson. The progress formula still occupies a central place in societies and minds all over the world. It remains the default, subconscious framework by which most of us understand our place in our species' history and our societies' trajectories through time, and thus by which policies are decided and enacted. It remains the foundation on which we are currently building the future.

To illustrate how the progress narrative still inflects our worldviews today, let's look at some examples a bit more recent than Jefferson's time.

In 1903 – a single human lifetime after Jefferson wrote his letter describing the Rockies as primitive wilderness, and a period nearly equidistant between the Declaration of Independence and the publication of the book you're reading – Orville and Wilbur Wright flew the world's first aircraft.[3] The wood-and-muslin biplane reached around 3 metres above sea level. Thirty years later, dozens of aeroplanes clad in metal were flying at about 13,000 metres above sea level. Just under thirty years after that, Soviet cosmonaut Yuri Gagarin became the first human to cross the Kármán line, the imaginary boundary between the Earth's atmosphere and outer space. That abstract line slicing the terrestrial from the celestial reaches about 130,000 metres above sea level. In other words, within just fifty-eight years, human beings went from flying 3 metres in the sky to over 130,000 metres. Sixteen years after Gagarin's flight, the United States launched the *Voyager 1* spacecraft. It is currently floating about twenty-one trillion (21,000,000,000,000) metres from Earth. If one made a graph of humanity's leap in flight capability, from life on the ground at the dawn of our species over 300,000 years ago up to the present, it would look like this:

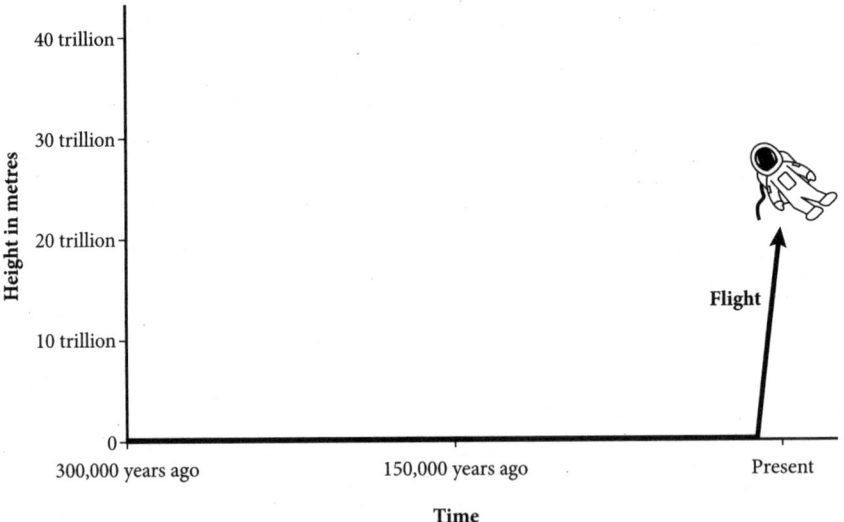

Figure 1 Flight through time

We could chart the development of many other technologies, plot them on a similar graph, and come out with nearly ninety-degree upward lines starting around the twentieth century.

WHAT IS PROGRESS?

In the twenty-sixth century BCE, the world's tallest building, the Great Pyramid, was built by Egyptians in Giza. At its tallest, it was around 146 metres. About 3,800 years passed before it was displaced as reputedly the world's tallest building by Lincoln Cathedral at about 160 metres, built by the Normans in England in 1311. But in the early twentieth century, skyscrapers began erupting from the ground all over the world, eventually reaching heights into the hundreds of metres, dwarfing the old world's pyramids and cathedrals. Where once monuments to the deified dead or to an invisible god reached highest into the sky, suddenly their status was usurped by monuments to finance and commerce, or to penthouse palaces for earthly lords. The tallest building at the time of writing is the Burj Khalifa in Dubai, hitting 828 metres (or a little more than five and a half Great Pyramids stacked atop each other). A chart of building heights (see Figure 2) would look like a fairly flat line for the first 300,000 years of human existence, then a higher line starting about 4,000 years ago made up of squat pyramids and stout temples, until the sudden, rapid ascension of familiar skyscraping towers of modern skylines starting around 1900.[4] While this graph may not go from 0 to 100,000 as fast as flight, it's still a razor-sharp incline, and both flight and buildings explode within a mere decade of one another.

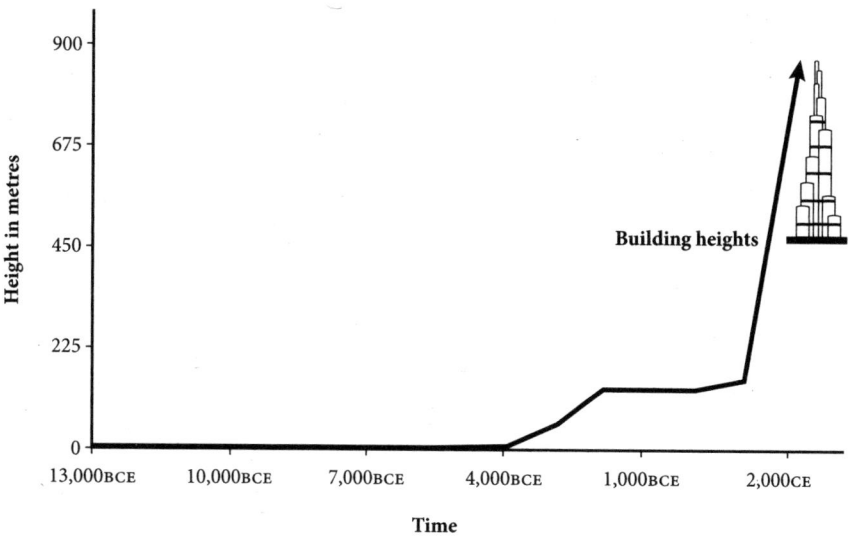

Figure 2 Building through time

Flight and construction engineering aren't the only feats of technology we've seen recently blast off into the stratosphere. A black powder that would eventually be used in weapons was first invented in ninth-century China. Use of gunpowder artillery in Europe began with the first English cannons rolling into battle using combustive chemicals packed into iron tubes to toss heavy balls around. This started sixteen years after Lincoln Cathedral was built. Such artillery didn't change much until the twentieth century when, mirroring flight and buildings, firepower exploded upwards in scale. Weapons technology reached its peak – literal and figurative – with the Russian *tsar bomba*. The *tsar bomba* was detonated in a 1961 test, just six months after the first human touched space. Its mushroom cloud reached a height of 65 kilometres, more than halfway to outer space, equivalent to about seventy-eight Burj Khalifas (or more than 460 Great Pyramids) stacked atop one another. Achieving such a massive explosion required huge amounts of energy to be accessed. A shell from a French 75mm 1897 modèle gun released about 2,600,000 joules of energy; the *tsar bomba* released 209,000,000,000,000,000 joules.[5] The *tsar bomba* was about 1,500 times more powerful than the atomic bombs the United States dropped on Hiroshima and Nagasaki combined, or, perhaps more stunningly, about ten times more powerful than all of the combined munitions used during the Second World War.[6] Again, a chart comparing the energy released from premodern cannons with that of the *tsar bomba* would look similar to the sharp upward growth lines of buildings and flight.[7]

These are just some of the starkest graphs showing the sharp increases in technological sophistication that started in the twentieth century. Other material shifts can be seen in public infrastructure – with widespread growth in electrification, indoor plumbing and urban sanitation occurring through that century – and in medical triumphs like the widespread deployment of penicillin, broad implementation of vaccines, and internal imaging. There are other indicators people traditionally point to when describing this rapid change: human populations skyrocket. Life expectancies increase. Per capita wealth rises. What demographers often call 'well-being indicators' all start to go up for many people during the twentieth century. Essentially, the amount of energy and resources that each person has access to increases, and that in turn translates for many into increased wealth and health. This vast increase in technological sophistication, human infrastructure, well-being indicators and wealth around the middle of the twentieth century is called the Great Acceleration, and it is often attributed to an innocuous-sounding process called economic growth. It is no surprise that such a phenomenon as 'growth' is tied to the kind of narrative formula of progress this book is interested in. No less a body than the United Nations Sustainable Development Group proclaims today on its website, 'Sustained and inclusive economic growth can drive progress, create decent jobs for all and improve living standards.'[8]

There is good reason for some to view the twentieth century as an era of progress. In the Global North, like the United States, the United Kingdom and Europe, there was an increase in not just wealth, but wealth *equality* during this period, which itself is a major contributor to increasing well-being indicators and the building of history's biggest 'middle' class. In the realm of social justice, there were important gains in minority rights and women's rights. All sorts of new regulations were passed in the mid to late twentieth century aimed at protecting public well-being, or at least reducing harms. It was during these dizzy decades that the twentieth-century idea of infinite economic growth and perpetual progress entwined and became the foundational default assumption of most people.

Does this mean progress is real? Is it true that humanity as a whole has moved from dark savagery to enlightened modern civilisation? Even if we home in on only the twentieth century, is it fair to say this period has been marked primarily by 'progress'?

It's worth taking a moment to complicate the answers to these questions and consider the possibility that they are more often closer to 'no' than 'yes'. So many of us have had notions of progress deeply forced into the architecture of how we understand the world, an alloy melted into the beams that bear the weight of our knowledge. In order to understand the story and phenomenon of progress, it is important to quiet the nagging voice that continually asks, 'But haven't we made progress?'

Today, when someone wants to make the case that the world has made progress, they typically deploy quantitative data and, from that, create data visualisation images like those earlier graphs.[9] Quantitative data and graphs can be good for answering certain questions, but when approaching the question of progress, they are less useful. To make the case for progress, you have to wade through historical complexities, moral dilemmas and the foggy realm of the human heart. After all, claims of progress imply an agreement around fundamental philosophical questions about good and evil, about what constitutes a meaningful life, about what should be the priorities of a society, that are far from settled.

When someone is trying to convince you that humanity has made progress in a grand narrative, in most cases they are doing so because they have an ideological agenda. The graph is the perfect tool for ideological propaganda because it erases complexity and encourages audiences not too look too closely at the data on which the graph is based, or the methods used to obtain and interpret that data.

One recent example of this is a narrative of progress based on poverty data. Many of those propounding a narrative of progress like to claim that economic growth has lifted all boats and poverty has been dramatically reduced. One statistic often deployed to support this grand narrative suggests that the percentage of those living in poverty has fallen from 90 per cent of the world's population in 1820 to just 10 per cent in 2015. But as economic anthropologist Jason Hickel has argued, this statistic is grossly misleading.[10] The data depends on redefining 'extreme poverty' as earning power of $1.90 per day, but the purchasing power of this amount is 'lower than the consumption level of enslaved people in the United States in the 19th century.'[11] To arrive at this claim that poverty has been greatly reduced and at the impressive-looking graphs, proponents of this claim, including billionaires like Bill Gates, simply reduce the amount of daily per capita income that constitutes 'poverty', creating the illusion of progress. In fact, poverty is not falling in most places, it's rising, and it will likely continue to rise as energy and resource limits are exceeded and wealth continues to be concentrated in ever smaller groups.[12]

'Data' is not an objective mass of numbers and lines. The numbers themselves are gathered by fallible people, from often incomplete sources, and within set ranges. Those numbers must then be interpreted, and all data interpretation is inevitably skewed by the biases, limitations and agendas of the interpreter. This does not mean that there aren't true facts or falsehoods and it doesn't mean we shouldn't try to understand the world through data gathering. What it means is that we have to approach data and simplistic visualisation tools with scepticism, because very often such neat graphs and charts have been constructed to present a particular picture, not to give us raw information or rigorous interpretations of how the world works. When they want the line to go up, it goes up. When they want the line to go down, it goes down. It just depends on where they put the numbers on the X and Y axes and what they choose to measure, or to exclude.

When such narratives are invoked, it is very rare that these invocations are accompanied by a sophisticated definition of what constitutes progress. They rarely include research on what actually makes people better off or what worthwhile goals of society should be. Such definitions and goals are simply taken for granted. So let's not take them for granted. Let's break down the question, *But haven't we made progress?* with more sophistication than can be found in a graph.[13]

One of the first major issues with these sorts of graphs is that they extrapolate very recent data to include all of human history. But the evidence they provide to support that claim rarely, if ever, covers the span of time or space they are making claims about. Further, they attempt to use quantitative data – sometimes cherry-picked or manipulated – to make claims about phenomena that simply cannot be made into a numerical abstraction, but which may be vitally important to the question of health and well-being, like whether a person feels intimacy and belonging, feels a sense of meaning and purpose, feels contentment or an absence of fear. Another issue with this sort of data interpretation is that it tends to create aggregates and averages that represent a fictional being. Such aggregates are meant to encapsulate sometimes many millions of people (or more!) in a single average. The averages very rarely reflect the actual realities of the physical beings they are attempting to describe.

One claim in particular illustrates these issues well, and that is the claim that life expectancy has increased through history, thereby indicating universal progress. This claim seems straightforward: life expectancy is the average number of years a person can expect to live, and, some claim, that number has moved from low in the past to high in the present. It seems like a simple number that you can plot on a graph and see the line go up over time. If you glance at such a graph, it can be easy to simply accept it and move on. The conventional narrative is that, in the past, human beings could only expect to live into their forties or fifties, at most. Then, the twentieth century saw a massive increase in per capita GDP and suddenly people began living much longer, into their seventies or eighties.

But the simplicity of this narrative can be misleading. To make such a claim, commentators will include an average for large populations from birth to death. The problem with that methodology is that it does not take infant mortality into account. Lots of children dying in childhood skews life expectancy data. What has decreased child mortality are interventions in medical practices and public sanitation, like germ theory in medical science and publicly funded water and sewage treatment and waste collection. These have led to much greater hygiene in both hospitals and other public spaces, making natal and public healthcare far safer. The reduction in infant mortality has made graphs of life expectancy show an increase. Of course, it is a good thing that there is more widespread practice of good hygiene and that fewer people die in infancy. To use this to prove a grand narrative of human progress would require showing that it is, first, universal, second, a major break not just from the near past but the deep past, and third, permanent.

None of these criteria hold up to scrutiny. First, there are the geographic disparities in life expectancy. Even today, there are wide margins in life expectancy from region to region, with the lowest country, Chad, at fifty-three years and the highest, Monaco, at eighty-seven years.[14] Some kind of geographic disparity has without doubt existed throughout human history, so extrapolating an average across all of humanity doesn't give us a very revealing picture of whether progress has occurred for all of humanity. But the bigger issue is proving progress with the dearth of data from tens of thousands of years ago when lifestyles and risk factors were quite different from today.

There is also, and has always been, the confounding factor of class. Those in more affluent economic classes will enjoy higher life expectancies than those in lower classes. Elite ancient Romans, who enjoyed more leisurely conditions and had better access to good hygiene, food and medicine, could expect to live for around seventy years, with some individuals living to be over one hundred.[15] Meanwhile, according to medical historian Valentina Gazzaniga, a study of two thousand ancient skeletons from common graves showed poor Roman workers had an average life expectancy of just thirty years due to the prevalence of disease, accidents, and injury from hard labour.[16] Such inequality is not a thing of the past. Using the Gini coefficient, the tool economists use to measure inequality, with 0 being perfect equality and 1 perfect inequality, British archaeologist Ian Morris puts the first century of the Roman Empire at 0.43. What does that mean in real terms? In Morris's calculation, 'the Roman elite (comprising something like 10 per cent of the population) was extracting wealth from the rest of the Romans at roughly 80 per cent of the theoretical maximum possible rate of exploitation.'[17] Even during Republican Rome, enslaved people would have made up somewhere between 15 per cent and a third or more of the population.[18] While some of the most equal modern societies, like Iceland, Denmark and Norway, have Gini coefficients nearly half that (0.26), a sobering fact is that, at the time of writing, the United States has a Gini coefficient the same as Ancient Rome (0.43, or even higher by some measures); given where trends are going, it will likely be higher by the time you read this.[19]

The reality is that human beings likely have a 'natural' lifespan into their late seventies.[20] That is, the human body is fit to live well naturally into seven decades. What prevents humans from living this long are factors like disease, violence, famine and vulnerability to ecological instability. All of these conditions increased in number and severity with the rise of dense cities: close proximity to domesticated animals, which brought about zoonotic diseases like smallpox; intensive agriculture, which made people sedentary and tied them to food sources vulnerable to climatic instability; and high rates of warfare, all of which began to expand rapidly around 5,000 years ago. Measuring recent life expectancy against that rather recent history – rather than in relation to all of human history – skews the graph to create an illusory image of progress in longer-living humans, and avoids difficult questions about geographic and class differences.

So what is the problem with the methodology that leads to such skewed graphs, which are often at the base of narratives of progress? First is the fact that all commentators making claims of progress over human history based on life expectancy have insufficient data to make such a claim. This is because they typically start their period of measurement in the very recent past, within the last few decades or centuries, since data for life expectancy is simply more abundant and reliable in this period. They cannot obtain large datasets of the life expectancy of people living 300,000 years ago because the data does not exist, or does not exist with the ubiquity and reliability that one would need to make a graph or a strong statement. Even data reaching more than a few hundred years into the past becomes thin and open to broader interpretation. To make a graph or statement about all of humanity through all of history requires extrapolating far beyond what such information can show.

As the data is limited temporally, so it is limited spatially. Data about life expectancy is geographically fragmented even for recent periods. Some regions will have much more robust and reliable data than others, and there will be a large gap in life expectancy across regions. When such a gap exists, the data visualisers have only one recourse: to aggregate and average the data, creating a meaningless abstraction of a lifetime. These temporal and spatial limits mean that what one can say about life expectancy across time and across all human beings is extremely limited. But this does not stop many from extrapolating a grand narrative about progress across all people and all history from twentieth-century data. They set goal posts that are rhetorically meaningful, but not analytically meaningful.

Another underexamined issue with the life expectancy graphs as evidence for progress is that commentators tend not to question whether greater life expectancy necessarily yields more positive outcomes. Even if it were true that humans are living much longer today than they have at any other point, they still take for granted that more years is on its own an unquestionable good. This is because many such commentators put forward a worldview based on the belief that *quantity* of human life is more important than *quality*, both in years and in number of humans living at any given time.

It should be self-evident that more years do not mean better years, nor does more people mean better lives for those people. We know this when we apply the principle to a species other than *Homo sapiens*. For example, there are many animals that have longer life expectancies when they live in captivity, like a zoo or rehabilitation centre. Yet we understand that their quality of life may be significantly diminished if they are not allowed to live a 'natural' life in the wilderness, regardless of how many extra years of living they have. In a similar way, human life may be extended in quantity even as it severely deteriorates in quality. We may have more years, but they may be worse years. Better years are impossible to quantify and turn into a graph. Similarly, population densities do not predict greater happiness, health or wellbeing, and very often correlate with (and often cause) the opposite result. More people does not mean happier or healthier people.

Life expectancy is just one of the measures that are commonly used to make the case for the progressive movement of history, and general improvement over time. Other oft-cited indicators include reductions in poverty, increases in per capita gross domestic product – the average wealth that individuals control – and growth in the accessibility and usefulness of certain conveniences, like faster transportation or automated washing machines. In each of these cases – most of which we will return to later in the book – the reality is more complicated. Reductions in poverty have been greatly overstated, and perhaps even exaggerated to the point of falsehood. A major recent study found that increases in GDP do not correlate or have any meaningful causal relationship with increases in human or animal well-being.[21] As mentioned, reductions in infant mortality are primarily associated with very recent improvements in public hygiene, sanitation and medical hygiene practices, which are certainly good in themselves, but have little to say about the thrust of human history and more to say about the benefits of very recent science-based practices. As for technological conveniences, their accessibility varies greatly across the world, and though they make some things easier, they make others harder and worse. Faster transportation is one of the greatest causes of premature death and illness and the greatest driver of climatic warming, while washing machines are a major source of poisoned water. Plastic bags are sometimes convenient; they have not made the world better.

Perhaps a more convincing narrative of progress today is in social justice issues, the belief that there have been steady improvements in fairness, equality, and opportunity for more people. Social justice narratives begin to reveal *why* a lineage of 'progress' is used instead of simply pointing to a positive change. Many claim that with the abolition of slavery and the enfranchisement of women, and many other new laws like civil rights legislation and changing gender norms, there has been great progress. Today, this kind of progress is most often depicted by representational politics, the idea that the representation of a minority group or of women in positions of power and privilege are indicators of progress. But is progress really the right frame to apply to these changes? Again, that a good thing happens does not necessarily indicate a pattern of progress.

In 1964, Malcolm X said: 'If you stick a knife in my back nine inches and pull it out six inches, there's no progress. If you pull it all the way out that's not progress. Progress is healing the wound that the blow made. And they haven't even pulled the knife out, much less try and heal the wound. They won't even admit the knife is there.'[22] Consider for a moment this quote and what it means with regard to the Atlantic slave trade. Huge numbers of African people were living out their lives, often in peace and equanimity, when out of nowhere, armed men appeared. These men kidnapped them and put them on ships where they either died or suffered extreme fear and illness. They then arrived in an alien land in harsh conditions, and worked brutally for long hours under the constant threat of corporal or capital punishment doled out by masters who feared no consequences for their violence. Meanwhile, the slaves' families were broken up, left impoverished, and in many cases their cultures, languages and land were soon imperilled by invading colonial forces. Even if the severity of this mayhem is grudgingly reduced – after great public agitation and a civil war – given the starting point in Africa prior to enslavement, we cannot reasonably call that 'progress'.

The abolition of slavery cannot be counted as evidence of a long moral arc, a momentum of human history, indicating a pattern of progress. Those who were captured lived a far different trajectory from the European one they were thrust into: they suffered a sudden, history-breaking apocalypse, the traumas of which their descendants still endure, with many trapped in the limbo of a dystopian system. Even if certain conditions have changed for the better, their enslavement lasted 246 years, while their relatively emancipated condition (such as it is) has lasted for only 156.[23] To return to Malcolm X on the 'freeing' of African Americans: 'How can you thank a man for giving you what's already yours? How then can you thank him for giving you only part of what's already yours? You haven't even made progress, if what's being given to you, you should have had already.'[24]

And for many Black people in America, conditions have only changed to some extent. The Atlantic slave trade was evil and a world in which it is no more is a world improved, but its abolition is not evidence supporting a grand narrative of historical progress. There are a few reasons for this.

One is that for all or most of the people ensnared in the slave trade, their lives were undoubtedly better before capture and their ancestors' lives were better for generations upon generations prior to enslavement. To place people in the worst possible calamity and then take some of that evil away, as mentioned, is not an example of progress. In the same vein, slavery as an institution is an evil that, to our knowledge, is probably only five thousand or so years old. As we will see throughout the book, for the vast majority of human history, slavery was non-existent or rare, sporadic rather than institutionalised. Then it became commonplace and inescapable. Returning to a situation in which slavery is rare, or even a situation in which slavery is totally abolished, is simply returning to a more natural state of being human, not advancing to some higher state from a primitive one.

But I should hasten to point out that we *do not currently live in* a condition in which slavery is rare or has been abolished. There are three times more enslaved people today in absolute numbers than there were at the height of the Atlantic slave trade – 40 million people, or around one in every 200, are enslaved (this is not including 'wage slavery', in which people are forced to participate in a labour market for a pittance or even for wages that are then stolen, which is much more widespread).[25] As Kate Hodal writes for the *Guardian*, 'there isn't a single country that isn't tainted by slavery', with enslaved people still manufacturing clothes, growing food, and mining minerals used in technology, even in richer countries. The UK has an estimated 13,000 enslaved people. Even though there are prohibitions on slavery, the institution persists across the world.

The second reason is that progress narratives depend on an arc bending in a particular direction, which presumes the existence of a future end state. But there are no such stasis points in time: things change and rarely settle permanently. There's no good reason to believe that slavery will remain 'abolished' in the United States and Europe and elsewhere. There have been many times throughout history when slavery was suppressed, even abolished in part, for extended periods, and then resumed. The Achaemenid Empire, founded by Cyrus in what is now Iran, likely diminished slave labour considerably for as much as two centuries before it returned for many centuries more. In the ancient Mauryan Empire in India, Ashoka may also have temporarily reduced slavery by abolishing slave trading, but this change too was not permanent.[26] There are many reasons to believe that some form of forced labour will return to many of the places where it has so recently (in historical terms) been abolished – as it was recently in the wake of Muammar Gaddafi's overthrow in Libya[27] – though it may not look like it did in the past, or may not be racialised in the same way. This is because climate change and ecological collapse are very likely to cause political fragmentation that nullifies legal and cultural precedents like abolition, and bring about agrarian and manufacturing crises and scarcities in which people are forced into labour. If market economies continue, there is little reason to assume they will not return to trade in indentured human beings.

But what if we home in from the very large scale of all of human history to more recent times? What if we just look at progress in social justice within the scope of a period from the recent past to the present: have things generally been getting better?

The United States has by some way the highest incarceration rate and imprisoned population in the world,[28] with 2.1 million prisoners. That is about 400,000 more than the next highest country, China (about 1.7 million), even though it has less than a quarter as many people (328 million to 1.4 billion). The proportion of Black Americans in the United States is one-fifth (13 per cent) that of White Americans (64 per cent), but Black people represent the *same* proportion of the prison population (about 40 per cent) – meaning the incarceration rate of Black people is five times higher.[29] 'More than half of all black men without a high-school diploma go to prison at some time in their lives,' writes Adam Gopnik for the *New Yorker*. 'Mass incarceration on a scale almost unexampled in human history is a fundamental fact of our country today – perhaps *the* fundamental fact, as slavery was the fundamental fact of 1850.'[30] About twice as many Black men are under state and federal criminal supervision – in prison, under probation – as were enslaved at the height of antebellum America.[31]

Those prisons can be hellish. Forced unwaged or underwaged prison labour, sometimes in extreme, dangerous conditions like fighting wildfires, remains a stark reality for many.[32] 'Prison inmates are picking fruits and vegetables at a rate not seen since Jim Crow,' reads one recent report on the practice of 'leasing' convicted federal prisoners to do farm labour.[33] Forcing people in chains to work in fields or at wildfires is not meaningfully different from, or morally superior to, nineteenth-century chattel slavery. Is this really progress?

A callous observer might reply, 'Well, if they didn't want to be forced to work in brutal conditions, they shouldn't have committed a crime.' But the fact is that Black people are policed differently, sentenced differently and imprisoned differently from non-Black Americans, who are *also* over-surveilled, over-incarcerated and living in a virtual police state. The vast majority of inmates in local jails have not even been convicted, and only about half of inmates in prisons are there for violent crimes.[34] According to one study, Black men are sentenced to an average of 10 per cent more time in prison than White men for committing the same crime.[35] Though the rate of marijuana consumption among Black and White Americans is comparable, Black users are 3.7 times more likely to be arrested for possession.[36] A recent study at Northwestern University found that White Americans vastly overestimate racial progress.[37] And it's not just in the difference in treatment by the legal system: the fact is, the wealth gap between Black and White Americans has not shrunk in over half a century.[38] Although the Voting Rights Act of 1965 was a landmark policy change, rampant voter disenfranchisement remains a major problem among Black communities.[39] Social psychologist Jennifer A. Richeson wrote for *The Atlantic*, 'Some essential civil rights have advanced, though unevenly, episodically, and usually only following great and contentious effort. But many areas never saw much progress, or what progress was made has been halted or even reversed.'[40]

The representation of marginalised identities in power, business or media is often hailed as indisputable proof of progress. But this representation does not yield improvements for most people who share those identities any more than having monarchs and emperors of a certain race or gender has improved conditions for workers and peasants of the same race and gender during history's millennia of slavery, serfdom and conscription. Such representation is more often used as a tactic for blocking egalitarian policies than for achieving them.

Former President Barack Obama offers an illuminating illustration of this. President Obama has been one of the deftest wielders of progress narratives in modern history. It comes as no surprise that his 2020 autobiography is titled *A Promised Land* – the 'promised land', as this book will detail, is a millennia-old cliché representing one of the original progress narratives. In a way, Obama's career perfectly represents the hollowness of representation politics and the cynical deployment of progress rhetoric.[41] Instead of widespread liberation for many, his own success stands in to embody progress: '[Obama] admits that his campaign deliberately "helped to construct" this association in the public's mind between *the election of Barack Obama to the presidency* and *the fulfillment of America's promise and the end to people's troubles*. The route to the "promised land" was through his presidency.'[42]

But even despite Obama's presidency, the Promised Land remains out of reach for most. Or, in part, it remains out of reach *because* of Obama's presidency: aside from the financiers his administration bailed out with trillions of taxpayer dollars, the top 1 per cent of wealth holders who captured 95 per cent of income gains under his administration (one of the largest transfers of wealth to the already-wealthy in history) and his own family, whom he enriched thirty-fold while in office, President Obama did not lead anyone to the Promised Land.[43] Far from Black people being liberated, conditions changed little for marginalised groups and Obama's administration even found opportunities to brutalise them. In 2015, Black Lives Matter protests erupted against police terrorism *and a lack* of racial progress during his administration. Obama's response was the violent suppression both of Black Lives Matter protestors and primarily of Indigenous Water Protectors, who protested against the expansion of oil pipelines in Standing Rock land in North Dakota.[44] The aggregate wealth of Black Americans *decreased* under Obama's administration. As *Jacobin* reports:

> Between 2007 and 2016, the average wealth of the bottom 99 percent dropped by $4,500. Over the same period, the average wealth of the top 1 percent rose by $4.9 million. This drop hit the housing wealth of African Americans particularly hard. Outside of home equity, black wealth recovered its 2007 level by 2016. But average black home equity was still $16,700 lower. Much of this decline . . . can be laid at the feet of President Obama.[45]

In a perhaps too on-the-nose example, Obama's Presidential Center, currently under construction in Chicago, is set to displace a community of working-class Black people.[46]

Obama's administration increased drone strikes tenfold in the Middle East, deported a record number of people, and constructed the border detention centres that would become infamous under his successor, President Trump.[47] A report by the Manhattan Institute (a 'free market' think tank, to be sure) argues that Obama's presidency failed to deliver any of the promised progress to Black Americans, stating, 'During an era of growing black political influence, blacks as a group progressed at a slower rate than whites, and the black poor actually lost ground.'[48] It was Obama's administration, as well, whose handling of the overthrow of Libya's dictator led to slave markets being reopened there.[49] Obama even dramatically increased fossil fuel exploitation and carbon emissions during his administration, and then bragged about it to a roomful of oil moguls.[50]

This elite trickle-down racial progress is not new, and Obama's presidency was not the first time a person of a minority identity has done harm to marginalised people. Despite claims that Kamala Harris was the first vice president of colour, the first person-of-colour to hold that role was Herbert Hoover's Vice President Charles Curtis, a member of the Native Kaw Nation.[51] Progress narratives often erase past instances of what they're defining as 'progress' to appear new, as if they actually are a step from a worse past to a better present, and thus an even better future; and Native Americans are generally erased from mainstream history as part of their ongoing genocide.

Curtis used his position of power to harm other Native Americans, even his own nation. As a Congressman, he wrote the Curtis Act of 1898, which broke up tribal self-government and the communally owned lands of five tribes – the Choctaw, Chickasaw, Creek, Cherokee and Seminole – in what soon thereafter became Oklahoma. The law stole 90 million acres from the tribes and granted authority to the federal government to choose tribe members.[52] And in 1902, the Kaw Allotment Act dissolved the Kaw Nation, which led to its ultimate extinction: the last full-blood member of the Kaw Nation, William Mehojah, died in 2000.[53] As an enrolled member, Curtis and his family were allowed to claim 1,625 acres of land in Oklahoma.[54] Curtis officially supported the move.[55]

When it comes to gender equality, the idea of trickle-down progress is just as pernicious as in racial matters. Margaret Thatcher, Britain's first woman prime minister, did no more to help women than Queen Elizabeth I, or any other queen, has ever done to help liberate or elevate other women.[56] Her government eroded public services, while unemployment rose, the economy was wracked by recessions, and poverty and inequality both increased.[57] These conditions hurt most Britons, of course, not women exclusively, although austerity measures do typically do greater harm to women.[58] But perhaps more egregiously, the pay gap between men and women *grew* under Thatcher's government, again illustrating how those of a particular identity can use their power to harm others of that identity.[59] While women have enjoyed some great successes in expanding equality – like the right to vote, own property, control reproduction and leave unhealthy marriages – much of women's liberation has meant only that a privileged few women get to dominate while most of the rest still must toil for a small elite, and many of these successes are being eroded. And, as we will see often in deconstructing the progress narrative, the systematic curtailing of gender equality or women's rights is relatively new, with the rise of a certain kind of society we call civilisation. Creating oppressive conditions and then improving them slightly does not indicate a long-term trend of progress.

* * *

Whether or when progress can be shown to have occurred depends entirely on when measurements start and end, where the lens is focused, how those measurements are conducted, and what data researchers choose to include or omit. Sometimes what appears to be progress in technology is actually just the solution to a new problem caused by some other technology that falls outside the scope of a given study, while progress in justice and civil rights may be but a slight improvement on conditions that were not natural or eternal, but were newly created horrors at the beginning of the time period under consideration. Grand narratives of progress begin to break down when you look closely at them. It soon becomes clear that they instead often serve more cynical purposes.

This faith, that history moves progressively, is at the foundation of my own upbringing. My parents have engaged in lifelong battles to improve the world they inherited, with an earnest urgency one sometimes finds – for better or worse – in US Midwesterners, which I could not help but absorb. They've protested in Washington, DC, they've volunteered for local and national candidates for office and for local and national organisations. They obviously did not believe progress is natural or inevitable. They believed you have to fight for it. And they instilled in me a deep-buried belief in the possibility of societal progress: of the value of participating in collective action movements to do whatever one can to make life better for other human beings and for all the other creatures that share the one planet known to house life. Some of my earliest memories involve attempting to decipher protest signs in the garage ('Keep your laws off my body'?) or accompanying my mother to envelope-stuffing gatherings for a local politician. I began attending rallies and protests in my teens, seeing both the heart-fluttering power of hundreds of thousands of voices ringing together and the cold terror of kids being beaten bloody by riot police. Growing up among wild habitat in northern Michigan, I felt more at home among forests and hills than anywhere else. Seeing first-hand the destruction of beloved places, I have felt deeply committed to wilderness protection. After years of participating in activism and becoming disillusioned with its lack of success, I have lately been focusing on scholarship and public writing. It is this combination of experiences and a desire to contribute to meaningful change that turned my eye towards the idea of progress itself.

Studying progress as a cultural artefact and its connection to material systems, like economies and ecologies, is not straightforward. It is interpretive and, given the importance and complexity of the idea, must be approached from multiple angles. I set out to study this phenomenon first through the lens of my area of expertise, geography. From the Greek γεωγραφία, combining 'earth' and 'writing', integrated geography is the study of how ecological systems interact with human systems through space and time. It is an interdisciplinary field, and as such I have approached the subject with the tools of multiple disciplines. Progress as a narrative formula that shapes the identity and priorities of cultures can only be understood this way, by understanding not only the culture itself, but also the ecological context in which that culture exists and is shaped.

As I began going through documents, reading books on the idea, thinking through my own relationship to these assumptions, it became clearer that the narrative infrastructure that I took for granted did not necessarily reflect history and reality. What if, I wondered, these progress narratives were not just wishful thinking, not just a human tendency to yearn for hope and paradise, but were active forms of propaganda? What if this narrative was itself a rhetoric deployed for political ends, like conquest and control, as a perverse obstacle to actually achieving progress? And, further and more subtle, though we can easily identify such propaganda when political opponents deploy it, is it possible that seemingly virtuous ideas of progress can be just as corrupt, just as wedded to systems that ultimately undermine their own stated goals?

* * *

Of course, we today can look at Jefferson's letter and know that his simplistic narrative is rather naïve. Jefferson's understanding of history was limited by the absence of what we think of today as systematic professional history. His depiction, as we will see, does not reflect how our species has actually moved through space and time. Perhaps he can be excused such an error based on simple ignorance. But, regardless of its accuracy, his depiction should not really be considered 'progress', even if we take that to mean only the advancing betterment of human life, or the extent to which technology and society can satisfy human wants and needs – or allow us, in his words written elsewhere, to pursue happiness. Moreover, for that error he has no excuse of ignorance. After all, Jefferson's progress was bought with the blood of millions of Native people, the eradication of billions of animals, the destruction of uncounted natural habitats and cultures, and the prohibition of ways of life that *do* maximise human health and well-being. He was personally involved in all of this, a wealthy slave owner and founding president who knowingly engaged in active, deliberate genocide and land theft. We twenty-first century people have evolved beyond this, surely, and can avoid falling into the same narrative trap.

But the fact is that twentieth-century growth, and now the twenty-first-century growth that so many among us yearn for, and who are eager to call it progress, has had consequences in some ways just as bleak and destructive as Jefferson's eighteenth- and nineteenth-century growth. Setting aside debates around past instances of progress for now – we will come back to those later – it is very clear that whatever genuine claims can be made to improvements in the twentieth century, most or all of these are rapidly disintegrating in the twenty-first, *as a direct result* of that success.

Thanks to the explosion in fossil energy deployed in the last hundred years, atmospheric greenhouse gases are causing temperatures to rise more rapidly than at any other time in human history, with cascading impacts and feedback loops like increased droughts, wildfires, floods, sea-level rise, the melting of permafrost, glaciers and ice caps, ocean acidification, forest diebacks and desertification, heatwaves, and more. As a result of the widespread expansion in use of synthesised chemical compounds in the Great Acceleration, other forms of airborne, waterborne and soilborne toxic pollution are proliferating at increasing rates.[60] As a result of skyrocketing human populations and the necessity for increased food production and housing during the last hundred years of growth, deforestation and habitat destruction are driving massive terrestrial extinctions, while overfishing, pollution and hotter temperatures are bringing about the collapse of marine life. Since the 1970s, 60 per cent of vertebrate populations have gone extinct.[61] Two-thirds of tropical rainforests have been destroyed.[62] Fish numbers have halved.[63] The result of this? The extinction rate is hundreds of times greater than it would be without such pressures.[64] Waste, too, is a growing problem: due to the invention of plastic and widespread single-use plastics, plastic waste proliferates across every region on the planet, reaching into the remotest parts of the globe, like the poles and sea floor; plastics are even found in human breast milk and blood.[65]

In part as a result of the destruction of the biosphere, and entirely as a result of the processes that drive that destruction, human welfare is worsening in a number of key areas: democracy is declining, economic inequality is increasing poverty and slums are a growing feature of urban life, global life expectancy is falling, maternal mortality for millennials is 300 per cent higher than for their parents' generation, new infectious diseases are rapidly increasing in reach and number, social trust in institutions and individuals is decreasing, the rate of economic instability is growing and social-economic mobility falling, along with other general well-being indicators in aggregate. In short, nearly everything is dying or degrading.[66] By the time you read this book, such conditions will undoubtedly be worse. That is not a pessimistic assumption, but simply a realistic assessment based on all trend lines in these areas. What had appeared to be indicators of progress in human well-being during the twentieth century – greater equality in wealth distribution, rising well-being indicators, or environmental regulations, for instance – may be little more than a brief blip, at best: a quick flash resulting from all the energy that also ignited the *tsar bomba*, skyscrapers and space shuttles.[67]

By coincidence, on the day I wrote the first version of this paragraph – Monday, 14 November 2022 – I went in search of the current number of living human beings on the planet, and on that day, the number passed the 8 billion mark.[68] It was 3 billion fewer when I was born in 1986. Between 2018 and 2023, 400 million people have been added to the total count – *net*, meaning that the number accounts for those who have been subtracted. That's adding more than the total population of the United States to the Earth in five years. The last time the total number of human beings declined is estimated to have been during the plague pandemic of the fourteenth century, which killed tens of millions.[69] This means the world has seen nearly 700 years of a compounding increase in the total number of human beings. As I write in the early 2020s, there are 4 million cities and towns that house just over half of us.[70] The roads on which many move could wrap around the planet 1,300 times.[71] To feed us, about 37 per cent of the planet's land is covered by farms, many of them intensive, and between 1 and 3 trillion individual creatures are hauled out of the oceans annually and slaughtered.[72] Two-thirds of the world's humans have access to the internet: mass digital networks of information, communications and markets.[73] More than three-quarters can read words generated from among this planetary sea of human neurons.[74] The anthroposphere has never been so vast.

The biosphere has shrunk proportionately.[75] Biomass is the total amount of biological material that makes up the living bodies of organisms. The biomass of wild mammals accounts for just 4 per cent of all the mammals in the world. This means human biomass makes up more than eight times that of wild mammals, while domesticated mammal biomass is more than fifteen times greater than wild mammals (see Figure 3). In 2020, the total mass of all animals was 4 gigatons. The total mass of plastic was more than double that, 9 gigatons. That proportion will have grown larger by the time you read this. The mass of humanity's built environment – the stuff of Jefferson's as yet most improved state – is greater than *all life on the planet*, all plants, all animals, all bacteria, everything, which means all known life in existence.[76] The destruction of biodiversity is the sixth mass extinction that researchers know about in Earth's history, and the first since 66 million years ago when what was likely an asteroid strike shrunk the planet's biodiversity and ended the age of dinosaurs.[77] There is an unknown and unquantifiable probability that today's extinction event could end up being the worst in life's 3.7 billion years on the planet – in all the universe, as far as we know – and could make us extinct along with much else.[78] There is nothing more evil, nothing more horrific or important to avoid, than a situation in which the one planet in the universe known to house life is made inhospitable to most life.

WHAT IS PROGRESS?

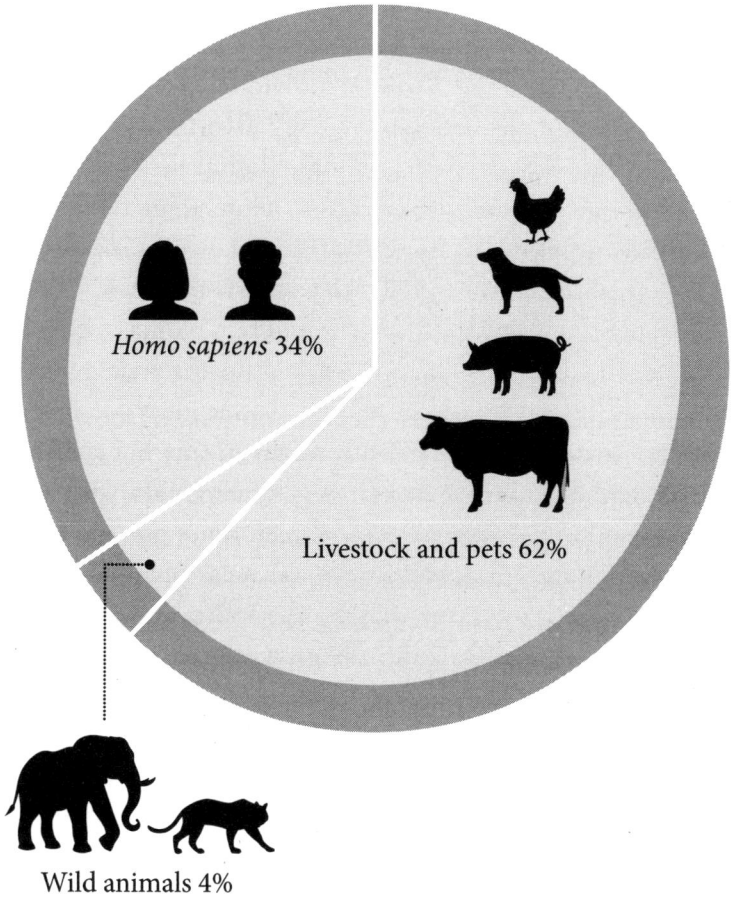

Figure 3 Mammal biomass on Earth[79]

Never before have so many lives, human and otherwise, depended on the immediate decisions of human beings. This exceptional era is the result of many complex processes, from war and conquest, to the global economy's material metabolism, like a huge god's gut, that involves processes like the globe-spanning *extraction* of raw material, the *production* of goods and services and the *excretion* of waste materials, to mass transit, to cultural processes like intellectual creation and discovery, and many others still. For all the progress cited in the contemporary version of Jefferson's narrative, we are all on a precipice whose near future looks far from any sane mind's idea of paradise. According to the best science today, it is within the range of possibility that the immediate future may bring the greatest destruction of life in all of history.

Progress, in this grand historical tradition, had always been the true north of my moral compass, the engine that powered the momentum of my life. This book focuses on the story that I have carried with me and have lately sought to extricate. I am of the West, raised and educated in its values, and so am sympathetic to that position, and yet it is the lineage of the West towards which I turn scrutiny in this book. It is sometimes painful to look at one's closest cultural ancestors and the crimes they committed and know that you are a product of those crimes. But it is better to have knowledge than not, though you must pluck out an eye to see, and better to live in a truthful story than one built on a lie. Living in a comforting lie is a coward's way of living, and there is no time or space for that now. The West is today the world's richest and most globally powerful region, unique in all of history for the quantity of its might. It is the great pioneer of the idea of progress, with the most available recorded sources engaged in the idea, and so it makes analytical sense to turn the lens of criticism in its direction. Such criticism does not come from any pre-existing hatred or discomfort with Western civilisation on my part. Just the opposite, this is a labour of love and desperation. To read it may be uncomfortable at times, but I hope by the end it will offer a clearer view of the world.

* * *

For all the fruits of the global economy's energy – and there are significant material benefits for many – they bring with them at least as many perils. To simply run the world's economies today entails, by necessity or not, inflicting pain and suffering on uncountable multitudes of people and other animals. The simple cruelty that is imposed on billions of people and creatures is difficult to examine in detail, which is perhaps one reason why so many are reluctant to confront it and so eager to embrace narratives of progress. The calamities of today are assuaged by believing those of yesterday were greater. But they weren't. The cruelties of today include heaping indignities on people daily, stoking fear and anxiety, and using deception to engage in psychological manipulation that undermines people's very sense of reality, even to the point that they blame themselves for the cruelty that is imposed on them. To run the present economy requires the destruction of individual plants and animals that have been living for hundreds of years, and of species of plants and animals that have been living for hundreds of millions of years. It necessitates the slaughter of well over a trillion living beings per year, and the killing of perhaps multiple trillions by accident, while both humans and other animals are held in cages against their will, complex minds wracked with fear and boredom. This economy is but the latest iteration of a lineage that has been built on war, genocide and torture, on mass murder of the most intimate and bloody form.

But *Homo sapiens* has what is known as a prosocial nature, a strong inborn desire to find camaraderie, communion and compassion among other beings, human and non-human.[80] We are a species of ape whose psychological profile is highly social and, under many circumstances, generous and gentle. Acting out all the evil that is required – or if not required, expedient – to run this economy requires engaging in advanced psychological manipulation, of ourselves and others. It is not simply that rulers must achieve positions of power and then trick their prosocial followers to engage in their projects. Class harmony depends on a shared mission. Only a collective mass delusion that is believed in as strongly, or more strongly, by the leaders of this economy as by those whose own hands turn its wheels would allow such a project to continue so smoothly, for so long. There have been many aspects of this psychological manipulation, but the form that is more powerful, more ubiquitous, and older than all the others is *Progress*.

The story of progress will start in Mesopotamia, where this narrative formula was born, and follow its evolution through other cultures in Southwest Asia, to those in southern Europe and North Africa, throughout northern and western Europe to the Americas, touching upon the influence of these ideas in the Far East. We will investigate the thought of Persia and Judea and Christendom, the geopolitics of Rome and Vikings and Islamic caliphates. This journey will travel through the Enlightenment debates that surrounded Euroamerican colonisation and evolved into the great ideologies, movements and nations that shook the nineteenth and twentieth centuries. The book is organised chronologically and will delve both into the various forms progress narratives have taken and the cultural and ecological contexts that shaped them. There are three primary eras in which economies took particular shapes and the progress myths they developed took unique forms with them, from a mythical age to a secular age to an economistic age.

The narrative formula of progress has grown in its potency and scope for the past five millennia. Because of this power, the idea has been wielded in ways positive and negative; its wielding is at the foundation of civilisation itself. Who defines progress controls the direction of politics, economies and nations. Who controls the parameters of what progress means sets magnetic north for many people's moral, intellectual and political compasses. Given that this narrative formula remains foundational to the societies and economies that rule the world and set the course for the future we are building, the course driving us all towards life's greatest destruction, this narrative and its role in society is in need of scrutiny more than ever before. As thin fracture lines branch through the current order and accelerating ecological destruction promises to bring catastrophic political chaos, more than ever those on the side of diverse life need the tools to strengthen our wavering hearts, yearning for hope against those who would promise us future salvation in exchange for present obedience. This book, in closely investigating every major version of this potent form of propaganda, aims to equip you with those tools.

Before Progress

In Jefferson's telling, North America before European settlement bore the image of a primitive past, simple, savage: what he calls 'barbarism'. This will be a familiar narrative to readers in the twenty-first century. The philosopher Thomas Hobbes rendered this idea most memorably with his description of human life in its original 'state of nature' as 'solitary, poor, nasty, brutish, and short'. In this original state, Hobbes claimed, there is 'no culture of the earth; no navigation, nor use of the commodities that may be imported by sea; . . . no knowledge of the face of the earth; no account of time; no arts; no letters; no society; and which is worst of all, continual fear, and danger of violent death'.[1] This Hobbesian argument, that prehistoric and Indigenous cultures described under the 'savage' and 'barbaric' designations are more violent, less civilised, less sophisticated, and more indolent than the sort of agrarian, urban and industrial European ones that Jefferson considers the height of civilisation, is alive and well at the time of writing.[2] The ideologies that lubricate the engines of global commerce and governance are held together by this general assumption. Though they may not admit it, many still carry around with them this narrative structure of native savages and enlightened colonists. Many still see a teleological aspect to this: that Indigenous colonised peoples represent a figment of the past, while those who conquer and replace them represent a more modern, more developed people. They may use terms like 'less' or 'least developed country' or the more politically correct 'developing country', they may frame their intentions as more benevolent than colonists of the past. But they remain the same. Does this formula on which so much is based bear even a grain of truth?

Savagery and Civilisation of the Ohio River Valley

When English colonists arrived in North America in 1607 – 46 years before Hobbes's book was published and 217 years before Jefferson's letter – they found sophisticated societies with towns and houses protected by wooden walls, whose food and resources were supplied by a combination of agriculture and foraging, who were governed by complex political treaties and had sophisticated systems for distributing wealth and power. The Powhatan peoples they first encountered, in what became modern-day Virginia, were ruled by a paramount chief (named an 'emperor' in some sources) who tended to be male and ruled over a complex hierarchy of other chiefs, reflecting on a smaller scale the kind of complex bureaucracy of classical empires. The Powhatan likely consumed the resources in the areas they inhabited to depletion, moving on and only returning once game, fish and soils had regenerated. Even as they practised somewhat gendered divisions of labour, women were still able to inherit wealth and become chiefs, and enjoyed considerable sexual autonomy compared with English women of the time, free to turn down suitors and – with their husbands' permission – take lovers.[3] American anthropologist Helen Rountree documents that the Powhatan would bathe in the river each morning before dawn both for cleanliness and hardiness, in preparation for a day of outdoor work, while the Elizabethan English may only have bathed a few times each year.[4]

Further north, the peoples of the Eastern Woodlands nations were quite different from the Powhatan. In contrast with the hierarchical and depletive societies of the mid-Atlantic, they engaged in a more sustainable relationship with the ecological systems they inhabited and operated their societies on more egalitarian principles between members – men and women, adults and children – and with extensive sharing of resources. They, too, had complex political treaties between tribes, they practised advanced silviculture, conducted layered and democratic political processes, and maintained a connection to the land that went back thousands of years, sustaining relatively high biodiversity and health and culture.

Almost immediately upon their arrival in North America, English settlers began shooting Natives and attempting to secure land for commodity development like tobacco farms, initiating a process of intermittent wars, alliances, peace and eradication. Alongside hostility and weapons, diseases endemic to Eurasia but not seen in the Americas, like smallpox and measles, decimated the densely populated Indigenous villages early on in the colonisation process. England's King James cheered this process. By 1620, he had established New England, claiming the land 'henceforth be nominated, termed, and called by the Name of New-England, in America . . . We do by these Presents, for Us, our Heyrs and Successors, name, call, erect, found and establish, and by that Name to have Continuance for ever.' Prior to his declaration, the king praised God for spreading the pandemic that wiped out potentially millions of Natives in the region, stating, 'And also for that We have been further given certainly to knowe, that within these late Yeares there hath by God's Visitation reigned a wonderfull Plague, together with many horrible Slaugthers, and Mur[d]ers, committed amoungst the Sa[v]ages and brutish People there.'[5]

As Euroamerican settlers pushed westward, their frontier inched into the Ohio River Valley, an area encompassing parts of contemporary Ohio, Kentucky, Indiana and West Virginia. There they found a complex, generally peaceful society of settled villages thriving on an abundance of resources gained from a combination of foraging and agriculture, even after the apocalypse of disease had passed through.[6] Women were central to agricultural production in these societies, developing advanced methods like no-till cultivation that enabled them to harvest continually without fallow periods or nomadism, and to achieve greater yields than even intensive European agricultural practices. This resulted in the building of a series of permanent villages along these rivers, from which the Native peoples gathered hundreds of aquatic species of fish, reptiles and invertebrates.

Historian Susan Sleeper-Smith has chronicled these societies and their conquest by Euroamerican settlers in great depth and detail. An early European colonial account of the area describes it as containing 'exceedingly heavy virgin forest, some of the heaviest hardwood forest I have ever seen as I have twice visited the Tropics (Central America) – covering almost the entire floodplain on the Indian side'.[7] These forests included sycamores up to 30 feet in circumference and nearly 200 feet high – one was big enough to hold twenty-nine men side by side in its hollow. Even with Native agriculture in the region, the forests and individual trees were immense and abundant. 'The original 20 million acres of primeval forest that once covered Indiana,' writes Sleeper-Smith, 'was nearly enough to encircle the globe one and a quarter times in a mile-wide band.'

Through a detailed accounting of the physical facts of these societies, we see a picture of an agrarian paradise combining great material prosperity with relative ecological stability, achieving an abundance of wealth in which production was mostly innovated, maintained and governed by women. When Europeans arrived, they found societies with elaborate dress and decoration, impressive structures, ample leisure time and entertainment, and complex politics and culture. The image we have of the scrappy, impoverished Native American and decadent, advanced European would have been reversed: the Euroamerican settlers were often poor, bedraggled soldiers and traders, the Natives settled and prosperous. Like the encounters between the Powhatan and English in Virginia swinging between periods of peaceful trade and brutal violence, so when the Euroamerican frontier reached the Ohio River Valley region, the Indigenous tribes and colonial settlers traversed knife-edge relations. European fur traders settled down in Indigenous villages and interacted frequently. Miamitown, the largest trading village north of the Ohio River, became a dynamic zone where 'manners, formality, and social etiquette were a captivating blend of Indigenous and European customs,' notes Sleeper-Smith. 'This was a lively, informal world where Indians and traders were social intimates and trading partners.'[8]

But peace did not last. As Euroamerican states and markets expanded, demand for land and resources grew, and the new United States government – free of legal, moral and economic shackles imposed by faraway European empires on its expansion – sought them with ever-greater ferocity. We can view the American Revolution of 1776 as a righteous rejection of foreign monarchic rule, but also as a necessary precondition for intensifying the extractive, genocidal economy on the continent, given that the English crown – and the French as well – had at one point sought to restrain expansion of their colonies in North America. The new country's first president, George Washington, engaged in underhand tactics to utterly destroy the Ohio River tribes and the complex societies they had built there. His secretary of war, Henry Knox, reassured Charles C. Scott, the general who led the militias charged with carrying out this endeavour, that his actions 'were not bound by any moral codes of army behavior'.[9] The Natives were not to be afforded the basic recognition of common humanity usually extended to enemy soldiers and civilians. Scott unleashed 750 volunteers, many chosen for their particular hatred of Natives, to decimate villages, tear up gardens and orchards, burn down houses, capture women and children, and kill anyone who attempted to flee. President Washington explicitly ordered the capture of women, knowing the central role they played in maintaining food supplies for Native warriors who had previously been successful in repelling American invasion. Destroying the breadbasket that fed this resistance by capturing those who maintained it would ensure Washington achieved a swift military victory.

What compelled Washington to pursue this scorched-earth campaign of kidnap and terror, even while trade relations had been peaceful and prosperous? The answer, primarily, was his failure to persuade these societies to give up control of their lands to US citizens, and his personal imperative to capture them exclusively for his own profit.[10] Washington was the new country's wealthiest man, and much of that wealth came from both the cultivation and speculation of land. Washington's Indian Wars of the last decades of the eighteenth century succeeded in expelling Natives from the land. Perhaps more importantly for the growth of this young nation, his campaigns destroyed any evidence of Indigenous prosperity in order to build a new mythic narrative of Euroamerican superiority and Native

American savagery, supporting the progress narrative formula that Jefferson evoked in his 1824 letter.[11]

What happened to those vast virgin forests upon Washington's victory? 'Today, scarcely enough old-growth forest remains to cover the Indianapolis Motor Speedway at the same one-mile width' described earlier.[12] In their quest to transform the land to enable more short-term, high-intensity productivity, settlers destroyed 20 million acres before 1870, clearing forest for farms, often simply burning the wood in huge heaps, day and night, for months at a time, 'turning the sky sallow'.[13] It is a pattern of destruction repeated across the world today in rainforests throughout the tropics, like the Amazon, the Congo and Borneo.[14]

The descendants of the first inhabitants of North America enjoyed great natural wealth. Through the work of many generations, many Native Nations achieved stability and accord with the abundant life around them: the trees dense, the rivers quivering with life, the communities tightly woven. One day a horde of hungry, stunted, tattered brutes intent on their death or capture amassed on the border of their land. These savage hordes murdered, enslaved, raped and, ultimately, burned it all down. This is not the myth, this is the reality.

The central claim of the modern progress narratives that First Nations America represented an earlier stage of human development – in any sense of the word: moral, economic, technological or sociological – is simply not true. The view of Native Americans as primarily composed of small bands of roving hunter-gatherer-warriors wrapped in animal skins is largely a caricature – and even if it were accurate, would not justify their harm, dispossession, or misnomer of 'primitive'. Prior to the Europeans' arrival, North America had been extensively populated, with towns, complex food systems often utilising agricultural techniques more sophisticated than those in Europe at the time, watercraft and building technology, treaties and economies. 'First Nations' isn't a rhetorical flourish: these peoples were and *still are* organised into nations, even if those nations were not always organised as states, kingdoms or empires (though, as noted, some were). Indigenous societies, in North America and elsewhere, no more represent a primitive 'state of nature' – and indeed perhaps less so – than Manhattan; the current one, that is, not the Indigenous island at the same site that the Lenape called Manahatta, from which the Dutch forced the original inhabitants.[15]

In a letter to his friend Peter Collinson in 1753, Founding Father Benjamin Franklin noted the colonies having 'little success' in attempts to 'civilize' Native populations, because 'in their present way of living, almost all their Wants are supplied by the spontaneous Productions of Nature, with the addition of very little labour, if hunting and fishing may indeed be called labour when Game is so plenty'.[16] Consequently in the eighteenth century, while Native people rarely assimilated with the settler population by choice, European settlers frequently defected to Native societies. Hector de Crévecoeur, a French migrant farmer, described the colonists' impulse to join Native communities in 1782:

> there must be in their social bond something singularly captivating, and far superior to anything to be boasted of among us; for thousands of Europeans are Indians, and we have no examples of even one of those Aborigines having from choice become Europeans! . . . Without temples, without priests, without kings, and without laws, they are in many instances superior to us.[17]

Even Euroamerican settlers captured by Natives as hostages frequently decided their captors' community was more attractive than their own, and resisted being rescued. As Franklin attests,

> When an Indian Child has been brought up among us, taught our language and habituated to our Customs, yet if he goes to see his relations and make one Indian Ramble with them, there is no perswading [sic] him ever to return . . . when white persons of either sex have been taken prisoners young by the Indians, and lived a while among them, tho' ransomed by their Friends, and treated with all imaginable tenderness to prevail with them to stay among the English, yet in a Short time they become disgusted with our manner of life . . . and take the first good Opportunity of escaping again into the Woods, from whence there is no reclaiming them.[18]

Given the choice between Native and colonial ways of life, the former society – rich in leisure time, generous and egalitarian in its distribution of resources, abundant in communion with people and wildlife – won out.

Euroamerican success in colonisation was not a natural process of ever-improving health and well-being, it was the success of a conqueror who was randomly lucky, with the great advantage of biowarfare resulting from the Natives' lack of immunity, and deploying cultivated ruthlessness. Even after the great dying from smallpox that wiped out a huge proportion of Native Americans, their survivors remained highly organised and enjoyed sophisticated lifestyles, with a complex and sustainable relationship to the other life on the continent. If smallpox cleared the original path for conquerors from Europe, and more advanced military technology gave European colonists an edge in combat, there was yet one critical cultural difference between Indigenous Americans and Euroamericans that led to the colonists' successful domination. What conditioned minds to embrace this expansionist attitude was a faith in progress and its iteration in this period, what came to be called manifest destiny (an idea on which a whole chapter will focus later). As Indigenous studies scholar Traci Brynne Voyles writes, for settlers to take Native land required a powerful sense of entitlement:

> Remaking Native land as settler home involves the exploitation of environmental resources, to be sure, but it also involves a deeply complex construction of that land as either always already belonging to the settler – his manifest destiny – or as undesirable, unproductive, or unappealing: in short, as wasteland.[19]

This primary difference emboldened colonists to extract biomass without an awareness of its finitude: to cut down and burn forests without sense of limit, to trap fur-game, to shoot birds from the sky, to fish the seas, lakes, and rivers with no regard for their ability – or inability – to regenerate. Not only did they believe the land rightly *theirs* for the taking; it was also, counterintuitively, worthless savage wilderness that could only be improved by destruction. By creating an airtight logic and justification for the greater good – civilising an otherwise savage mankind – the progress narrative empowered settlers to engage in treachery, regularly breaking treaties with Natives, to rape and pillage, to over-exploit and overpopulate the land, to enslave the people already there and bring others enslaved from elsewhere. The logic of these economies – to extract from the people and ecologies without regard to their limits or personhood – is largely, though not entirely, absent from Native societies at this time. The sort of relationship with the land that allowed numerous Native Nations to thrive in unbroken cultural lineages for many millennia likely yielded an important military disadvantage against those unencumbered by such restraint, who could suck in an abundance of resources with only short-term considerations. Given the subsequent decline in liberty, equality, ecology and human well-being, the European-driven depopulation and settlement of the Americas was by virtually every reasonable measure, even the narrowest definition as increasing human health and happiness, the perfect inversion of genuine progress towards human betterment, despite the backwards narrative represented by Jefferson's letter. Indigenous human life – including the countless diverse variations found outside the Americas – had been organised into systems that, in general, were better designed to maximise human well-being and nurture the long-term integrity of life than those of the Europeans who arrived later. The sad tragedy of Euroamerican replacement is that Native Nations were better equipped to enshrine life, liberty and the pursuit of happiness than the republic that took their lands and justified that theft by claiming to usher in those very outcomes.

If these primitive savages among the Native Nations in what became North America were neither primitive nor savage, and in many important ways were much less so than the societies that sought to conquer and replace them, what about the peoples who were their ancestors, and the ancestors of all peoples? Even if Hobbes and Jefferson were – knowingly or otherwise – deceitful about existing Native peoples, could they still have been right about ancient human societies? That we humans lived in a wild, violent, dark infancy that we have only recently grown up from?

The Vital Science of Human Ecology

What we know now is that *Homo sapiens* evolved in southern and eastern Africa at least 300,000 years ago. Populations of humans began spreading outside of Africa at least 60,000 years ago, and possibly as long as 84,000 years ago.[20] As they did so, they may have caused local extinctions of large animals – including other human species like Neanderthals – either by overhunting them or outcompeting them for scarce resources. It is likely that human presence combined with other environmental factors led to these extinctions.[21] They tend to coincide with both human arrival on a continent and environmental instabilities such as climate variations. But, like the Native American Nations mentioned earlier, the peoples who settled on all the habitable continents eventually – after the initial disruption caused by the arrival of an invasive predatory mammal – developed a relationship with the ecological systems they inhabited that maintained high biodiversity, as in the Ohio River Valley. Although it is difficult to know much about the cultures and societies of those peoples who populated the earth tens of thousands of years ago, of that fact we can be absolutely certain, given the vast evidence of biodiversity persisting for so many millennia around the world.

Human beings are like every other animal in that we must extract energy from the environment in order to live. Cell metabolism evolved to transform nutrients and energy into motion and replication. Chlorophyll-based life, like plants, evolved to directly metabolise sunlight into its bodily growth. Animal life evolved to consume and metabolise those plants and each other. Some species of fungi use radiosynthesis, capturing energy from gamma radiation rather than solar radiation. Some extremophile bacteria, meanwhile, use chemosynthesis, extracting chemicals at undersea thermal vents for their energy. Most other organisms extract energy from organisms in the environments in which they live, and incorporate it – 'incorporate' literally, from the Latin *incorporātus*, 'to make into a body'. This process is what I will call 'concrete energy capture' – 'concrete' meaning tangible, physical and material, and 'energy capture' meaning the extraction of energy and mass captured from outside an entity and brought within and turned into energy and mass.[22]

Great diversities of plant and animal life, alongside fungi, bacteria and viruses, evolved to create vast, intricate networks of life covering the world: regionally they are called biomes, and globally the biosphere. From these networks of organisms combined with physical elements like water, air and earth emerge ecological systems. 'Ecology' is the study of these systems, from the Greek *oikos* (οἶκος), or house. This discipline draws from data and theory from many other physical sciences – physics, chemistry, biology, hydrology, geology and more – to understand the dynamics of the world's many networks of mass, energy and biomass, the living matter that makes up organisms.

Human beings, or in Carl Linnaeus's Latin-based binomial nomenclature, *Homo sapiens*, meaning 'wise man', add an additional complexity to the study of ecological systems, since this species has become a powerful force in every ecological system in the planet's biosphere. *Human* ecology, then, is a discipline, or a confluence of disciplines, that examines the ecological relationships that human beings develop with each other and other species. Given that there is a whole array of disciplines focused on the mechanics of human beings – from their minds to their immediate social groups in psychology to their broader cultures, societies and geopolitics in the social sciences – this megadiscipline brings social sciences into engagement with the physical sciences involved in the study of ecological systems. *Homo sapiens* is an animal that, like any other, lives within an ecological system, and their living – which involves concrete energy capture, the production of goods and waste, reproduction, and lots of other activities in between – creates a certain kind of relationship with the other mass, energy and biomass existing in the same ecological systems.

Ecology identifies certain forms of relationship between organisms that capture energy from one another. Three of the most important kinds of relationships are mutualism, commensalism and parasitism. In mutualistic relationships, two organisms benefit one another. The classic example of mutualism is the clownfish and sea anemone. The fish provides nutrients and clears parasites off the anemone, which in turn provides protection from predators. Humans have had mutualistic relationships with many species. One of the more well-documented is the Hadza people of Tanzania who work with honeyguides, a type of bird. The bird's calls signal to the humans that they've found beehives. The humans respond with their own call, one they have developed over generations, which the bird recognises. The bird then leads the humans to the hives. The birds can't break into a hive while the bees are protecting it, so they call the humans, who use smoke to tranquillise the bees and break into the hive for the honey. The birds then eat the remaining beeswax and larvae.

In parasitism, one organism captures energy from another to the other's detriment. There are many examples of this, like vampire bats and ticks sucking blood from hosts and weakening them, or internal worms feeding on nutrients within the body of another organism, sometimes even leading to the death of the host.

In commensalism, one organism captures energy from another without doing either harm or good to the other. Many species – like birds, squirrels, ants and flowers – have a commensalistic relationship with trees, making their home on the tree without harming it. Barnacles on whales or harmless bacteria on human skin are other examples of commensalism.

This framework can be usefully adapted for understanding phenomena on a larger scale, beyond the interaction of two species or organisms. Rather than an organism-to-organism scale, we can apply this conceptual framework to an organism-to-ecological-system relationship. An organism – or an aggregate of organisms, like a society or hive, of a given species – can have one of these three sorts of relationships to the biome it inhabits.

We could say that human systems – economies, societies, cultures – form similar sorts of relationships with the ecological systems they inhabit. The human societies that migrated out of Africa and spread across the world eventually developed relationships with the local ecological systems that were commensalistic or mutualistic: they used naturally regenerative energy capture, meaning they extracted energy – in the form of inorganic mass and biomass – from ecological systems without doing much harm to the ecological system, or at least not catastrophic harm.

Not all such commensalistic or mutualistic societies have been based on foraging wildlife for energy capture. Research by British archaeologist David Wengrow suggests that many of the epochal breaks or stages of development upon which numerous macrohistorical narratives depend are less profound than is sometimes suggested.[23] The 'agricultural revolution', for example, has been cast as a pivotal moment in human history so regularly and for so long that it is taken as conventional wisdom. The very nature of agriculture, the story goes, necessitates dense, complex social groups that yield hierarchical political economies like monarchies and empires.[24] But some form of agriculture has been practised more widely for far longer than the 10,000-year date often used as shorthand for the agricultural revolution, and did not yield hierarchy even when cities arose. By contrast, some foraging communities have developed hierarchical societies and resource inequality.[25] As in the case of the Miami and other Ohio River Valley tribes, sporadic agrarianism was compatible with high biodiversity, and thus commensalistic.

Why and how did humans develop commensalistic – or in some cases, even mutualistic – relationships with their ecological systems? As to *why*, there were probably multiple reasons. For one thing, it is rational to not destroy the ecological basis on which your health and well-being depend. After eliminating large competitors, it may be that some human groups decided it was simply inefficient to continue destroying smaller or smarter or more elusive competitors (like small primates, wolves or birds). Another likely reason is that many animals, including humans, may have a natural instinct towards biophilia, the love of living things, and genuinely care about biological health and diversity. Most readers who have been lucky enough to experience it will recognise this from the personal feelings of awe and wonder experienced when seeing wildlife and wilderness. Conversely, those without will experience suffering from 'nature deficit disorder', a genuine condition recognised by medical researchers and psychologists, whereby nature's absence causes reduced well-being, lower mood and measurable negative health impacts.[26] More evidence can be found in the widespread development of animism, a belief system that renders other life and ecological elements like rivers, mountains and lakes as divine entities worthy of worship. Derived from the Latin *anima*, literally 'life, soul, spirit', animism applies to a large category of beliefs in philosophy and spirituality that have persisted across thousands of cultures and geographies throughout human history. Most societies for most of human history have likely been animistic, if the oral traditions of the longest-lived societies are any indication. The ubiquity of animism throughout human cultures and history also starts to answer the question of *how* humans developed sustainable, commensalistic relations with their ecological homes.

Animism and a Theory of Everything

Taking another example from North America, the Abenaki are a First Nation people who live in Ndakinna, now commonly known as northeastern North America. They've lived in this region for thousands, possibly tens of thousands of years.[27] They have lived among bears and lynx, mountain lions and wolverines, eagles, owls and corvids, foxes and wolves, divers and racers. They have lived among oak and beech, pine and spruce, maple and fir. They have lived on rivers of snowmelt and ledges of pink granite, beside the realms of other peoples and nations. Historically, if you grew up Abenaki, the belief system in which you were raised imbued all of these things – the animals, the trees, the granite – with a form of personhood. In place of a mass of objects, the world was studded with countless subjects; instead of only one species – *Homo sapiens* – containing a soul, all these creatures and things would have been *ensouled*, endowed with a living spirit.

In animistic worldviews, other ensouled beings are little different from human persons; humans are of the same order of life as the trees and animals in the biomes they inhabit. An animistic world is vibrant and full of life, the forest a bustling city of creatures with unique and mysterious minds, a universe populated by millions of interiorities. The Eswazi anthropologist Jason Hickel illustrates this condition well with his description of the Achuar people of South America, who live 'in small circular clearings in the middle of jungle, with dense walls of trees rising up all around them . . . dark, brooding, pulsing with the noises of frogs, toucans, snakes, monkeys, jaguars, millions upon millions of insects, plus a universe of mosses and mushrooms and curling, roping vines'. One might be inclined to imagine that the Achuar feel isolated, cut off from other human communities. 'But the Achuar see the jungle quite differently. They see people all around.'[28] Hickel also points to Austronesian Indigenous groups, who made deliberate decisions to stop over-exploiting the places they newly settled, often being the first *Homo sapiens* ever to set foot on those many Pacific islands. Although Austronesian peoples started by entering new islands between two and four thousand years ago and decimating native species there as they went, they soon changed tactics:

> They learned that to build a thriving society within a bounded island ecosystem requires a completely different approach to ecology. They had to swap the ideology of expansion for an ideology of integration. They had to learn to pay attention to other species – learn their habits, their languages, and their relationships with others. They had to learn how much they could safely take from any given community, and how to give back in order to ensure its continuation. They had to learn not only to protect but to *enrich* the island ecosystems on which they depended. . . . Societies that took these steps ended up thriving in the Pacific islands.[29]

In Australia, dozens of cultures coexisted for 50,000 years or more on the continent, developed agrarian and foraging economies, built dams and wells, preserved surplus yields in permanent structures, and, like most Indigenous groups, manipulated the landscapes they inhabited – all while maintaining relatively high levels of complex biodiversity.[30] In other words, they shaped and extracted from their environment but did not destroy it.

Lisa Brooks, an Abenaki scholar, writes about how many different Native creation stories have long coexisted, in stark contrast to the homogenising belief systems later brought by the missionaries of organised religions with an aim to convert: 'There are hundreds of indigenous nations on this continent, and hundreds of creation stories. . . . All are considered equally valid, and have been relayed between nations in both formal and informal contexts for millennia.'[31] Native creation myths depict a particular place and ecology. Rather than a sterile, impersonal history, Brooks writes, these narratives constitute 'the map of the land, the map of how to be human in a particular place, the key to survival in a specific environment'. Animist creation narratives are imbued with practical knowledge, accumulated over countless generations, often containing information about seasonal food supplies and reliable water sources, or cautioning against driving species to extinction. Each myth is inflected with local and regional particularities to reflect the unique ecosystems in which they arose.[32] It is not one myth to rule them all, but a myth system carefully crafted for every particular culture, history, ecology and locality.

Most of the thousands of Indigenous cultures that have existed, and particularly those that have survived millennia of disasters, strife and daily toils, have been animistic in these similar ways. It makes good evolutionary sense. If your means of livelihood come largely from wild sources and the whims of the weather, from the health of the soil and water in the realm you inhabit, then a belief system – and the society that lives by it – which respects and understands those sources will last far longer than one that doesn't. Belief systems that stay the hand of the overzealous hunter and provide rules for how to maintain human systems in mutual reciprocity with natural systems have proved far longer-lived and more resilient to environmental changes than ones that don't. Potawatomi botanist Robin Wall Kimmerer describes the legendary monster of the Anishinaabe people called the Windigo, a grotesquely oversized human being with yellow fangs and a heart of ice, who brings ravenous hunger in the midst of winter famines. In the midst of such harsh conditions, a person may become possessed by the Windigo and resort to hoarding food and cannibalism to survive while their community fellows perish. The cost of accepting the Windigo's unholy survival tactics is banishment from the spirit world, cast out to roam in isolation for eternity and suffer insatiable hunger. These beliefs serve to enforce a value system that puts the survival of the community in hard times before the greed of any individual. As Kimmerer notes, 'It is a terrible punishment to be banished from the web of reciprocity, with no one to share with you and no one for you to care for.'[33]

Animistic beliefs align with *Homo sapiens*' highly prosocial nature. As such, animistic beliefs perform several important functions: they help to maintain a healthy relationship between human systems and natural systems; they help to make humans effective stewards of the land they inhabit, and thus to make their societies more enduring; and they help cultivate mutual respect between the members of a given society.

Such beliefs have long been dismissed as primitive or unscientific by the societies that have invaded, displaced and sought to eradicate them. But many modern Euroamerican fields of inquiry – from archaeology to biology to zoology – are beginning to vindicate long-standing Indigenous intellectual systems. One example of this has been noted in the journal *Nature*: 'DNA analysis of the Australian outback's only palm tree, *Livistona mariae*, indicates that it originated from seeds brought from the north of the country – a finding backed up by a recently unearthed Aboriginal myth.'[34] The millennia-old oral teaching that recorded the migration of this tree species across the continent was simply accurate natural history. Cultures have cultivated advanced navigation techniques, including celestial and marine navigation. Polynesian explorers traversed the huge expanses of the Pacific in canoe-like vessels thousands of years before European explorers crossed the Atlantic (which is just over half the size of the Pacific). In what is today southern California, the hundreds of Native tribes who have inhabited the area for generations, like the Chumash, Hupa and Serrano, managed the fire ecologies of the region for many thousands of years, conducting controlled burns and adapting to wild ones. When settlers arrived in the area and displaced or annexed tribal lands in the nineteenth century, they stopped the controlled burns, which led to bigger and more devastating fires, significantly harming both the human and animal populations in the area. Only recently has fire management begun to catch up with Indigenous science and consult the remaining Nations in the region.[35]

Indigenous peoples' knowledge of their ecologies tends to be deep and broad, spanning the topography, hydrology, biology, weather and physical contours of the continent. Members of these cultures have grown up learning a vast, complex curriculum of ecological knowledge, from edible plant and animal species, to tracking, to sophisticated tool making, on top of all the social dynamics of living in and among shifting societies.[36]

It makes sense that animistic belief systems would reflect the reality of the world more accurately than recent intellectual inventions: they've been cultivated and refined over tens of thousands of years. But animistic worldviews are no longer the norm, they're held by only small and shrinking proportions of the world's population. The dominant belief systems today – even those, like Judaism or Hinduism, that we consider ancient – are newborns, at most a few thousand years old. Instead of reflecting the *reality* of the world, as animistic beliefs do, the divine in modern religions is more suited to reflect an imagined world: invisible and relegated to some other dimension. Far from primitive in the production of knowledge, commensalistic cultures have to develop sophisticated knowledge systems for navigating and shaping an infinitely complex natural world. Indigenous cultures that practise commensalism have developed a science of existence, an all-encompassing knowledge system. They exhibit an understanding of everything from human behaviour to anatomy and from epidemiology to species dynamics in the ecological systems they inhabit, from seasonal changes to appropriate ways of moving through the world.

Joy and the Meaning of Nature

One foundational presumption that most progress narratives rest upon is the notion that 'nature' or 'wilderness', that zone of ecological relations that is believed to exist outside of some imaginary human boundary – marking 'society' or 'urban' or another abstraction meant to delineate human from natural – is full of suffering. Progress, this narrative states, occurs to the extent that humans separate themselves from the harshness of nature: the more effectively human structures can insulate humans from the vicissitudes and violence of the weather or other animals, the greater progress has occurred. Perhaps the most famous iteration of this belief, or at least the most enduring in the Western canon, is that of Thomas Hobbes. His description of life in nature as 'nasty, brutish, and short' has the benefit of sticking like a catchy tune in the mind, and was clearly used for the political motive of justifying the rule of a centralised state.[37] The idea remains persistent today. An animal ethicist recently drew criticism for proclaiming that animals who live in the wilderness must be suffering because nature is brutal. Humans, he argues, should therefore intervene in wildlife and create 'predator-free' zones, rendering domestic animals tranquil so as to decrease suffering in the world.[38] But such notions are as incorrect today as they were in 1651, the year Hobbes's *Leviathan* was published.

Animals in general, and animals in the wild in particular, experience a range of emotions that sciences in the Western tradition have been either innocently ignorant of, or have cynically ignored – and in some cases probably both – for centuries. Traditional ecological knowledge, the kind of science that has been practised for many millennia longer in cultures around the world, has known for most of its existence that animals experience many emotions. That animals experience a complex range of thoughts and feelings is self-evident to those who live with domesticated animals and can readily see them expressed every day. Whether dogs, cats, horses, cows or chickens, it is very clear that they enjoy a rich cognitive and emotional experience of the world. In captivity, they display feelings like joy, pleasure, fear, pain, confusion, relief, boredom, excitement, contentment, focus, suspicion, love, anger, disgust, loyalty, protectiveness, determination, and more. Such feelings are no less common in the wild, and may be even more abundant and varied. It is only clearer with domesticated species because those of us who live in highly regimented societies that have banished most wildlife have far more opportunities to observe them. If we lived as closely with wildlife as many other cultures have done throughout history, we would see just as readily that wild animals too possess a rich cognitive and emotional experience of the world.

Western ecological and biological sciences today are beginning to record this fact. Most large mammals, for instance, have a similar neuroanatomy to humans, with many of the same brain regions performing the same functions. They also engage in complex abstract thought and experience a complex emotional inner life. Humpback whales, for instance, have been observed saving seals from predation by orcas for no apparent reason other than altruism. Sperm whales are thought to share information collectively; at the height of nineteenth-century whaling, pods under attack taught one another to swim upwind to escape humans' wind-powered ships and their harpoons.[39] For some species of mammals, tool use is common: the extremely endangered orangutans of Indonesia not only use tools to capture food, like dipping sticks for honey, but have also been observed hanging up their favourite tools to use again in future.[40] When their habitats are destroyed and groups become separated or diminished — which is happening at an extremely rapid rate thanks to the activities of global commodity corporations — this knowledge, too, is lost.[41] They didn't have a spear-wielding instinct leading them to unthinkingly sharpen sticks, any more than humans do. They observe, experiment, think through problems, and invent. Virtually every order of creature displays feats of cognition that are surprising researchers. A recent study of African grey parrots found the birds spontaneously giving up currency tokens — which they had learned to trade for much-coveted walnuts — to other parrots, selflessly helping their friends without any apparent assumption of reciprocity; they have also, perhaps regrettably, been taught to use video calls on smartphones and tablets, and have been seen independently calling each other and maintaining connections over months.[42] Magpies have been observed helping each other to remove tracking tags applied by researchers, sometimes within minutes of being released.[43] Creatures with very different neuroanatomies from mammals also show high intelligence: cephalopods like octopuses can figure out complicated puzzles and show evidence of abstract thought, even recognising individual humans (and holding grudges against those that annoy them). Given that they dislike excessive brightness, captive octopuses have been known to squirt water at lightbulbs to short-circuit them.[44]

Frogs show play behaviours, ants orchestrate mock battles, and honey-bees can do basic maths.[45]

It is true that some wild animals suffer. To be a living creature capable of positive emotions – joy, pleasure, love – also entails suffering. It is not however necessarily true, contrary to the opinion of the ethicist mentioned earlier, that wild animals suffer more than domesticated ones. Given how meat corporations regularly treat 'livestock', it's likely that wild animals suffer a good deal less than most domestic ones.[46]

What is joy? This is a question central to concerns around progress, for joy is a critical element of human well-being, and if progress means increasing human well-being, then attitudes towards joy, how it is nurtured and sustained, become of central importance. It is easy even – especially? – for some thinkers like Hobbes to develop myopia in answering this question, gathering only the data found in their own personal experience of joy in order to understand it in others.[47] But even a rudimentary reflection on happiness and suffering will reveal they are far more complex than the oversimplifications that many deploy: to increase 'positive' emotions or sensations and decrease 'negative' ones, as if tipping the balance of an equation. Joy and suffering are themselves deceptive words, containing within them myriads of texture. One may find the greatest happiness in the midst of pain and peril, or find deep sorrow in a palace of luxuries. Harmony between inner nature and outer circumstance is one reliable means of ensuring greater contentment and joy. A fish that can swim, a bird that can soar, a cat that can hunt. Even through toil, struggle and suffering, it is this congruity between inner and outer natures that can reliably yield contentment and joy. Balance in the diversities of life is another: a world of many minds is wondrous.

Given that many animals do similar things to those that humans do, and for similar reasons, and while we cannot take a look inside their brains and read their neurons for definitive proof of their emotions any more than we can with humans (and, to be sure, we *can't* with humans either), there is good reason to believe that, if they have similar brain structures that evolved in the same ways for the same reasons as human brains, many of them have the same or similar thoughts and emotions for the same reasons: affection towards both their kin and larger social groups, curiosity at new phenomena, excitement and exhilaration, and sometimes fear, boredom, ennui, anxiety and grief. They are kin, even as they are fundamentally different and mysterious.

One of the foundational ideas that narratives of progress are often built on is the idea that as humans have 'escaped', as they put it, from wilderness by building more elaborate infrastructures that banish plants and animals that are not immediately useful; that they have thereby grown increasingly happy, healthy and safe, and that their suffering has decreased. It is a version of the animal ethicist's argument, applied to humans. The more domesticated we have become, the more tranquil our lives have become. But there isn't really any more evidence for this assumption than there is for the argument applied to wild animals. In fact, as with domesticated animals, and as we shall see in more detail later, it's likely that the reverse is true: that we have become more miserable the more we have insulated and distanced ourselves from the wild. Indeed, a recent study bore this out, finding that small-scale, low-income societies closer to nature report the highest life satisfaction in global opinion polling among all other societies.[48] To understand why this is the case, and to carry on with our story, we need to establish a theoretical framework that can clarify some of these complex phenomena.

The Science of Hierarchy

So far, we have primarily looked at forms of *concrete energy capture* – through foraging and farming – that humans perform within the ecological systems they inhabit, and how beliefs like animism have shaped, and been shaped by, those strategies of energy capture. But there's another kind of energy capture that is unique to social species like human beings, what I term 'abstract energy capture'. With *concrete* energy capture, an organism or system extracts physically existing material directly from another organism or system: a fish eats algae, a company of humans catches fish. By contrast, *abstract energy capture* is a process whereby organisms extract energy from other organisms not by physically taking it and chemically incorporating it to sustain their bodies, but instead by controlling or directly benefiting from the secondary energy output of the organism being extracted from. What I mean by 'secondary' energy is the energy that is expended in the form of the extracted organism's time, labour or attention. For example, a human child depends on the time, labour and attention of carers. A fetus and a nursing infant may extract *concrete* energy from a mother's body directly, but, once weaned, they are dependent on *abstract* energy capture from the family or community in which they are raised – as well, of course, as the concrete energy capture that the family or community engages in. This kind of abstract energy capture is natural and inevitable, particularly given that humans are born underdeveloped and incapable of surviving independently of a direct carer for several years. Other, less natural and less desirable forms of abstract energy capture will be explored later.

Like the relationship between society and ecological system outlined earlier – concrete energy captured through mutualism, commensalism or parasitism – we can use the same framework in understanding the relationships between individuals and groups within a given society. Members of the aforementioned societies that developed commensalistic or mutualistic relationships within their ecological systems, also tended to develop mutualistic or commensalistic relationships within those societies. When a society has a primarily commensalistic relationship with its ecological system, that relationship tends to be mirrored within the society, in the relationships between its members. This is not a hard rule, there are exceptions, but it does seem to be a consistent tendency. This means that members of commensalistic groups tend to capture *abstract* energy from one another in a way that is mutualistic, variably reciprocal and linear, and primarily non-harmful. For example, one person may be particularly good at tending and nurturing children, another at hunting or protecting the group or scouting out new opportunities, another at weaving textiles or stories or songs, another at cooking up new recipes, another at delegating tasks. Each person may contribute more or less, one may depend on the labour of others, and others may depend on the labour of one, but no individual depends disproportionately on many others unless great need demands it: infancy, old age, infirmity or disability, in which case the rest of the group takes care of that individual to the extent possible.[49]

Though such societies may not have always been perfect utopias, the disparities they exhibited between classes, genders or other differences in personal autonomy, as detailed in the earlier examples, were likely much smaller than in the sorts of societies we will examine later. Gender and wealth inequalities in many Native American cultures, for example, were minimal and complicatedly interconnected. As the late American anthropologist David Graeber notes, the Iroquois nations' 'main economic institutions . . . were longhouses where most goods were stockpiled and then allocated by women's councils'.[50] Laguna Pueblo scholar Paula Gunn Allen argues that the central roles occupied by women in many Native American cultures – including as political leaders and heads of the family – were downplayed by European colonists, who were primed to interpret their encounters through the misleading patriarchal lens of their own cultures. She writes that within these woman-led societies, goods tended to be distributed among people evenly, and corporal punishment was absent as a means of social control.[51] Recent evidence suggests that in many cultures, even gendered labour divisions were rarely rigid, with between 30 and 50 per cent of women participating in hunting activities.[52] According to a study in *Science*, not only was sex egalitarianism high in many prehistoric societies, but such an arrangement 'may have been a survival advantage and played an important role in shaping human society and evolution'.[53] This is a brief example of a commensalistic or mutualistic sort of abstract energy capture, but we will see more varied kinds as we go along.

For most of human history, energy capture – whether concrete or abstract – was predominantly non-harmful. It is very likely that commensalistic societies dominated the world for a vast majority of the sixty or seventy thousand years that human beings have lived on nearly every continent on Earth, in a colourful multitude of forms.

This relationship between nature and culture was relatively stable, if wildly diverse, for all that time. The narrative formula of progress was unnecessary and likely not a major part of any of these cultures. The many cosmogonies and stories that made up animistic belief systems reflected the cycles of nature. The eternity of paradise, the accumulation of humanity as a dominant force in the world and the binary of less or more 'developed' simply did not make sense for most such cultures.

But around 5,000 years ago, everything changed.

I Heaven

Overview

Date: 3000 BCE to 1400 CE
Progress: theological and mythical
Parasitism: contiguous and regional
Agents: city-states, kingdoms, empires

Context

The story of progress really begins in Mesopotamia, in the basin of the Tigris and Euphrates rivers, a realm fringed on its east by the Zagros mountain range, marking the border between present-day Iran and Iraq, and on its west by the Mediterranean. For uncounted millennia, the region that we think of now as generally arid was covered in trees and fertile soil. It was primarily because of this lush tranquillity that Mesopotamia was home to the first major, systematic agricultural societies. Farms, towns and cultures coexisted in the region for nearly 7,000 years after the first evidence of settled farming emerges there before the tale of progress begins. It may well have looked a bit like the Ohio River Valley, with farmers growing vegetables and cereals in verdant rows, shepherds leading small flocks into mountain meadows, well-dressed citizens plucking fish and shellfish from splashing rivers, and towns with ordered dwellings, temples and communal halls, all verged by – instead of American sycamores – hulking cedar forests.[1] The rise of settled, systematic agriculture – the domestication of plants and animals edible and malleable to humans – between 8,000 and 10,000 BCE, and of metallurgy – the shaping of metal into tools – a millennium later, are often presented as among the most important shifts in human history.[2] While they are both undoubtedly important, when we are considering humanity's place in the biosphere and our relationships to one another, a different change eclipses them.

Around 3,000 BCE, the world shifted. With relative rapidity, these Mesopotamian cities began to coalesce into the world's first empires and market economies. The Early Dynastic Period covers nearly six hundred years in which city-states throughout Mesopotamia began to expand. British archaeologist David Wengrow considers the expansion of Uruk to be the first historical instance of imperial colonial activity, 'in which entire communities budded off from an urban metropolis, re-establishing themselves in distant locations, yet maintaining distinct identities and regular trade relations that linked to their cities of origin'.[3] Uruk and the trading activity that surrounded it also represents the first known practice of labour-intensive, value-added goods produced by a centralised group of regimented workers weaving and sewing in a proto-factory setting, forming a thriving textile industry. Wengrow describes one of the earliest records of profit-generating industry, occurring through a supply-chain originating in southern Mesopotamia and moving north into Assyria and as far as Anatolia:

> Assyrian (north Mesopotamian) traders were thus able to generate substantial profits by using relatively small amounts of Anatolian silver to purchase bulk quantities of tin and textiles from production centres further south. These commodities were then traded back over the Taurus mountains into central Turkey, where they were in high demand, in return for large shipments of silver (acquired at a highly favourable rate of exchange), which could be reinvested in the purchase of new merchandise to support further commercial ventures of the same kind.[4]

This rise of centralised labour, serving a large-scale market, administered by consolidated monarchies and oligarchies with expansionist aims, all fuelled by intensive environmental harvesting, was a vitally important break from the past. For hundreds of thousands of years prior to around 3,000 BCE, most societies in most parts of the world developed commensalistic relationships with the ecological systems they inhabited. They either extracted according to the carrying capacity of their area, within the bounds of naturally occurring energy, or they migrated. They were not necessarily Edenic utopias – though some may fit that description – and many may have even caused the extinctions of major competitor species, like other large predators and grazers. But these commensalistic societies extracted energy from their ecological zones without over-extracting to the point of biodiversity collapse, and they cultivated relationships with one another that were variably mutualistic, with abstract energy being shared reciprocally.

Those examples of societies we've already looked at, the animistic egalitarians in North America and elsewhere, represented methods of energy capture that are very ancient to human beings. The new types of societies that arrived with the first empires captured energy with very different strategies. We could call them 'parasitic'. In ecological relationships, a parasite is an organism that extracts nutrients and energy from a host to the detriment of the host, like the leech, cuckoo or intestinal worm. These sorts of societies were extracting and expanding rapidly, to the detriment both of the ecological diversity in their region and certain members of these societies. Specifically, parasitic societies harmed the fertile soils by overproduction, they harmed the once lavish forests and the species that called them home by cutting down the trees, and they harmed the classes of workers, slaves and soldiers deployed to toil and make war and die on behalf of small groups of elites, new semi-divine rulers at the apex of their economies. Instead of operating within local carrying capacities, they expanded beyond the naturally occurring energetic limits of a given ecological system.

When the limits to their extraction of resources are exceeded, the parasitic system must either suffer a crash or must invade and take the energy of a more distant ecology or society. The boom-and-bust process of parasitic economies is evident whether or not they are administered by imperial bureaucracies. Suddenly, cultures that persisted for tens of thousands of years became or were replaced by others lasting only decades or centuries before their eventual collapse. It was not only ecological instability, but also political and economic instability that accompanied them.

This terminology is not necessarily meant to place a value judgment on such societies; 'parasite' can be a loaded term, after all, weighted with negative connotations, for good reason. But in this case, 'parasitism' is simply the most appropriate and clarifying terminology to describe the relationship between these sorts of human systems and the ecological systems they inhabit, and (as we will see) the relationships between groups within those societies. It's also a good time to point out that this sort of society does not represent all of 'humanity' any more than the use of guns or cars represents all of humanity. Parasitism represents a very particular kind of system and subject. It was a divergence from most of human history and most human cultures. It is a great anomaly.

If a society is primarily parasitic in its relation to the ecological system it inhabits, there is a good chance the relations within that society will also be parasitic. In this case, a few individuals extract disproportionate amounts of abstract energy from other members with little expectation of reciprocity while doing harm to those extracted from. In the more populous versions of parasitic societies, a diagram of their internal and external relations looks something like Figure 4.

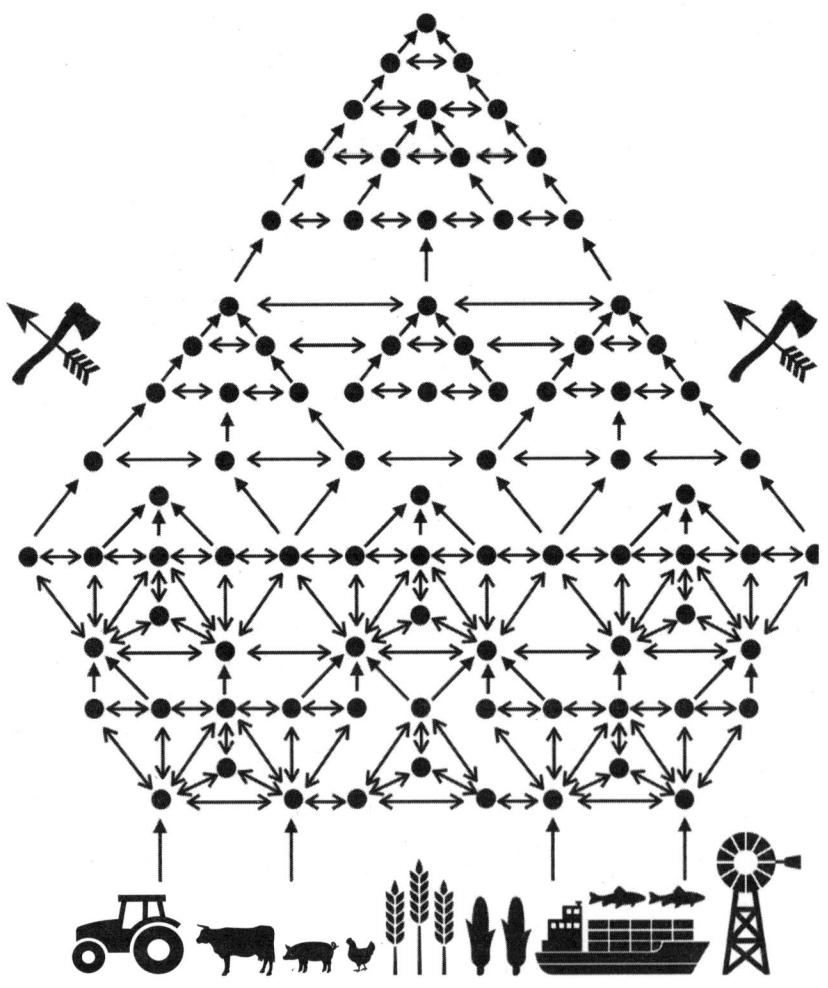

Figure 4 Complex parasitic society

In this diagram, concrete energy capture occurs through a variety of modes of production: foraging, systematic agriculture and synthetic augmentation. Abstract energy capture happens through layers of dynamically interrelated hierarchies and networks.

What this means is that societies began to extract more from their ecological habitats than those systems could support over the course of a couple of generations. They did this through practices like intensive, large-scale agriculture, irrigation and grazing. This required those societies to replenish their stock of biomass and energy by, instead of moving out of the area, expanding the zone from which they captured biomass and energy. In order to expand that zone, these societies required centralised and hierarchical social and political organisation to coordinate actions like war and the conquest of new frontier territory. Because they were extracting from new areas where people were often already living – peoples who had themselves, in many cases, cultivated richly productive lands – they needed to be able to coordinate militaries that could seize such land. We could call this particular form of parasitism 'contiguous parasitism'. What 'contiguous' means in this context is that this expansion tended to occur like a bull's-eye rash spreading out with a city-state at the centre; territorial boundaries expanded overland.

One example that we will come back to later but that is illustrative of this principle is Rome. Starting as a city-state in 753 BCE, Rome gradually expanded over the following centuries, reaching its height of land coverage nearly 1,000 years after its founding. The city-state practised intensive agriculture and centralised into a series of kingdoms lasting more than two hundred years before citizens overthrew their monarchy and established the Republic. To continue their expansion, they developed regimented militaries and won a series of battles, taking captive slaves and maintaining a large agrarian peasantry. After 500 years of Republican rule, Rome became formally imperial and fully conquered the Italian peninsula by 7 BCE. The imperial period intensified their expansion, setting up extensive contiguous colonies throughout Europe and maritime routes to North Africa, extracting resources from these areas through the taking of taxes and captives by force, diplomacy and coercion.

We could see this history through flows of energy. The city's initial expansion occurs with a centralised monarchy; as it grows and brings more resources into its economic metabolism, those resources fund a larger ruling class of oligarchs – senators, generals, merchants and so forth; eventually, energy becomes too dispersed in private hands. It needs to be concentrated again – in the hands of a new monarchy – in order to fund public works, militaries, and more expansion. After reaching its height in the second century AD, Rome began its inevitable collapse and the gravitational agglomeration of energy dissipated; it was broken into two empires, Western and Eastern, and former colonies, and finally into a politically fragmented Europe and Italian peninsula. (The latter only briefly, before energy began to consolidate in monarchies and oligarchies throughout Europe, but we're getting ahead of ourselves.)

These forms of energy capture, both concrete and abstract, were often violent. They were contrary to the instinctual tendencies of most humans. Human beings, like some other social predators, birds and mammals, tend to have certain in-built intuitions about how their groups should work. Wolves, for instance, show a distinct aversion to inequity and unfairness; sadly, there is evidence that dogs are losing this instinct due to their cohabitation with humans.[5] Those social intuitions include the value of merit-based fairness in the distribution of resources and decision-making authority, cooperation with those in one's in-group, the right to exercise personal agency, and some kind of biophilia or altruism, the care for life outside the in-group and even one's own species.[6] The dark side of these traits is the human capacity to band together to do great harm, to feel hatred towards members of out-groups, and to use their effective capacity for extracting energy without limits. The concrete energy capture that emerges with parasitism requires a violation of biophilia, by creating huge tracts of agricultural land that evict other life, and the mass, mechanistic slaughter of both domesticated livestock and wildlife. Such societies frequently engaged in the eradication and extirpation of competitive predators and large mammals and birds in order to build these mass farming operations. Rome was particularly effective in this, denuding large swaths of Europe and Africa. The abstract energy capture that accompanies parasitism, meanwhile, violates or manipulates intuitions around cooperation, fairness and freedom: suddenly, humans have to engage in behaviours that are directly contrary to those intuitions, or that depend on mangled interpretations of them. Parasitic societies begin, for instance, to engage in large-scale invasion and organised warfare; they engage in the enslavement of other humans, rent-seeking and debt-holding, and theft; they engage in forcing people into hard labour or battle and selling people into bondage; and they engage in genocide, the wholesale, large-scale eradication of entire peoples and cultures.

A schism arises in the relationship between the parasitic society and the external resources that it aims to extract, whether those resources are ecological, like forests and fish, or human, like another society's stored wealth or human bodies. We could see the division in energy capture *within* these societies as taking on a similar binary. That is, between on one hand a majority consisting of workers and soldiers who extract from outside and bring resources into the societal metabolism; and on the other, a smaller group, an authority, which consists of administrators who organise, direct and extract from the majority. The parasitism of extractors from natural and Indigenous ecologies was mirrored in the parasitism of the categories of labour that developed within these new sorts of societies.

Another way to understand this simple class division is with host-parasite terminology. With individual organisms, the parasite feeds off – extracts from – the host. In a society, the host class is made up of those from whom a parasitic class extracts energy and other resources. But, in many cases, the host class themselves have a parasitic relationship with external sources of biomass, energy and (often forced) labour – the aforementioned ecological systems and the Indigenous cultures that had been cultivating them for millennia. But these extractors who are parasites facing outward, become 'hosts' in the relationship to those who are tasked with *ordering* the processes of extraction, production, consumption and excretion. The latter are, in this relationship, the parasitic class.

We will see many examples of the parasitic class in the following chapters, and their favourite bedtime story, the progress narrative formula, in which they are often the heroes. If the host class is made up of the majority of a population, they are the body on which the parasitic class feasts. Though not in the literal way of consuming the biomass of this class – except in rare circumstances – but by feasting on the souls of the host, in whatever colloquial way we might understand a 'soul': they harvest the living time of this class, consuming the output of their minds, their attention, their internal peace, their creativity, their hopes and dreams, their imaginations.

During this period of contiguous extraction, the parasitic class is made up of hierarchies of managers, landed gentry, military leaders, spiritual advisors and monarchs, like kings, queens or emperors; they use a variety of social tools to extract energy from the host class. Some of these tools include debt, conscription, enslavement, taxation, market profit and wage slavery.

It is important to emphasise what a fundamental change this represented in the basic ways human beings related to each other and to the rest of existence, the other animals, and the plants and inanimate materials that make up human experience. The notion of 'profit', that you take out more than what you put in, encodes expansion in one's very value system and in the incentive structure of institutions. This taking more than you give creates an imbalance whose effect can be seen in every other realm: in taking more from other people than you give in return, in taking more energy and biomass from ecological systems than you return to those systems. This is a critical imbalance that represents an affront to nature.

The priorities of a member of the host class would likely consist of familiar goals: cultivate a healthy family, spend leisure time with friends, work in meaningful toil, and feel yourself an important part of a tight-knit, safe community. These often conflict with the demands of the parasitic class, whose priorities – other than making high returns on their investments (or profit, taking out more than they put in) – are to compel members of the city-state into the maximum amount of work, conscription into the military, and harmonious cooperation with each other and obedience to the authority in peacetime. What sorts of ideas would be useful if host and parasite, majority and authority, are to be united in a mutual, collective endeavour, which may consist of collective actions like conquest, deforestation, economic profit or public works?

Smoothly maintaining the social organisation that effectively engages in parasitic energy capture requires either subverting, overruling or manipulating those in-built pro-social intuitions. A sort of mass, collective delusion must occur if the parasitic and host classes are to collude in maintaining successful surplus energy extraction. Narrative formulae of progress have been the most consistent, potent, adaptable and enduring tools for achieving this mass delusion. The first progress narratives at the start of this current line of parasitic human ecologies also emerge in Mesopotamia, along with the first economies in that lineage. This section will look at myths in Uruk and Babylon, some of the world's first empires, before moving to the birth of monotheism in Persia and then Judea, to the polytheistic empires of Rome and Greece, to the kings and queens of Christendom, to the Norse raiders, and finally to the Islamic Golden Age. Though these myths do not always provide fully formed narratives, they are the bricks and mortar that set the foundation for more potent progress narratives to follow.

Myths of Mesopotamia

The Epic of Gilgamesh is the world's earliest known literary endeavour. It also may be the first piece of propaganda used to justify a new parasitic relationship to ecological systems. It is probably the first recorded iteration of the progress narrative formula. The tablets that bear the original myth are incomplete, with dozens of lines missing. But from the remaining text we can interpret some important information that tells us much about the beliefs and priorities of the people of the region. The tale begins with a lengthy hagiography of Gilgamesh, a demigod and king of Uruk.[7] He is described as 'awesome to perfection' and as dominating the natural world, a 'raging floodwall who destroys even walls of stone', digging wells, opening mountain passes and traversing the ocean 'seeking life'. The story then depicts Gilgamesh converting a wild man to the sedentary and 'civilized' life in his own walled city. The man is described as 'a primitive, a savage fellow from the depths of the wilderness'. Gilgamesh converts him with the help of his 'voluptuous' woman servant, Shamhat, whom he orders to have sex with the savage. The man, named Enkidu, is further seduced by the bread and beer Shamhat has brought, of which he drinks seven jugs and 'became expansive and sang with joy!' Whereupon he 'splashed his shaggy body with water, and rubbed himself with oil, and turned into a human'. From there, the formerly savage, now civilised man chases off lions and wolves 'so that the shepherds could eat'. As a servant of civilisation, he becomes a defender of domesticated energy capture.

Following this transformation, the narrative turns to Humbaba, the protector of the Cedar Forest who is described as a terrible god. The narrative is primarily interested in Gilgamesh and Enkidu slaying this guardian of the forest and taking the timber as property to build a city. The first known instance of human-caused deforestation occurred in this region. Vast forests had covered the region until agriculturalists began cutting them down, a process that began potentially six thousand years before *The Epic of Gilgamesh* was chiselled into clay, and brought about complete deforestation by the time the words were hardened in 2100 BCE.[8] Given that it took so long, at least on a human timescale, for the deforestation to be complete, there is good reason to assume that it was not necessarily done in a deliberately depletive way. It may be, for instance, that newly agrarian societies took wood at a rate that was imperceptibly depletive to them, the forests appearing vast and inexhaustible. They may have even been careful about not taking too much wood, while being unable to accurately calculate the rate that would be needed to avoid eventual exhaustion. This is partly why it makes sense to date the rise of progress and parasitism to this later date, during the Early Dynastic Period, when it becomes clear that societies are not lamenting the destruction of the forests but celebrating it, as in *The Epic of Gilgamesh*.

On one hand, there is no evidence of such an attitude when deforestation begins several thousand years earlier, in part because there are so few records of any kind before this. On the other, the myth may be based on a much older oral tale that recorded the actions of a living ruler, but it is only with its recording that we have significant evidence of an attitude that celebrates the destruction of wilderness to build cities and introduce private property to markets and trade. Given its language, which unequivocally exalts Gilgamesh, and then celebrates his taming of Enkidu and slaying of the forest guardian Humbaba, the story is obvious political propaganda. The myth is a clear attempt to both explain the established order of expansive, ecologically destructive kingdoms through a creation myth, and to morally justify them through its glorification of Gilgamesh. The mythic figure of Gilgamesh served a dual purpose: he could both represent the legacy of the empire as a founding father of Uruk and also act as a symbolic stand-in for subsequent monarchs who sought to uphold the state's conquest.

The Epic of Gilgamesh was not the only Mesopotamian narrative wielded to solidify the beginnings of this lineage of parasitism. Along the Tigris and Euphrates rivers, the Babylonian creation myth, the Enuma Elish, was scratched into seven clay tablets.[9] They are incomplete, with large pieces lost. In 1849, an English archaeologist recovered the existing tablets in what was then the ruined Library of Ashurbanipal, in the ancient city of Nineveh.

Though the tablets date to 1200 BCE, they are based on older stories that date to at least the reign of Hammurabi, beginning in 1792 BCE; like Gilgamesh, they may also be based on even older oral traditions. The tale tells of the first two gods, the freshwater god Apsu and the saltwater goddess Tiamat. From their flowing together the younger gods were born. These children were loud and tumultuous, and so Apsu planned to kill them. Hearing of his intentions, one of the young gods swoons and kills Apsu. Enraged at the murder of her mate, Tiamat summons forces of chaos, eleven great beasts, to wage war on the young gods. One god arises to contest her: Marduk. Much of the myth is the tale of Marduk's war on Tiamat and his ultimate victory. After Marduk vanquishes Tiamat and her forces of chaos, he brings order to the universe. The other gods praise him, hailing him 'Bestower of planting', '[Founder of sowing]', 'Creator of grain and plants', 'who caused [the green herb to spring up]!'[10] Again, we see a clear attempt to glorify the domestication of energy capture, as with Gilgamesh taming Enkidu, who in turn defends the shepherds, and both heroes vanquishing the wilderness of the forest. The myth continues:

> The gods give Marduk dominion over creation:
> O Marduk, thou art chiefest among the great gods, . . .
>   Henceforth not without avail shall be thy command,
> In thy power shall it be to exalt and to abase.
> Established shall be the word of thy mouth, irresistible shall
>   be thy command;
> None among the gods shall transgress thy boundary. . . .
> O Marduk, thou art our avenger!
> We give thee sovereignty over the whole world.[11]

In these lines, a monarch is born whose dominion is the world itself. The myth speaks not only of Marduk subduing Tiamat and chaos, thus bringing order to creation, but also frequently of abundance and prosperity. Aside from the titles clearly indicating Marduk as a bringer of agriculture, he is given other titles, like 'Creator of Fulness and Abundance' and 'the Founder of Plenteousness'. Towards the end of the myth, the epilogue proclaims: 'Let a man rejoice in Marduk, the Lord of the gods, That he may cause his land to be fruitful, and that he himself may have prosperity!'[12]

If the fruit of Gilgamesh's progress is the taming of the wild, Marduk's is the bringing of order and prosperity to his kingdom. Priests and civic leaders in Babylon read the story to the public each New Year to celebrate 'the progress of the cosmos from initial anarchy to government by the kingship of [the god] Marduk'.[13] Marduk was among a pantheon of gods commonly adopted by aspiring rulers who sought to consolidate expansionist political economies in Mesopotamia, as in Uruk and Assyria mentioned earlier. It was in Babylon that the Marduk cult was strongest, helping it develop from a city-state to the nucleus of a contiguously parasitic empire. The myth promoted a sense of forward momentum in Babylonia, 'resonating with an audience who understood Babylon as a city breaking with the traditions of the past to create a new and better future', its progressive nature encoded in its future focus. The creation myth in the Enuma Elish was central to an emerging understanding of societal and historical progress, the movement of society from a lower form to the higher one ruled by a contemporaneous parasitic class.

Babylonia represents a typical example of the sorts of empires of the classical world that depended on expansion and the use of forced labour and conscription to extract energy through this form of parasitic energy capture. The precise proportion of enslaved people in the population is not known, but wealthy households regularly held many people enslaved, and slaves in Babylonian society rarely left servitude.[14] Myths like Gilgamesh and Enuma Elish could modulate and mitigate any potential backlash against such conditions and orient conscripted soldiers' struggle within a narrative, part of a lineage of divine struggle for land, order, prosperity and domination. The promises made by this progressive myth system were useful tools for rulers to justify to peasants, soldiers or slaves their place at the bottom of an internal hierarchy of energy capture and at the top of an external one directed towards conquered peoples, and to validate the rulers' own place at the top of both. The progress narrative formula, in whatever details it comes cloaked, can serve not only to justify these hierarchies but simultaneously to erase them, giving the host and parasitic classes a shared grand mission in the pursuit of which they put aside potential conflict and work together.

This mythology was combined with newly systematised civil law, commonly known as the Code of Hammurabi, named after the Old Babylon Empire's deified sixth monarch – a document that opens by praising Marduk. Tightly entwined with the myth, this codified law also served as powerful psychological and social mortar holding together Babylonian civil society. Archaeologist Ian Morris calls the social contract that maintained such systems the 'Old Deal', writing that it represented 'a simple idea: that nature and the gods required that some people should give commands while others obeyed them, and as long as everyone played their parts properly, all would be for the best in the best of all possible worlds'.[15] In other words, this belief system was based on the assumption that both material progress – increasing the amount of market goods, lands and slaves – and spiritual progress – spreading the dominion of Marduk and his earthly champions, the kings and emperors – could only be achieved with rigid adherence to established hierarchies of energy capture. Characters like Marduk and Gilgamesh, and the promises of progress priests and kings made on their behalf, represented a critical element of this first Old Deal social contract, one that remained remarkably consistent through millennia and across many distinct geographies that relied on contiguous parasitism. But it represents only the first bricks in what would grow into a mighty tower.

The Ends of Zoroaster

The Persian religion of Zoroastrianism has roots reaching back as far as 1500 BCE. It is one of the world's oldest religions, and in the lineage we're concerned with likely the first monotheistic one. It may have served as direct inspiration for many of the themes of later religions, including Judaism, Christianity, Buddhism and Islam, and of Greek philosophy.[16]

While the religion introduced numerous important themes central to many of the faiths that followed – a singular God and an evil demonic figure in constant conflict, a messiah and prophet, and human free will – one particular Zoroastrian invention is important to this discussion: end times, and the study thereof: eschatology. The progress narrative formula includes the belief in a utopia that exists in the future, or currently in some parallel world that will only be revealed at a future time. This can be seen clearly in modern secular progress narratives. Political theorist Francis Fukuyama argued for the 'End of History' notion, that history has led in a linear fashion to an apex of human organisation and political economy – 'liberal democracy' – such that the class and ideological conflicts of history had come to an end in 1989. Although this claim as an empirical fact has been challenged, even mocked, in many rebuttals that point out that history has not ended but has in fact continued, one might nevertheless interpret Fukuyama's argument as calling the bluff of progress narratives that perpetually situate paradise in a never-arriving future. Fukuyama's argument, essentially, was that the paradise promised by liberalism has arrived, undercutting the narrative that depends on holding that paradise out as a prize to those in the present, never necessarily intending for it to happen. Many faiths, too, project an eschaton into the future, a day of judgment and the arrival of a new and perfect paradise. Prophets occasionally foresee in these traditions that this paradise is imminent; so far all have suffered the same fate as Fukuyama's End of History idea. In Christianity, inspired by the Zoroastrian tradition, the version of this principle is the New Earth of Revelation:

Then I saw 'a new heaven and a new earth,' for the first heaven and the first earth had passed away, and there was no longer any sea. I saw the Holy City, the new Jerusalem, coming down out of heaven from God, prepared as a bride beautifully dressed for her husband. And I heard a loud voice from the throne saying, 'Look! God's dwelling place is now among the people, and he will dwell with them. They will be his people, and God himself will be with them and be their God. "He will wipe every tear from their eyes. There will be no more death" or mourning or crying or pain, for the old order of things has passed away.'[17]

Like the Mesopotamian progress myths, Zoroastrianism served as a useful tool for social cohesion on the part of rulers in ancient Iranian kingdoms and empires, just to the immediate east of Mesopotamia. As in most cases, it is open to debate whether rulers were true believers in the faith or, rather, were simple opportunists using the faith for their own ends. It is likely that progress propaganda is deployed both by zealots who believe as much or even more in their own myths than their subjects do, and also by pure pragmatists who deploy the myth for political expediency, or by a complex and shifting mix of both positions in the same mind.

In the case of Zoroastrianism, there remains debate about whether it was actively instituted as the state religion of the Achaemenid

Empire, which reigned for two hundred years beginning in 550 BCE in Persia and stretched west to Greece and modern Libya. Whether its founder, Cyrus II, was an adherent of Zoroastrian ideas also remains up for debate.[18] But Zoroastrianism was a foundational part of the Sasanian Empire, a successor kingdom to the Achaemenid that lasted nearly twice as long. Providing a framework for understanding of the world through a binary of good and evil, a conflict which leads to an end times in which good triumphs over evil, would have served a similar purpose to other progress narratives: motivation for both rulers and those whose labour fuelled expansion and extraction. Zoroastrianism, like Marduk and the succeeding myths, gave a divine right to the Sasanian dynasty. Although the history of Zoroastrianism is woven through with myth, legend and competing accounts – as are the histories of most contemporaneous faiths – there is good reason to believe that some of the important tenets of divine kingship and eschatology were only solidified with the unification of the Sasanian Empire, rendering it a thoroughly imperial tradition.

The end times promised by the Zoroastrian faith by this period – which begins in the third century CE – will look familiar to contemporary Abrahamic adherents. When the process of judgment begins, the dead are resurrected in a period lasting fifty-seven years. Although souls are reunited with their bodies, and people living from all periods of history mingle together, none crave the needs of the body. Instead, each is judged and sent to either paradise or hell, but not for eternity. Instead, they await a final battle between Ohrmazd, the supreme God, and Ahriman, the demon who rules over sinners in hell. Ohrmazd banishes all evil from the world, initiates a process of cleansing all but the most heinous of sinners through a fountain of molten metal, and all live together in perfect bliss.

The Dominion of Abraham

It would be reasonable to see these early millennia of contiguous parasitism as a process of developing ever more effective means of maintaining smooth-running states, markets and militaries, via the implementation of innovations in political propaganda narratives. Given that these societies and their myths emerged in one region with substantial trade and communication, such narratives almost certainly did influence one another and were adapted to the cultures in which they were deployed.

One of the most enduring forms of mythic progress arrived west of Mesopotamia, with the creation myth of Judaism. The Old Testament's first chapter, Genesis, serves as the cosmogenic foundation of Judaism, Christianity and Islam. The Babylonian Enuma Elish myth may have directly inspired some of the elements of the Old Testament, which contains similar themes and narratives, an original void and global cataclysms like the flood myth. Given that some of the earliest sources of the Old Testament originated with the Judean exile from Babylonia, it would make sense that there was mythic cross-pollination. Whether inspiration or coincidence, the two myths served a similar function: to reproduce the social and ecological relations that could maintain an extraction- and expansion-dependent human ecology and political economy. If the tale of Gilgamesh has embedded within it an ethic of a kingdom dominating nature by building its cities out of once verdant forests, and if Marduk represents a god breaking from the past to build a better future, Genesis was an innovation that provided even more explicit justifications for the sundering of humanity and ecological systems and for human domination of other life:

> And God said, Let us make man in our image, after our likeness: and let them have dominion over the fish of the sea, and over the fowl of the air, and over the cattle, and over all the earth, and over every creeping thing that creepeth upon the earth. And God blessed [Adam and Eve], and God said unto them, Be fruitful, and multiply, and replenish the earth, and subdue it: and have dominion over the fish of the sea, and over the fowl of the air, and over every living thing that moveth upon the earth.[19]

In this story, the God of Abraham granted humans immaterial souls separate from nature and granted them supremacy over nature. The 'dominion' mentioned in the Genesis myth provided a divinely derived mandate for humanity's exploitation of nature. Whereas the biocentric values likely common for hundreds of preceding millennia granted intrinsic value to other life, the dominion idea helped believers feel vindicated in transforming complex, interdependent ecological systems into discrete units – livestock, productive fields – possessing only instrumental value, meaning non-human life was valuable only to the extent that it served human needs and reflected God's glory. The importance of this shift in the psychological means by which humans relate to their natural environments cannot be overstated. It is the shift from a biocentric ethic to an anthropocentric one, in which human beings were no longer just one of many sorts of persons inhabiting the world but had been moved into the centre of the universe. 'God created man in his own image,' says Genesis; or rather, the authors of Genesis made this particular God in their image, revealing a deeply narcissistic divinity, empty and nihilistic, stripped of any meaning deeper than the human soul, mind and face. It is from this belief that human supremacy becomes fully sanctified: humans are not only apex predators navigating complex webs of life, but seize this divine directive to climb atop a new biological hierarchy and rule over all other life as masters. As they dominate, we must now worship humanity through the middleman of God.

Nature suddenly ceased to exist as an irreducible reality. God, instead, became the atom of existence, the source and totality from which nature is merely projected. Parasitic energy capture in these early imperial economies took place through vast agricultural projects, which pared biodiversity down to a few productive species in service to the human body alone. As such, the people playing extractive roles no longer had any need to understand the intricate mechanics of ecological systems. As long as they understood the will of God and the many practical agrarian lessons imparted in theological texts, they could maintain their ecological and political supremacy.

The book of Genesis provides the theological basis for dominion, but it also contains the promise of progress, which takes the shape of frontiers, or living space set aside for God's chosen people, to be found in new land. It models practically, for contemporaneous and future adherents, the rewards that may come from putting one's faith in promises of divine progress, showing in detail the benefits of obedience to this particular god. For example:

> Now the Lord had said unto Abram [later Abraham], Get thee out of thy country, and from thy kindred, and from thy father's house, unto a land that I will shew thee: And I will make of thee a great nation, and I will bless thee, and make thy name great; and thou shalt be a blessing: And I will bless them that bless thee, and curse him that curseth thee: and in thee shall all families of the earth be blessed.[20]

The promised land was not *only* The Promised Land, what has since become a metaphor for the aspirational place at the end of all progress rainbows, but a real place: Canaan. And Canaan was not an empty territory granted by God to the Israelites to fill; it was already populated by another culture, now identified as a target of conquest.[21] That is to say, this early form of progress myth was explicitly used as social and moral justification for opening a frontier of conquest into Canaan. The Old Testament goes on to describe in great numeric detail the uniting of the Hebrew tribes into one vast army hundreds of thousands strong, assembled and united by this belief system for the purpose of conquering others and taking their land. The narrative formula remains at the basis of this myth – that of a primordial past, a triumphant present and a glorious future, in which the chosen are separated out from the heathen, and a paradisal land is granted by the one true God. The Promised Land notion would become a central motif in future progress myths, even those with a more secular orientation.

But as is fundamental to the progress narrative formula and economic parasitism, the original promised land did not suffice. Limits are reached and expansion must continue. New rulers emerged to pursue new targets of conquest, and God promised a new land. Abraham's grandson Israel was guaranteed another bounty in Egypt where his own son Joseph had been made governor:

And God spake unto Israel in the visions of the night, and said, Jacob, Jacob. And he said, Here am I. And he said, I am God, the God of thy father: fear not to go down into Egypt; for I will there make of thee a great nation: I will go down with thee into Egypt; and I will also surely bring thee up again.[22]

The promised land is central to the progress narrative driving events in Genesis, and continues in Exodus with Moses leading the Hebrews out of Egypt and into the Levant. As historian Clifford Backman notes, at this stage Abrahamic myth was largely based on an earthly progress narrative, 'one designed to secure for the Jews security and prosperity in this life, rather than spiritual rewards in a life hereafter'.[23] This is a clear member of the lineage of Gilgamesh and Marduk – also bequeathers of prosperity through their earthly agents – and Zoroastrianism. The promise of a better future in the physical world could sustain, at least according to the legend, even decades-long wandering in a desert, which was an instructive lesson for believers: if you mirror such faith even through the travails of deprivation and collective suffering, you too may be granted by God a land of plenty, one day.

The Age of Athens

Given the tradition of oral histories that many Indigenous societies have used to record the timeline of their existence, histories that sometimes span tens of thousands of years, it is likely that *Homo sapiens* has been compelled to maintain histories for most, or all, of its existence. While history always moves unidirectionally, there is a split in the understanding of history according to whether it is seen to be moving in a more cyclical way or a more progressive way. Cyclical histories are those that are understood to move perpetually in ups and downs, forward in a circular path, according to cycles. Progressive histories tend to take the shape of a gradual ascent to a permanent plateau. Central to the latter is the contour of a society's timeline that ascends, step by step, in a compounding way from savage or

I HEAVEN

simple or primitive to civilised or complex or sophisticated, parenthetically sealed by a beginning and an end.

The little scholarship that there is on progress narratives tends to trace their origin to the European Enlightenment, starting in the late seventeenth century CE. Scholars treat earlier European histories, like those of ancient Greece, as more cyclical. But sociologist Robert Nisbet diverges from this view and convincingly argues that the origin of a progressive historiography – the writing and recording of a given history – in Western history can be found much earlier, in Greek antiquity as early as the eighth century BCE (or well over two thousand years earlier than the Enlightenment and predating the Old Testament myths by at least two centuries, though not as old as the mythical versions in *The Epic of Gilgamesh*).[24]

Nisbet contends that the secular version of the idea that society emerges from a primitive past into sophisticated civilisation and gradually moves towards an idealised golden age comes from one of the earliest thinkers of the Western canon, Hesiod. Hesiod's history leads, like Enuma Elish, from a chaotic past to a stable present. These ideas then develop through other thinkers such as Protagoras, Aeschylus, Sophocles, Plato, Aristotle, Epicurus, Zeno, Lucretius and Seneca.[25] Hesiod's eighth-century didactic poem *Works and Days* – a work of verse that imparts practical lessons – marks the beginning of the European epistemic tradition of *secular* societal progress.[26]

This tradition continues with the ancient Athenian historian Thucydides, who begins his *History of the Peloponnesian Wars*, written in the fifth century BCE, by declaring that although the Greeks had once been like barbarians, they had ascended by the progressive momentum of history to the pinnacle of civilisation.[27] Like others of his class, time and culture, Thucydides saw Athenian civilisation as the apex of history, perhaps even as an 'end of history' for the Periclean Golden Age. There is legitimacy to seeing Athens as one of the major founding sites of Western knowledge production, and the fact that this knowledge compounded in some progressive ways should indeed be celebrated. But though much praise has been heaped on the citystate's proto-democracy, still about half the population was enslaved, while women held roles subordinate to men and rarely with any political representation. It served the Athenian authority's material political ends to place itself at the height of all human history. It did so not by simply exalting in its own superiority, but by creating a duality between those fit to rule and those most fit to be ruled, or to extract and be extracted from; between the host and the parasite, between the savage and the civilised: the binary central to the progress narrative formula.

Swiss-Israeli archaeologist and classicist Benjamin Isaac observes that classical Greece developed theories to rationalise both territorial expansion and the domination of other peoples. Athenians believed themselves to be endowed with exceptional racial 'purity', in contrast with other peoples whom they considered to be lesser due to being 'mixed' and degraded:

> the essence of this is first encountered in the treatise Airs, Waters, Places, which insists that the inhabitants of Asia are soft because of their good climate and resources. They are less belligerent and gentler in character than the Europeans, who are more courageous and belligerent. Aristotle then claimed that the Greeks, combining the best qualities of both groups, were therefore capable of ruling all mankind – an early, if not the first, text to suggest that Greeks should achieve universal rule. No less important: these ideas were taken over, suitably adapted, by the Romans.[28]

Such ideas were useful in justifying the enslavement of these lesser peoples who 'live best in a symbiotic relationship with fully human masters', and solidifying the dichotomy in relations of abstract energy capture between the host class and the parasitic class.[29] These racial hierarchies served dual purposes. Externally, they provided a rationale and an ethical imperative for conquest by framing it as the act of civilising otherwise savage races; internally, they provided a means of ethnically classifying and distinguishing those most fit to work in forced labour, whether agricultural slaves or house servants or otherwise, from those fit to hold positions of political authority – in other words, property-owning male citizens. Such a race theory, and such a binary between savage and civilised, can exist only in connection with a progressive view of history in which some races are less developed, through time, and others are more developed, a theme that will play out in similar societies over the next two and a half thousand years. Along with the Athenians' many positive contributions to Western history, this secular racial theory, consistent with the progress narrative formula, would remain among the most enduring.

Peace of Rome

Lucretius was a Roman poet who influenced later and now more famous poets like Virgil. He left only one work, *On the Nature of Things*, likely written in the first century BCE. This work offers a prime example of a Roman narrative of progress, and mostly follows the standard formula. His narrative of humanity's progress begins with the notion of a savage whose cunning in forging weapons enabled him to build civilisation, which then birthed kings, property and gods. Lucretius departs from a strictly linear path for this historical development. As he sees it, society descended into mob rule for a while, but civilisation reasserted itself, triumphed, and a new phase appeared with 'genuine government and law, with settled principles existing where once only arbitrary, monarchical power had prevailed'.[30] He designed his narrative with the aim of having a 'reassuring effect' on his fellow Romans.[31] This narrative will be a familiar outline to Americans today who see the United States as a republic rising out of monarchical power, arriving more than two thousand years after the Republic in which Lucretius lived – and self-consciously modelled on it.

Lucretius's preference for an oligarchy ('republic') over a monarchy may be reasonable, and his rough sketch of Rome's history not far off, but, nevertheless, his narrative formula and historiography would prove useful to the new Roman monarchy, which arrived soon after his time. While Lucretius himself favoured republican government, the ideas of progress and the intellectual lineage to which he contributed were soon co-opted by Roman political elites for the renewed assertion of monarchical power and territorial expansion. Six years after Lucretius died, Julius Caesar seized Rome, marking the tipping of the city-state's symbolic – and soon material – balance of power from republican institutions towards a new class of emperors. Rome's economy had long been parasitic in the sense discussed here, expanding far beyond the city-state's original borders while ruled by an oligarchy of senators and generals. It underwent a far greater expansion after the foundation of the Republic than it had during its original monarchic phase. But its slip into empire accelerated the process considerably, with emperors feeling an imperative to expand their territories and grow the economy to maintain their legitimacy and the loyalty of their militaries and citizenry. This parasitism had a major effect on surrounding peoples and ecological zones. South African researcher Ashley Dawson notes that 'the Roman Empire was probably responsible for the greatest annihilation of large animals since the Pleistocene megafauna mass extinction', the time when human beings first started migrating out of Africa and settling in all the habitable continents. 'As was true of the Sumerians,' that is, the Mesopotamian culture from which the city of Uruk and *The Epic of Gilgamesh* came, 'Rome annihilated most of the large animals it could get its hands on and reduced most of the lands it conquered to desert.'[32] Dawson further illustrates Rome's parasitism vividly:

> First Egypt, then Sicily, and finally North Africa were turned into the granary of the empire in order to provide Rome's citizens with their free supply of daily bread. Deforestation caused by the Romans' agricultural enterprises spread from Morocco to the hills of Galilee to the Sierra Nevada of Spain. Like the Sumerians, the Romans failed to engage in sustainable forms of agriculture, seeking instead to expand their way out of ecological crisis.[33]

Julius Caesar's nephew Octavian, taking the imperial name and title Caesar Augustus, laid the foundation for nearly two centuries of

I HEAVEN

relative peace *within* the empire – the frontiers remained rife with battles – that has come to be called the *Pax Romana* (Roman peace). Historian Ali Parchami describes it like this:

> Ideologically, the Pax Romana was predicated upon the universality of Rome's political and military authority, as well as its laws, institutions, customs, and cultural mores. It was a conceptualisation which claimed that Providence had ordained the Romans with the task of conquering and civilizing the world. The mythological stories surrounding the foundations of the city of Rome suggest that from the earliest times the Romans firmly believed in Providence and in their national destiny.[34]

Rome had been a republic for nearly five hundred years, even if it was fundamentally a parasitic political economy. Augustus's was the first in a lineage of dynasties to rule since the kingdom was overthrown. This was a major break from the past, and so Augustus had to justify it to conservative elements of Roman politics. He did so in part with an aggressive propaganda campaign that styled his administration a Golden Age of expansion, order, prosperity and progress, engaging with important thinkers like Livy and Horace.

The Romans were quite reflective and conscious of how their economy played a role, though they may not have called it 'parasitic', in the despoliation brought about by their conquest. In 72 CE, Rome invaded what is now Scotland. The Caledonians residing there were led by a legendary leader whom Roman historian Tacitus called *Calgacus*, a Celtic word that can be interpreted as 'bearing a blade'. Calgacus built a confederation among the disparate tribes of Caledonia to repel the Roman invaders. When the Romans and Caledonians met at the Battle of Mons Graupius in 83 CE, Tacitus chronicled the battle while his father-in-law Agricola led the invasion. Tacitus imagined Calgacus giving a rallying speech to his soldiers on the brink of battle in which the swordsman proclaimed, 'To robbery, slaughter, plunder, the Romans give the lying name of empire; they make a desert and call it peace.'[35]

Even if Tacitus was conscious of how Rome's enemies might view imperial expansion, within Rome such growth was inextricable from a sense of historical and religious progress. We may look back at this sort of narrative as recognisably secular. The Pax Romana, after all, would be invoked nearly two millennia later by Austrian aristocrat Richard von Coudenhove-Kalergi as an apologia for the 'Pax Americana' and US military supremacy in the twentieth century, which emulated Augustus's reign by bringing centuries of, in his words, 'peace and progress'. It's no coincidence that this Pax Americana rhetoric accompanied the world's greatest period of economic expansionism via the Great Acceleration.[36] But even if they may be familiar to secular readers today, the Pax Romana and Roman life and power were suffused with religiosity. The polytheism of Rome inflected the cult of empire – granting emperors a degree of divinity – just as it did the daily life of most Romans. Virgil, one of the most famous Augustan poets, insisted that the gods willed Roman rule, and that the Empire 'should bring the entire world under its laws'.[37] The *Aeneid* is, in Parchami's words, 'littered with the notions of manifest destiny, providence, and divine will'. Virgil's Jupiter, Rome's syncretic version of the Greek supreme god Zeus, proclaims: 'To the Romans I assign limits neither to the extent nor to the duration of their empire; dominion have I given them without end.'[38] This attitude is indistinguishable from Abrahamic dominion and is little changed from notions of 'infinite economic growth' casually invoked by pundits and politicians today.

While the Pax Romana did deliver real material and cultural benefits to some Romans – expanded infrastructure, more wealth, which bought more art and architecture, and a cessation, or at least a pause, of constant warfare – it was less a utopia than a grim and, for many, non-consensual bargain. For those who did not directly benefit from it, or even those who thought they did, still there was servitude: it remained a parasitic economy that disproportionately extracted from some for the benefit of a few others, further solidifying the dichotomy between host and parasite. In Britannia, Rome's most westward territory, 'Roman peace' was imposed at the end of spears and swords. 'Incorporation into the empire was violent,' notes Ian Morris. After initial revolts were suppressed, 'civilisation, secure borders and law and order would pacify and enrich even the most turbulent natives.'[39] But Morris notes eight wars in Britain in the 150 years following the country's submission to the Pax Romana, and that there were likely others that went unrecorded. In addition to wars, there are many recorded examples of the Roman army committing crimes against the native population even in peacetime, showing the limits to 'law and order'.

The aforementioned general Agricola did more than any other to establish the Roman colony in Britain. The Caledonians likely halted their northward advance and, as a result, Rome failed to establish a foothold there. But most of the rest of Britain submitted to Roman rule. In addition to his famous line, placed in the mouth of Calgacus, Tacitus made a less famous statement about the allure of the fruits of Roman civilisation. With Roman rule, he wrote, the indigenous Britons were 'gradually led into the demoralising temptations of arcades, baths and sumptuous banquets. The unsuspecting Britons spoke of such novelties as "civilisation", when in fact they were only a feature of their enslavement.'[40]

Kingdoms of Christ

The Roman Empire's first rise coincides with the life of Jesus of Nazareth and the birth of Christianity. The historical figure called Jesus has been interpreted as a freedom fighter resisting the extractive incursion of Rome into Judea.[41] Though it may be somewhat optimistic to interpret Jesus as equivalent to a modern revolutionary, the Gospels do clearly spell out virtues we might consider emancipatory; the tale of the cleansing of the temple, which depicts Jesus whipping and expelling those who extract from individuals through debt, is among the most famous. The New Testament is fairly unequivocal that hoarding wealth is an evil and fraternal poverty a virtue. Yet the church that grew out of his legend was soon put toward more elite purposes. Historian Theodor E. Mommsen suggests that early Christian thinkers successfully tied the material prosperity brought about during the Pax Romana, and its embedded narrative of progress, to Christianity: 'As the appearance of Christ coincided with a marked improvement of all things secular, so, the early apologists argued, the growth of the new faith will be accompanied by further progress.'[42] Eusebius and Tertullian, important early Christian polemicists, also argued that the material prosperity of the Pax Romana could be directly tied to the rise of Christianity. Such an association of Roman ambitions and Christian myth did not stop with the end of the Pax Romana in 180 CE. Two centuries later, emperors found it expedient to enlist the message and mission of Christianity – as they had done with their own polytheistic cults – and turn the fast-growing religion towards the ends of Rome's economy.

Emperor Constantine I – who ruled between 306 and 337 CE – converted the empire to Christianity for primarily political reasons: to improve the cohesion and controllability of the state and its subjects. He reasoned that a monotheistic empire would be more effective at both unity and continued expansion than a more pluralistic or polytheistic one. 'Constantine does not just Christianize the Roman Empire; he unifies it too,' writes historian Garth Fowden. 'And he expounds a worldview to which, though it was not absolutely original, he gave new force: one god, one empire, one emperor.'[43] The scriptural evangelism embedded in the notion of the New Testament's Great Commission helped spread the faith far afield:

> Then Jesus came to them and said, 'All authority in heaven and on earth has been given to me. Therefore go and make disciples of all nations, baptizing them in the name of the Father and of the Son and of the Holy Spirit, and teaching them to obey everything I have commanded you.'[44]

This principle of the mission began a process of missionaries venturing forth as seemingly benign pioneers, warming the waters for subsequent armies and settlers. The faith would be an outstretched white glove concealing the brass knuckles of empire. Constantine aimed to replace the cults of Rome with Christianity, 'beyond as well as within his own frontiers – an entirely new understanding of the Roman emperor's role'.[45] Constantine's ambition was vast, aiming to pursue an 'unprecedented opportunity' for building 'world empire' and the 'infinite expansion of the dominions he had inherited and conquered'. He sought to achieve this 'by the union of imperial impetus and missionary monotheism, Christianity and Rome'.[46] Simply swap Jupiter's edict of Rome's limitless dominion with that of some new god. The mission embedded in Christianity is one of progress: submission on Earth, no matter how excruciating, would yield eternal peace and joy in heaven.

The faith was well suited to Rome's parasitic economy and culture. Constantine's Christian empire fulfils the original promise of Abraham by becoming 'the definitive force of providence in history', which, Robert Nisbet writes, 'promises to the Christian the prospect of an ever triumphant and ever improving society'.[47] That is, Christianity wedded to the Empire guaranteed progress. A vital function of the Christian mission, and one that departed from the earlier, pluralistic attitude of the Roman Empire, in which diverse religions coexisted, was the eradication of competing belief systems both within Christianity and between this religion and others.

While Constantine could be said to have accelerated the process of Christianisation, it was non-linear: thirty years after Rome officially adopted Christianity, polytheistic Visigoth King Alaric sacked the city, leaving Christian hegemony in doubt. This was a temporary setback, however; Christianity was quickly and deeply imbued in the power structures of emerging European kingdoms, for which Constantine set the stage by instituting quasi-judicial powers for bishops through changes to marriage and divorce law, and by introducing new punishment for heretics.[48] Constantine's immediate successors swung between traditional Roman polytheism and Christian rule until the latter fourth century, when Theodosius I launched an aggressive campaign destroying pagan temples and purging the empire of heretics. He deployed a paramilitary force of Christian monks who attacked non-Christian sites of worship. Where these monks invaded, according to Roman thinker Libanius, 'utter desolation follows, with the stripping of roofs, demolition of walls, the tearing down of statues and the overthrow of altars'.[49] As they rampaged like a biblical plague, the monks 'eat more than elephants, and demand a large quantity of liquor from the people'.[50]

Though it started with Rome, Christianity left the Empire behind. Its political usefulness made it one of the primary tools fuelling the engine of extraction and expansion in new kingdoms that littered Europe in the shadow of Roman ruins.

If Constantine built the political foundation for a Christian progress myth at the heart of expansive economies in Europe, Augustine of Hippo gathered together the philosophical foundation. Born Aurelius Augustinus in 354 CE in the Roman province of Numidia – in what is today Algeria – Augustine became the most important figure establishing the intellectual basis of Western Christendom. In Augustine's primary text, *City of God*, Nisbet notes that 'all of the really vital, essential elements of the Western idea of progress are present: . . . the unfolding, cumulative advancement of mankind . . . the idea of time as a unilinear flow . . . the conception of stages and epochs . . . the idea of conflict of cities, nations, and classes as the motor spring of the historical process.'[51] He contributed to the concept of original sin and 'just war' theory, which is an ethical accounting to justify certain wars. In Augustine's case, he argued that war could be justified if it were used as a means of mitigating sin, or if rulers used war to maintain peace, a clear descendant from ideas central to the Pax Romana. This could reasonably be interpreted as an attempt to reconcile Christian pacifism with the intrinsically violent imperative of parasitic states to wage war and conquest. This was an essential moral and philosophical innovation if Christian monarchies like Constantine's were to maintain their legitimacy. But not only does Augustine clear the way for viewing Christianity as an engine of progress, he presents it as the sole engine: 'For over and above those arts which are called virtues, and which teach us how we may spend our life well, and attain to endless happiness – arts which are given to the children of the promise and the kingdom by the sole grace of God which is in Christ.'[52]

Vikings of Valhalla

While Christianity was spreading through central Europe during Rome's long decline, the northern fringes still held to pantheons of many gods. Early Christians deployed the term 'pagan' to distinguish themselves from non-believers. 'Pagan' derives from classical Latin meaning 'rural' or 'peasant', carrying a connotation that urban Christians first used to condescend to Roman polytheists – those in the countryside slower to adopt trends widespread in the core – but which remained useful when applied to polytheists outside of Rome. Anthropologist Jason Hickel writes that, 'As Christendom expanded through Europe it sought to repress [pagan] ideas wherever it encountered them.'[53] The Nordics were among the last Europeans to have large polytheistic populations, even as they were the first to settle on a continent west of the Atlantic Ocean.

The first known Europeans to attempt colonising North America were settlers from Norway. Leif Eriksson reached what is now Canada, spurred on by demand for timber, a coveted commodity for fuelling shipbuilding in maritime economies. There Vikings met the Beothuk and Inuit Thule peoples already long resident there and began trading and living side by side. This foray into North America was the furthest west Norse settlers went. Far more of them colonised nations throughout Europe; Vikings settled across the British Isles, Western Europe, and into the Mediterranean and Levant; they named Russia and left legacies in Iceland, non-continually in Greenland, and Vinland in modern Newfoundland. Viking raiders opened new frontiers to fuel their parasitic form of human ecology and political economy. What makes this example unique is that it is a decentralised form of parasitism, but one that still used a mythic form of progress narrative to justify their expansionism. Despite establishing the short-lived 'North Sea Empire'[54] – which existed for about forty years towards the end of the Viking Age (793–1066 CE) – Norse settlers did not expand contiguous territory from a centralised authority like an imperial bureaucracy. Scandinavian kings took tribute, but did so without exercising the sort of administrative control over conquered territories that Rome and other contiguous empires had done, or like the maritime empires of Britain, France, Spain, Portugal and others that would follow.

Nevertheless, they were without doubt parasitic in the same sense. Their concrete energy capture forms created a boom-and-bust economy dependent on linear, compounding, and ecologically unsustainable extractivism. They quickly deforested the places they colonised, particularly in Iceland and Greenland, and pushed native species to extirpation. Erik 'the Red' Thorvaldsson set up colonies in Greenland that devastated the walrus populations there, overhunting the animal to population collapse for ivory traded in Europe, which played a significant role in the collapse of the Greenland Norse.[55] Abstract energy capture was also generally parasitic: Norse settlers pushed Indigenous peoples out of their land – in this case, driving the Sámi people in Scandinavia to the Arctic fringes – and took captives from the places they raided and forced them into slave labour.[56] With few exceptions, they set up strictly hierarchical societies with feudal extractive relations between members. These forms of energy capture directly imperilled some of their lost colonies: Norse Greenlanders relied on agriculture and livestock, and on this hierarchical social structure. Their way of life was dependent on the order imposed by a handful of chiefs and bishops (of those who had already converted to Christianity) who owned most of the land, controlled access to the boats and trade, and ruled over the majority of settlers. All of these factors contributed to their downfall. American geographer Jared Diamond summarises this dynamic:

> There were many innovations that might have improved the material conditions of the Norse, such as importing more iron and fewer luxuries, allocating more boat time to Markland journeys for obtaining iron and timber, and copying (from the Inuit) or inventing different boats and different hunting techniques. But those innovations could have threatened the power, prestige, and narrow interests of the chiefs.[57]

Norse settlers were driven not only by material imperatives of resource extraction, nor did they succeed only with the military advantages of technologically advanced longships. They were also aided by cultural traditions that made them 'adventurous and aggressive and scornful of death', in the words of Kevin Crossley-Holland, which 'must have given added momentum to the impulse to raid and trade,

I HEAVEN

conquer and colonise'.[58] Norse pagan myths of cosmic battle and combat death rewarded in the afterlife – the purgatorial mead hall called Valhalla and the end-times battle Ragnarök, similar to those in Abrahamic and Zoroastrian myths – propelled marauders from medieval Scandinavia across far reaches of the globe. While access will certainly be granted to a promised land of cosmic paradise for obeying your battle commander and conquering other peoples – or the opposite, damnation in a dreary hell for those who have failed to die in battle – there is also the promise of end-times judgment, demise, and rebirth. During Ragnarök, the gods and giants and all the warriors of Valhalla will do terrible battle, 'the nine worlds will burn and the gods will die', as Yggdrasil, the tree of life, burns and 'men and women and children in Midgard [the middle-land of human habitation] will die'.[59] But after this conflagration of blood, storm and fire, a new fertile land rises from the tumultuous sea and a breeding pair of humans survives to repopulate the earth.

When combined with promises of material wealth and fertile homelands, Norse myths helped galvanise berserkers, making them some of the most feared warriors in Europe. But soon after the Vikings set forth from Scandinavia, they adopted a new god to replace their traditional pantheon: the God of Abraham. While there is evidence of Islamic Vikings from Norse funeral garments bearing the word 'Allah' and Islamic coins circulating in the far north of Europe, Christianity is the Abrahamic religion that captured Norse populations throughout Europe.[60]

Like other former pagans abandoned in Rome's fall, Christianisation of the Norse can be traced to many causes, but the presence of political leaders motivated by social control and pursuing imperatives of parasitism was paramount. As Christianity had served Constantine in these matters, so the Norse saw profit in adopting this religion that promised an afterlife of paradise for all converts willing to submit to a rigid but practical power structure, one that could unify disparate populations. As historian Caitlin Ellis has written, 'coming into contact with Christian kingdoms which were more politically centralised arguably led to greater unification of the Scandinavian realms.'[61] Figures like Harald Bluetooth of Denmark, Rollo of Normandy, and

Olafs Tryggvason and Haraldsson were important agents in this process, and by the twelfth century, all the Scandinavian nations would be officially Christian. Prohibitions on pagan practices soon followed. The Christian Norse, descendants of pagan Vikings, participated in the Crusades, raiding with promises of Heaven comforting them in the way Valhalla once had: as Ellis notes, 'Instead of an afterlife of feasting in Valhalla as a reward for dying in battle, those who died on Crusade would go straight to Heaven.'[62]

Christianity also helped to accelerate the Norse conquest of Indigenous lands, and coincides with their becoming more aggressive colonists. Norse pagans had always maintained a complex relationship with the Indigenous Sámi peoples, sometimes intermarrying, but Christianisation transformed theirs into a more consistently hostile relationship.[63] In the seventeenth and eighteenth centuries, missionaries confiscated and destroyed large numbers of the Sámi's sacred noaidi-shaman drums, punishing and sometimes executing those who used them; the brutality employed was enough to earn this period the name 'the burning time'.[64] The missionaries sought the 'complete destruction' of the 'old worldview', in the words of historian Francis Joy.[65] By the end of the eighteenth century the Sámi came fully under the formal control of the Christian Church.[66]

With the Christianisation of the Norse, European Christendom was complete. Christian kingdoms during this period consistently used Christianity as a political and social tool, and they clearly did so in the interest of maintaining and expanding parasitic economies and energy capture. British archaeologist Ian Morris describes in his materialist history of Britain, *Geography is Destiny*, the various monarchies that spread their grip over the island after the fall of Roman rule there. He portrays in great detail the ways monarchic power welded to aristocratic power and channelled through markets consumed energy and resources from the land and people, all girded by violence: 'Ever since their arrival, the richest Saxon chiefs had got that way by killing and robbing people. Fierce warriors signed up to serve even fiercer ones, and as the latter turned into kings, they rewarded the former by making them lords and giving them estates.' This was a process certainly not unique to Britain, but there remains a lot of recorded history on the process, so the picture there is quite clear. 'When not fighting, kings and their war bands would travel between their properties, consuming all the meat and mead on site then moving on to the next spot.' Markets began to develop by 650 and offered kings and lords new vehicles for extraction. In an almost perfect copy of the profit process described at the start of this section in Uruk nearly 3,000 years earlier, kings and lords in Britain would hire agents deployed at farms and 'confiscate the lion's share of its output', taking it to market in exchange for more durable goods. 'Continental merchants wanted food and drink (and slaves) to sell in the cities back home; Anglo-Saxon elites wanted Continental ornaments, clothes and weapons to distinguish themselves from their poorer peers. Everybody gained, except the slaves.'[67] Squeezing peasants and land for more than they gave back was central to profit-making in Mesopotamia as it was in medieval Britain and elsewhere throughout Europe. There, Christianity was an integral part of the process. In addition to the crude unification role of progress propaganda pioneered by Constantine three centuries earlier, the Church provided much-needed administrative expertise and clout. 'Fortunately for kings,' Ian Morris writes, 'the church was keen to provide' men who could read, keep records and organise large-scale trade. These literate advisors from Rome, the seat of church power, could exercise influence behind the Anglo-Saxon thrones, and often did shape the direction of kingdoms to the enrichment of the Church and continued spread of the faith. But as potent a force as this Abrahamic religion was for advancing European parasitism and progress narratives, another was nearing its birth in the East.

Empires of Islam

At the beginning of the seventh century CE, the Arabian Peninsula was populated by disparate tribes and kingdoms practising a wide variety of religions and methods of energy capture, and often at war with one another. The Quraysh was one of the most important tribes in the region, growing wealthy by maintaining important trade routes and engaging in enterprises like mining. Whereas many of the tribes were nomadic, the Quraysh had begun to settle around the middle of the west coast of the peninsula, on the Red Sea. There they maintained the burgeoning city of Mecca as a safe haven for markets, exempt from the frequent raids and wars that broke out between the scattered tribes of Bedouin and other clans across the peninsula. One result of this sedentariness and new wealth was the stratification of their society, based on hierarchies of abstract energy capture. A new oligarchy arose, whose economic agenda came to dominate the city's priorities and push out other values. Already at this time there were attempts to envision the unification of the Arabian Peninsula into a single state. But the diversity of religions, economies and ways of life, spread over a great distance of many difficult-to-inhabit miles, made this unification unlikely. What was needed was the sort of intervention Constantine had deployed to hold together his stretched empire in Europe three centuries earlier.

Abu al-Qasim Muhammad ibn Abdullah was born in this Arabia in 570 CE into the Quraysh clan. Though not much is known of his early life, Muhammad married a wealthy older woman and worked as a merchant until the age of forty, when he began secretly assembling followers of the faith he would found and, within three years, begin preaching publicly: Islam. Based on Judaism and Christianity, Islam (meaning 'submission') is the third major Abrahamic faith, and carries many of the central themes of the others. Like Christianity, an aggressive evangelism is embedded in the faith. The result of this is its great efficacy in uniting disparate peoples under a single belief system and government. Muhammad spent the next twenty years building a following and interjecting his new faith, sometimes by force of arms, into the various tribes and societies spread around the peninsula. In his book on the role of monotheism in major empires, *From Empire to Commonwealth*, historian Garth Fowden argues:

> Muhammad shared Constantine's ability to perceive the historical moment, but was both more radical and luckier. On the

political level he was able to set in motion a sequence of conquests that resulted in world empire. And on the cultural level he did not merely choose one religion rather than another and then rewrite history accordingly. Instead he gave history new impetus by proclaiming a new revelation and a new religion, while cleverly drawing on the momentum built up by earlier monotheist prophets.[68]

Soon after Muhammad's death, an Islamic empire began rapidly spreading beyond Arabia across the southern Mediterranean. From this expansion came the Islamic Golden Age that endured from the eighth century to the thirteenth.[69] Perhaps more than any that came before, the Islamic society that arose during this period mixed mythic forms of the progress narrative formula with secular ideas – many based on their own sorts of progress narratives – that presaged those that would follow in the European Enlightenment.

At the beginning of Ramadan in 859 CE, an Arab woman named Fatima al-Fihri founded a mosque that became the world's first degree-granting university. Today, the University of al-Qarawiyyin in Morocco remains the world's oldest, longest continually running university. During the Islamic Golden Age, Baghdad became a global centre of trade, scholarship and invention. Scholars developed algebra, recorded, preserved and translated classical texts, invented new writing systems, and contributed new theories in law, philosophy, psychology and logic. Medicine saw much new research during this time: Ibn Sina's *Canon of Medicine* remained an authoritative text for half a millennium.[70] Fields like astronomy, mathematics and chemistry received important refinements.[71] Al-Sufi developed a scientific description of the Andromeda galaxy and Al-Hajjaj ibn Yusuf translated Euclid's *Elements* from ancient Greek. Science writer Dennis Overbye reflects, 'Commanded by the Koran to seek knowledge and read nature for signs of the Creator, and inspired by a treasure trove of ancient Greek learning, Muslims created a society that in the Middle Ages was the scientific center of the world.'[72] Even so, Muslim scientists were not secularists; they were thoroughly committed to Islam. Frequently compared with Leonardo da Vinci, Abu Rayhan al-Biruni is often credited with the founding of anthropology and geology.[73] He also proclaimed, 'how much superior the institutions of Islam are, and how more plainly this contrast brings out all customs and usages, differing from those of Islam, in their essential foulness.'[74]

Embedded in Islam is a progress narrative akin to those of the other Abrahamic faiths. Most Islamic sects see Muhammad as the 'Seal of the Prophets', the final prophet, following from the other prophets through the Old and New Testaments, rendering the political leader the culmination of historical and spiritual progress. Muhammad founded this new branch of Abrahamic monotheism to unite disparate polities: to build and strengthen what became a parasitic, trade-based empire. Like its Christian cousin, Islam contains the promise of an off-world after-death paradise reserved for believers who correctly and fully submit to God, and those who are God's messengers and representatives on Earth. For secular readers today, it may seem difficult to square this sort of 'progress' with faith in what we may consider more material forms of progress, like the idea of economic growth or development delivering increases in health, peace and happiness. But it is worth remembering that for many people with such faith in the Abrahamic myths, the notion of after-death paradise *is* material in the sense that there is direct continuity between physical earthly bodies and the souls that endure to enjoy heaven – a heaven which is not a metaphor or abstraction, but a very real place capable of being experienced. The New Earth of Christianity is a physical place *on earth*, not in some parallel higher dimension. There is little psychological or emotional difference between faith in a future of eternal joy in heaven – whether on earth or otherwise – and faith in the secular future many contemporary narratives of progress promise: of enduring peace and pleasure in a prosperous society.

While the Abbasid Caliphs encouraged a high degree of learning during the Golden Age, these rulers were not benevolent philosopher-kings, but the same sorts of authorities that governed the parasitic empires in ancient Mesopotamia and the Mediterranean. The caliphs demanded absolute loyalty from their subjects, punished disobedience with public executions, and haunted palaces run by enslaved workers.

Male slaves were commonly captured, castrated, and shipped from Africa, India and the Caucasus Mountains, while female slaves were seized from surrounding localities by the caliphs' armies and held as concubines.[75] The Abbasid Empire also used slaves from East Africa to harvest salt from the marshlands of southern Iraq, work that historian Bernard K. Freamon describes taking place under 'horrific conditions'. As he writes, '[Most] of the imperial conquests accomplished by the Abbasids . . . were simply aggressive wars designed to subdue local populations and gain control of a new territory's land and resources. Conquest and imperial plunder was the norm.'[76]

In 1258 CE, the Mongolian empire sacked Baghdad, massacred its inhabitants, and destroyed its mosques and centres of learning, including its much-mythologised library, the Bayt al-Ḥikmah or 'House of Wisdom'. Rumours spread that so many books were dumped in the River Tigris that it ran black with ink and the bloated tomes enabled soldiers on horseback to cross. This initiated the end of the Islamic Golden Age, but scholarship continued. Ibn Khaldun was born in Tunis on the opposite side of the Islamic world in 1332 CE. Khaldun was a significant figure in introducing more secular economic and sociological theories that anticipated many of those that would emerge in the European Enlightenment three hundred years later. Like classical Greek scholars before him, and like the macrohistorians and social theorists to come, Khaldun's *Muqaddimah* sought to describe laws of societal change over large periods of time beyond a purely Abrahamic historiography, and he did so in a similar frame of progressive historical development:

> One should then look at the world of creation. It started out from the minerals and progressed, in an ingenious, gradual manner, to plants and animals. The last stage of minerals is connected with the first stage of plants, such as herbs and seedless plants. The last stage of plants, such as palms and vines, is connected with the first stage of animals, such as snails and shellfish which have only the power of touch. . . . The animal world then widens, its species become numerous, and, in a

gradual process of creation, it finally leads to man, who is able to think and to reflect. The higher stage of man is reached from the world of the monkeys, in which both sagacity and perception are found, but which has not reached the stage of actual reflection and thinking.[77]

Khaldun's notion that 'civilisation progresses gradually' by similar processes to biological evolution was developed more thoroughly in the nineteenth century and has reverberating effects even today.[78] Seeing the beginning of the Islamic decline in the Mediterranean, Khaldun was sensitive to the collapse of civilisations and developed a cyclical theory of social change as well: simple tribes lead to civilisations, but within four generations begin to decline as a result of a deterioration of *asabiyyah*, or social cohesion: '[a society's] group feeling decays because the people who represent the group feeling have lost their energy. As a result, the dynasty progresses toward weakness and senility.'[79]

But though his theory introduces some scepticism about a final perfecting of society, he writes explicitly of the scientific progress delivered by the Islamic Golden Age: 'Muslim scientists assiduously studied the (Greek sciences). They became skilled in the various branches. The [progress they made in the] study of those sciences could not have been better.'[80] In his *Wealth of Nations* published in 1776, Adam Smith would describe how the division of labour increases worker productivity. Khaldun 'presented almost exactly the same argument 400 years earlier', according to economist Dieter Weiss.[81] Khaldun developed a modern, proto-liberal economic theory, presenting arguments for low taxation, market regulations, economic growth and development, and describing a precursor to the Laffer curve (a model attempting to measure the relationship between tax rates and revenue). Meanwhile, he is credited with developing early versions of what we would now designate as Keynesian economic principles and what are considered their contrary, supply-side economic theories.

Khaldun anticipates Thomas Hobbes by arguing that societies depend on the presence of a strong centralised ruler or bureaucracy to prevent a war of all against all.[82] But, simultaneously, he presages a thinker often pitted against Hobbes, Jean-Jacques Rousseau, when he claims that the Bedouin are less tainted by civilisation, and that 'they are closer to the first natural state and more remote from the evil habits that have been impressed upon the souls (of sedentary people) through numerous and ugly, blameworthy customs.' He even prefigures Karl Marx, writing that 'Profit is the value realized from human labor.'

And, like the Athenians before him and the Europeans after, he espouses a racial hierarchy classifying humans into groups of 'savage' versus 'civilised' in service to the extractive imperatives of the parasitic economy he is writing in favour of: 'to the south, there is no civilization in the proper sense. There are only humans who are closer to dumb animals than to rational beings. They live in thickets and caves and eat herbs and unprepared grain. . . . They cannot be considered human beings.' The Islamic Arab settlers in Africa often treated black Africans with the same exploitative contempt that European colonists would later display. He goes on to write that, 'Negroes of the first zone . . . live in savage isolation and do not congregate, and eat each other.'[83]

Although many of the basic principles underlying the secular European Enlightenment, including the idea of civilisational progress, that we will examine in more depth in the next section can be found embedded in Khaldun's *Muqaddimah*, I do not wish to imply that this means Khaldun is superior to European Enlightenment thinkers, nor that Enlightenment ideas are all plagiarised from a North African scholar. Rather, what the resemblance shows is that such ideas create cohesive frameworks useful in the reproduction of parasitic economies across time, geography and culture. Many of the European thinkers who developed similar ideas likely did so independently, or as a result of different inspirations; they just happened to be as practically valuable and popular within elite realms in parasitic European market monarchies as they were in parasitic Arabic market monarchies. It is not necessarily the case that these thinkers are universally deploying such ideas with underhand or sinister intent. Rather, they are operating within this system and are mutually reproducing it, even as it is funding their research, so they are incentivised to uphold structures that maintain the status quo, as many thinkers do today with apparently benign intent.

Ultimately, Khaldun's role was not simply to make impartial theories about the world as it is. Enmeshed as he was in his religion and state and therefore acting as a champion of its fundamental ideas, his work, held together by the framework of societal progress, has the effect of bolstering and justifying the political and economic logic that governed his society. Khaldun reminds the reader of Muhammad's admonition: 'Whoever dies confessing that there is no God but God, enters Paradise.'[84] In one specific example, he praises the opportunity given to Mamluks – slave soldiers brought to serve in Egypt – to convert to Islam and become military leaders: 'cured by slavery, they enter the Muslim religion with the firm resolve of true believers and yet with nomadic virtues unsullied by debased nature, unadulterated with the filth of pleasure, undefiled by the ways of civilized living, and with their ardor unbroken by the profusion of luxury.'[85]

The values of Islam and of secular progress together aimed to create an ostensibly classless collective more effective at conquest; or rather to engender a shared faith that could, in its blinding brilliance, outshine the shadows – unfairness, unfreedom, indignity – cast by the hierarchies intrinsic to parasitic modes of economy. And like the other Abrahamic myths, Islam grants humanity dominion over other life: 'It is He Who made you vicegerents in the earth,'[86] and offers up animals for human use: 'It is Allah who created the cattle for you that you may ride some of them, and some of them you eat / and there are [numerous] uses in them for you, and that over them you may satisfy any need that is in your breasts, and you are carried on them and on ships.'[87]

The Islamic Golden Age was fuelled by the conquest of peoples and land: over 1,400 years, Islam and the notions of progress embedded in it provided an important vehicle for Arabic-speaking people to rapidly spread from small tribes in Syria and the Arabian Peninsula across southwest Asia, North Africa and southwest Europe. And it remains powerful today: Islam is the world's fastest-growing and second largest religion. The ideological cohesiveness of Islam, its promises of spiritual and earthly progress, and its adaptability in melding with secular theories in civil law, science and philosophy enabled Arabs to build one of the largest pre-modern empires. The relationship between parasitic economies and these ideas was circular, mutually reproductive: great wealth from the spoils of war funded the exchange and discovery of knowledge through trade routes and markets, and made possible the enlightenment in the Arab world as it would in the European Enlightenment centuries later, just as similar ideas – sometimes the very same ideas – justified, explained, and practically facilitated more conquest.

Saladin, one of the most famous of the Islamic emperors in the Western canon, defeated Christian crusaders in the Levant and went from that victory to found the Ayyubid dynasty. Despite their oppositional relation, these two statist religions, Christianity and Islam, were not – are not – so different. They were both extremely effective in imposing a parasitic economy on the Indigenous groups in their path of conquest, implementing policies that ranged from intense agriculture, sedentarisation and environmental despoliation, to church hierarchies and patriarchy. As Christianity conquered Celts, Britons and Sámi, so Islam conquered Moors, Somalis and Amazighs. But their very success made their resources insufficient, and as the world grew more interconnected, new forms of parasitism arose to continue energy capture and new forms of the progress narrative emerged to justify and reproduce them.

Coda

Five thousand years ago, a handful of city-states in Mesopotamia initiated a lineage of societies within the descendants of which most people still live. This lineage broke from most other cultures in how it related to the ecological systems on which it depended, and consequently, how the people within these societies related to each other. Such a shift could be understood as moving from more mutualistic or commensalistic – energy capture that does no harm – to parasitic – energy capture that does harm. This disruption in relations demanded a concomitant disruption in the intellectual and spiritual systems people used to understand the cosmos and organise themselves into functional institutions. In order for rulers and ruled alike to justify engaging in inhumane acts, like genocide, extinction and enslavement, a new intellectual framework would have to make these things seem not only to be necessary evils, but even to be heroic acts bringing the world into a more civilised state. From this we see a narcissism of monarchy and monotheism, a turning away from non-human life to an anthropocentrism that views humanity as central to all drama of life on Earth, and to dramas happening elsewhere than Earth: hell or paradise in an afterlife.

We also see a change in the balance of energy capture. In previous eras, extraction had primarily been carried out via the exploitation of nature; now it came via the exploitation both of other peoples – whether internally enslaved or externally colonised – and of the productive ecological zones cultivated for millennia by Indigenous foragers and horticulturalists, ransacked into strictly ordered farms and denuded pasture. It is in this period that quantity of human life becomes more important than quality of human life, or either quantity or quality of other-than-human life. When Rome ruled Britain between the first and fourth centuries CE, the population doubled, indicating an influx of resources sufficient to feed them. But quality of life did not follow, and likely deteriorated, given that life expectancy in post-Roman Britain rose with Rome's fall.[88] As anthropologist Marvin Harris illustrates: 'In the 4,000 years between the appearance of the first states and the beginning of the Christian era, world population rose from about 87 million to 225 million. Almost four-fifths of the new total lived under the dominion of the Roman, Chinese Han, and Indian Gupta empires.'[89] He describes the conditions in these empires as 'warrens full of illiterate peasants toiling from morning to night only to earn protein-deficient vegetarian diets. They were little better off than their oxen and were no less subject to the commands of superior beings who knew how to keep records and who alone had the right to manufacture and use weapons of war and coercion.'[90]

These societies used a form of 'contiguous parasitism', the spread of territories connected primarily by land, for their energy capture, and by mythic narratives for the cultural tools used to reproduce that energy capture. That is, they developed progressive ideas of time and a set of other narratives that supported and followed from this progressive notion of history.

In Mesopotamia, the cult of Marduk, the Enuma Elish and *The Epic of Gilgamesh* all introduced the idea that the world develops from a primeval beginning and progresses into civilisation; they glorified kings alongside a codified civil law that solidifies hierarchic rule, that one central god is a powerful motivator and supporter of aspiring rulers, and that nature exists to provide market goods. Zoroastrianism in Persia, meanwhile, introduced ideas of heaven and hell in the afterlife, developed what is likely the world's first fully formed monotheism and eschaton, an apocalyptic myth that puts history on a path toward an end point, which yields perfection. The Old Testament introduced explicitly divine dominion, the rule of man over nature, and the Promised Land myth that puts forth divinely ordained conquest. Athens and Rome introduced more secular versions of progress myths. The Athenians cultivated a racialised distinction between savages and the civilised. The Romans, meanwhile, introduced the Pax Romana, a hegemony in which the Roman state and military were tasked with imposing peace by force on far-spread colonies. Christianity proved, through the late Western Roman emperor Constantine, that it could be used as a powerful tool of unification and expansion. The Christian progress myth is one that incorporates many elements from the Old Testament and Zoroastrian ideas – dominion, end-times perfection, heaven – but adds an aggressive missionism. Embedded in the faith is the imperative to spread and conquer, to displace others and redesign societies around its tenets. This included the final Christianisation of Europe with the conversion of Norse pagans, who themselves were effective colonists, relying on the myth of battle-death and drunken paradise in Valhalla before adopting Heaven. Finally, Islam built on the Abrahamic progress myths, but took the missionism of Christianity even further: a faith was hammered together in a smithy for empire, designed deliberately to unify, spread and conquer. It also brought the most secular intellectual infrastructure into its faith for the practical purpose of governing a wealthy empire. The Islamic Golden Age paved straight and sturdy roads that Enlightenment thinkers in Europe would soon extend, even treading across oceans.

This view of traditional religious faith being put into operation as a form of social control is common: Karl Marx famously declared that religion is the opium of the people, or the sigh of the oppressed.[91] Popular historian Yuval Noah Harari has written, 'The crucial historical role of religion has been to give superhuman legitimacy to . . . fragile [social] structures. Religions assert that our laws are . . . ordained by an absolute and indisputable authority. This helps place at least some fundamental laws beyond challenge, thereby ensuring social stability.'[92] Meanwhile, quantitative historian Peter Turchin argues that religion may be seen as a social adhesive used by sprawling empires to maintain unity: 'World religions first appeared during the Axial Age and provided a basis for integrating multiethnic populations within first mega-empires, such as Achaemenid Persia (Zoroastrianism), Han China (Confucianism), and Maurya Empire (Buddhism).'[93]

Religious faith is effective as a political tool not due to an *a priori* or intrinsic dysfunction in human psychology, a 'god gene' or some other will-to-divine-subservience.[94] Instead, the rending of religion into a shape capable of summoning the collective delusion necessary to ease the trauma of conquest, genocide, slavery and ecological despoliation develops across time, culture and geography. It is so effective at social control and pacification because it can genuinely provide for deep-rooted psychological and emotional needs: faith in an ordered and just universe that cares about our toils; a sense of momentum and meaning in our work; clear boundary lines between in-groups and out-groups, good and evil; a promise of alleviation to our suffering, or comfort in the face of death's permanence and mystery. All progress-based faith does the same, whether theological or secular. The misuse of faith occurs alongside its reasonable use, and it does so by hijacking an instinct for progress. Instead of simple propaganda wielded by a small conspiracy of oligarchs, we ought to see this sort of psychology as endemic among both host and parasitic classes. The potency of this metanarrative is that it can be just as compelling, or more so, in capturing imaginations sufficient to organise continent-wide protection rackets, extraction frontiers and trade zones – the material promise empires typically base their legitimacy on – and compel others to do base violence for a higher purpose. The psychology of myth and religion as social adhesive or brainwasher of the masses is complicated, tapping into genuine needs, hijacking both prosocial and antisocial tendencies in cooperative, omnivorous mammals, and resting above all on progress: on the profound urge to feel hope, to maintain faith in the improvement that lies ever just ahead. That if we all join together, the high and the low, oars in hand, pulling as one, we will cross a river of flame or forgetfulness into some Celestial City.

II Nation

Overview

Date: 1400 to 1900
Progress: secular and rational
Parasitism: disparate and maritime
Agents: kingdoms, nation-states, empires

Context

While Islamic scholars were developing the foundations of many fields of inquiry still used today – and Islamic rulers were enslaving, raping and pillaging their subjects – European cities, tribes and nascent nations were at war after the fission of Roman collapse. As the polities of Rome exploded apart after the fifth century CE, they reformed into independent entities. Some cities across Europe developed more democratic means of governance, and some places warred over the voids of hierarchy formerly filled by Roman or Roman-aligned rulers. Emerging parasitic classes soon reasserted control and consolidation. Kingdoms arose, commercial dictatorships took over formerly, briefly democratic cities, aspiring emperors spread their boundaries, and, with the assimilation of polytheists and the expulsion of Jews and Muslims throughout Europe, Christianity took full hold of the continent.

Soon the contiguous parasitic realms of Europe bumped into one another, populations grew dense, the peace – or at least, order – imposed by martial laws, kings and religious conformity could not hold. With the rise of shipbuilding technology, and with the insatiable appetites of increasingly wealthy aristocratic and autocratic dynasties, a taste for faraway luxuries – the same sorts of tastes that fuelled the rise of the first parasitic economies in Mesopotamia, swapping lapis lazuli for silk and silver – compelled members of Europe's parasitic classes to invest in ventures seeking resources that could only be got from far afield. The causation goes something like this: royals have imperatives to grow their economies and treasuries because they want goods, because they want legitimacy as rulers, because they want to defeat their enemies, or to satisfy egos lusting for a glorious legacy. In order to grow their economies, they engage in rent, labour and tax extraction from their population, plus external extraction through war and conquest. To justify to themselves, their soldiers, their subjects and workers the need to participate in these actions, they need propaganda, one of the most potent forms being a grand narrative. They use the ideas within progress narrative formulae for this purpose, which in this case include Christian missions and theology promoting the conversion of 'pagans', the putative civilising of savage peoples, the cultivation of wilderness into agriculture, and more.

Although not as great a shift in human-ecological and human individual and class relations as that which occurred around 3000 BCE, the new shift in the *form* of parasitism would have more global and immediate consequences around the fifteenth century. Contiguous parasitism relies on a particular form of energy capture in which primarily centralised bureaucracies organise armies and industries to extract energy from both Indigenous human populations and native ecological systems in their immediate vicinity. As a central state, usually a city-state, achieves military and diplomatic supremacy it spreads its borders regionally. What this meant in terms of *concrete* energy capture was that agricultural projects, organised and maintained by large landholders and their typically enslaved workers, created large exclusionary zones in which native flora and fauna were extirpated, land intensively cultivated or grazed. Resources extracted from deforesting, mining and expropriation in these zones were pulled into the centre of the state, distributed to owners and rentiers, and recirculated through the economy and smaller elite economies. What it meant in terms of *abstract* energy capture was generally the rule of a small parasitic class of owners, citizens and politicians and a much larger class of peasants, slaves, soldiers and other workers – with some overlapping and swapping between the classes, like soldiers whose plunder made them large landholders. These sorts of economies relied primarily on land-based transportation and communication.

The shift around the fifteenth century in the West's parasitic economies meant that they began to rely more heavily on waterborne transportation and communication, utilising river systems and, more importantly, oceanic travel. Suddenly transit and communication could cross much larger geographic expanses. What was unique about this new sort of maritime economy was that the Atlantic Ocean no longer presented a vast expanse of hostile, uncharted wilderness. Instead, with the discovery of ocean currents and the development of tall ships, it became a highway for moving goods and people.

This meant that parasitic economies could establish faraway colonies, extract resources and move those resources at a commercial scale along sea routes. In this way, a more disparate form of extraction began to occur, distinct from that practised by the Norse, for instance, that we examined in the last section. Norse settlers used maritime routes effectively, but their activities were more decentralised and made use of waterways as a means of achieving land-based expansion. This new form of parasitism was more centralised and broader reaching, pulling resources from supply chains across the world into a home continent. As such, it necessitated more complex forms of bureaucracy and systems of exchange – most notably, capitalism – but also a range of new narratives, philosophies and intellectual systems to serve the purposes of justifying and systematising these new administrative infrastructures. Perhaps most importantly, all those involved in both organising the new systems and carrying out the necessary roles had to be convinced that they were thereby serving a higher destiny.

The narratives deployed as the mortar for the contiguous parasitic economies that dominated the classical and medieval periods in the West were primarily mythic. There is a consistent religious seam running through the bedrock of contiguous expansionism, even well outside those periods. It appears to simply be a characteristic of this mode of parasitism that its narratives take on a more mythical strain when conducted in a contiguous, regional way. In the same way, the disparate mode of parasitism that this section is concerned with carries with it patterns in the narratives used to make sense of – and promote in a propagandistic sense – this form of parasitism. The cultural reinforcements that began to attain dominance in the fifteenth century in the West became distinctly tinged with secular ideas. Whereas in the Islamic Golden Age secular concepts blended with Islamic doctrine, the European Enlightenment, which arose with this new form of disparate parasitism, saw complex entanglements between Christianity and secular ideas, and sometimes conflict between them. Christianity could be used as a pioneering force in the

New World, with missionaries establishing the first outposts before armed conquerors arrived, and could provide a divine justification for colonising, pilfering and enslavement. But church administration and contiguous imperial bureaucracies were not up to the task of efficiently capturing, moving and consolidating such great quantities of wealth from such faraway sources. New administrative forms would have to emerge.

Furthermore, members of the majority who would be cutting down trees, digging up silver and burning down villages would have to be, by definition, more worldly wise than their provincial counterparts in contiguous parasitic expansion, who could conquer relatively local dominions. They were, after all, sailing thousands of miles to worlds that no one from their cultures even imagined existing. A measure of learning and cosmopolitanism was necessary not just for those settlers making up the host class but for the parasitic class as well, the members of the authority uprooting from their familiar rural estates in Old World Europe to establish parallels in the Americas and the newly opened markets and colonies in Asia and Africa. And as commercial enterprises began to grow in importance, with their supremely secular imperatives of simply extracting resources and converting them into abstract wealth, numbers on a ledger, Christianity as an organising and unifying force could only serve to get in the way of profit. A new sort of belief system was needed.

Wealth of Europe

Capitalism is a word carrying immense baggage. It means many things to many groups of people, with all sorts of positive and negative connotations hanging heavily on it. The most fundamental definition of the term goes something like this: a system whereby material resources and labour are transformed into 'capital', in the form of commercial elements that may be either abstract or concrete. These may include in the former case stocks, bonds or debt packages, for instance, and in the latter physical property, raw materials or produced goods; or they may be both concrete and abstract in the

II NATION

form of human physical labour. Central to capitalism is the accumulation of profit, meaning an excess of value extracted from a transaction or production. Capital*ism* is the umbrella name for the administrative frameworks that move capital around the world. The conventional history of capitalism places its origin with the mercantilism that arose in southern European cities during the Renaissance (1300–1600 CE), a period commonly seen as a transition to the 'modern' world. However, as we have seen, many of the elements common to capitalism were present in economies of the classical empires of southwest Asia and southern Europe thousands of years earlier. Empires from Mesopotamia, to Persia, to Egypt, to Greece, to Rome had begun to build market economies producing capital – like property, debts, and currencies – all in the interest of growing profit, more than four thousand years before capitalism is understood to have emerged in Europe. This is because parasitic economies have been operating on many of what we now call 'capitalist' principles since around 3000 BCE. It was not capitalism but parasitism that was the origin of extraction of value from labour, abstraction of value and the centralising of growth in profit above all other concerns. Capitalism was, rather, an administrative innovation that accommodated a new form of economic parasitism to truly *global* networks of trade and extraction.

Even so, capitalism, or the variations on this administrative system that were percolating throughout European kingdoms, colonies and economies before their theoretical coalescence under that broader terminology in the eighteenth century, was an important cultural reinforcement for the introduction of parasitic forms of human-ecology and political economy across the globe. By some accounts, capitalism was imposed by European monarchies and commercial enterprises within the continent as a trial, using similar forms of extraction, enclosure (of land and property) and control of domestic populations before exporting it to overseas colonies.

One misconception lingers today, a legacy derived in part from the new set of Enlightenment myths that were soon to emerge, yet remains part of the conventional view of capitalism and its history. This, a view shared even by traditional anti-capitalists, is that capitalism was a major force in freeing European peasants from feudalism. Between about 1350 and 1500, European peasants from Florence to Paris, to England and Germany enjoyed increased wages, greater control over the land they worked, and more leisure time, with frequent feasts and festivals. Women's wages also increased relative to men's and food prices fell. On top of enhanced lives, social conditions and economic equality in this period, the biomes of Europe improved immensely as peasants developed a more reciprocal relationship to the land. In a sense, those peasants recalibrated their relationships both to each other and to the ecological systems they inhabited to be more commensalistic and less parasitic. The Gaelic concept of *Dùthchas*, for instance, was once widely used to describe the interconnectedness of living beings in Celtic regions of Britain; it is 'a haunting expression of the idea of unity existing between land, people, all living creatures, nature and culture' in the words of scholar Alan Riach.[1] Economic anthropologist Jason Hickel writes that this shift 'inaugurated a period of ecological regeneration [in which] Europe's soils began to recover. The forests regrew.'[2] Part of this change can be explained by the Black Death, the plague that burned across Europe in the late 1340s and decimated the population by 25 million. So many died, the reforestation that occurred where people had previously lived may have cooled the planet.[3] The massive reduction in availability of workers increased the bargaining power of labour, as the surviving workers could demand better pay without being undercut by competition. More than a century before capitalism is generally agreed to have emerged, peasants across Europe were rebelling against feudalism and overthrowing lords.[4]

Rather than capitalism freeing peasants from servitude, the reverse is closer to the truth. The majority of the population in England at the time worked on land that they owned and subsisted on the fruits of their labour in an often self-sufficient existence. The imposition of new regimes we would now call capitalism rendered the majority of the population landless and obliged them to work for employers to earn wages – an existence dependent on markets to obtain necessities and endowing landowners with the power to hire, fire and evict.[5] As historian Roxanne Dunbar-Ortiz writes, 'The first population organized

II NATION

under the profit motive – whose labor was exploited well before overseas exploitation was possible – was the European peasantry. Once forced off their land, they had nothing to eat and nothing to sell but their labor.[6] Many social and political relations shifted with this move. Historian Spencer Dimmock documents how in England, likely the first country to make the transition to agrarian capitalism and then industrial capitalism, peasants were often violently evicted from their land.[7]

The wealth gap between men and women in Europe grew around 1500 as rulers began to reassert power and agricultural work diminished.[8] Around the end of the fifteenth century, as lords and nobles began seizing back property from collectively owned peasant commons and communes, some European monarchies began stretching their fingers into new continents across the oceans to repeat the process there.

Perhaps the most consequential moment for the spread of these parasitic systems was the entry of European economies into the Americas, led by Christopher Columbus and financed by Spanish monarchs seeking new frontiers of extraction to fund their domestic regimes. Columbus made first contact in the Americas on the island he called Hispaniola, now the location of the Dominican Republic and Haiti, in 1492. A passage from Columbus's diary describes his first contact with the native Arawak peoples:

> They are very well made, with very handsome bodies, and very good countenances . . . They neither carry nor know anything of arms, for I showed them swords, and they took them by the blade and cut themselves through ignorance . . . They are all of fair stature and size, with good laces, and well made . . . They should be good servants and intelligent, for I observed that they quickly took in what was said to them, and I believe that they would easily be made Christians, as it appeared to me that they had no religion, our Lord being pleased, will take hence, at the time of my departure, six natives for your Highnesses that they may learn to speak.

Three days later, Columbus wrote:

> These people are very simple as regards the use of arms, as your Highnesses will see from the seven that I caused to be taken, to bring home and learn our language and return; unless your Highnesses should order them all to be brought to Castile, or to be kept as captives on the same island; for with fifty men they can all be subjugated and made to do what is required of them . . .[9]

He took prisoner the first Arawak people he stumbled on, hoping they would lead him to gold.[10] When other natives refused to trade their bows and arrows to him, he ordered two stabbed to death. Those he captured to take back to Europe froze to death on the way.[11] Though they found little of the gold they were commissioned to secure, Spanish colonists in the Americas pilfered 16 million kilograms of silver, more than tripling European reserves.[12]

Imagine yourself, a member of the Tule or Naso peoples, standing in a fertile clearing with your family and fellow villagers in what is today Panama, trading gossip about the strange, pale men who have been seen stalking around the nearby forest. Suddenly, from the south, an army of these steel-clad brutes arrives bearing long blades. One such beast marches before your group and holds out a large swatch of rolled-up skin. Reciting from it, he spits out a long string of words, commanding you to convert to his nation's religion and to recognise his rulers as your own, and threatening that:

> if you do not do this, and maliciously make delay in it, I certify to you that, with the help of God, we shall powerfully enter into your country, and shall make war against you in all ways and manners that we can, and shall subject you to the yoke and obedience of the Church and of their Highnesses; we shall take you and your wives and your children, and shall make slaves of them, and as such shall sell and dispose of them as their Highnesses may command; and we shall take away your goods, and shall do all the mischief and damage that we can, as to

vassals who do not obey, and refuse to receive their lord, and resist and contradict him; and we protest that the deaths and losses which shall accrue from this are your fault, and not that of their Highnesses, or ours, nor of these cavaliers who come with us. And that we have said this to you and made this Requisition, we request the notary here present to give us his testimony in writing, and we ask the rest who are present that they should be witnesses to this Requisition.[13]

Of course, it's all harsh, nasal gibberish to your ears. No one around you has any idea what the many, many words mean, except for an interpreter who may understand some Spanish, but who is equally baffled, comprehending neither the characters mentioned nor the demands made. Upon finally finishing the recital, the armour-clad men leave for a while, going in the direction they had come.

After some time in which you look around at each other in bewilderment, attempting to decipher this strange behaviour, suddenly the steel-clad army emerges from the trees again and begins wantonly slaughtering your people and taking women and children captive. This farce was called the Spanish Requirement (*requerimiento*, or 'demand') of 1513, a legal declaration made by the monarchy and read to Native Americans in the path of Spain's conquest. The letter would be read aloud in Spanish to unsuspecting Natives, who typically understood none of it, immediately before the conquistadors attacked. The document proclaimed explicitly that the invaders were justified in their conquest thanks to holy scripture, and that all those who did not convert would be exterminated and enslaved in accordance with God's will, despite the obvious practical difficulties of conversion for people who did not share a language or culture. The document gave the Spanish crown the legal and religious right, as they saw it, to moral immunity in committing virtually any crime against the Native Americans in their warpath. It was an elaborate theatre of political and spiritual absolution for committing some of the most horrific acts one group of humans has ever committed against another. The Spanish read this aloud to the tribes they annihilated, but they also sought to conquer the existing empires already in the Americas.

Today, as then, Spanish colonisation of the Americas is often also justified by contrasting the Spanish with the brutality of Mesoamerican empires, particularly the Aztecs whom they decreed heathen savages and whom they should thus overthrow and replace. The Aztecs, although their buildings possessed great beauty and size, were seen as barbarians not only because they were not Christian, but because they sacrificed people to their gods. The reality is that the Aztec Empire was indeed brutal towards both its subjects and the other cultures living at its frontiers, as all empires are, but its human sacrifices were ritualistic and selective. This is not to defend the practice, only to point out that the Spanish colonists sacrificed tens of thousands of Native people to their god, too, and possibly more so than any Mesoamerican empire did in such a short span of time. Whereas the Aztec mostly sacrificed young men, the Spanish were more indiscriminate.

Spanish clergyman Bartolomé de las Casas personally witnessed and recorded the means by which the Spanish began to settle and extract from the Americas in his *A Short Account of the Destruction of the Indies*, and the effects this had on the Indigenous inhabitants:

> The *Spaniards* first assaulted the innocent Sheep . . . like most cruel Tygers, Wolves and Lions hunger-starv'd, studying nothing, for the space of Forty Years, after their first landing, but the Massacre of these Wretches, whom they have so inhumanely and barbarously butcher'd and harass'd with several kinds of Torments, never before known, or heard (of which you shall have some account in the following Discourse) that of Three Millions of Persons, which lived in *Hispaniola* itself, there is at present but the inconsiderable remnant of scarce Three Hundred. Nay the Isle of *Cuba*, which extends as far, as *Valledolid* in *Spain* is distant from *Rome*, lies now uncultivated, like a Desert, and intomb'd in its own Ruins. You may also find the Isles of St. *John*, and *Jamaica*, both large and fruitful places, unpeopled and desolate.[14]

After enduring such horrors, Indigenous populations either began to flee to higher terrain, or made plans to attempt to expel the Spaniards.

II NATION

Possessing far less effective military gear, they were subdued by the Spanish genocidaires. De las Casas went into some detail about what the colonists did upon first seeing the islanders' attempt to defend themselves. His journals describe Spaniards overrunning cities and towns and killing people of every age and sex, not sparing pregnant women, instead 'ripping up their bellies' and tearing them 'alive in pieces'. He describes the conquistadors making bets between themselves on whether they could cut a man in two with one sword blow, or who could behead most elegantly. The Spaniards snatched 'young Babes from the Mothers Breasts, and then dasht out the brains of those innocents against the Rocks', while others they tossed into rivers 'scoffing and jeering'. The Spaniards gathered prisoners in thirteens to represent Christ and the twelve apostles, whereupon they tied them to a low gibbet and burned them alive. Others had their hands cut off only halfway, leaving them hanging by the skin, to carry messages to others hiding in the mountains.

But those who fled to the mountains did not long evade the colonists:

> I was an Eye-Witness of these and innumerable Number of other Cruelties: And because all Men, who could lay hold of the opportunity, sought out lurking holes in the Mountains, to avoid as dangerous Rocks so Brutish and Barbarous a People, [. . . the Spanish] bred up such fierce hunting Dogs as would devour an Indian like a Hog, at first sight in less than a moment.

And if a native fought back and killed a colonist in retaliation, the Spanish made it law to kill one hundred natives in response.

The Spanish conquest was only the first incursion of an economy that was fundamentally of a parasitic form. While the Spanish and Portuguese in South America were the first, soon the English, French and Dutch made the voyage to North America, encountering a continent full of diverse forms of human ecology and political economy, from more mutualistic in some places – like the civilisations of northeastern North America and the Great Lakes – to more parasitic in others, like those of the northwest coast and the raiding peoples of southern North America.

The settling of the Americas by Europeans was a long and complex process. It was often more complicated than the simplified picture many of us have in our minds today – of either a triumphant conquest by a superior race or a quick genocide by an intrinsically evil race (Spanish treatment of the Arawak notwithstanding). But the fact is that, as in the case of Columbus, the process was often brutally violent, whether in tactical cold blood, or by the tragedy of unruled emotion, or by the biowarfare of epidemics. When it was done in cold blood, those organising the colonisation of the Americas often took advantage of peoples who were neither physically equipped for well-ordered combat nor mentally equipped to confront the minds of those long committed to parasitism, capable of treating human and non-human animals alike as disposable objects.

After Columbus took the first slaves seized by a European empire in the Americas, it was only around thirty years before a transatlantic system for moving captured people emerged. It is important to note that the Atlantic slave trade does not constitute the invention of slavery, as it is sometimes treated today. Enslavement was endemic in Europe and Africa for millennia prior to this period. There were people working and being sold into forced labour throughout Europe and Africa even as the first Europeans captured the first Africans for the purposes of resource extraction in the Americas. The parasitic classes of Europe and Africa colluded with one another, trading people they saw as products and sources of energy capture.

Even so, the Atlantic slave trade was notably ruthless and represented an intensification of this business. Suddenly, there was an abundance of labour involved in transforming the Americas from largely intact and biodiverse ecological systems into vast tracts of homogeneous farmland. And there was a newly efficient method of administration – capitalism – to order this transatlantic commerce. Between 12 and 15 million Africans were captured, packed tightly onto ships, and sailed across the Atlantic. The journey was deadly; nearly 2 million did not survive it.[15] The number alone cannot capture the suffering embedded in abducting and trafficking people, destroying cultures and sundering families.

Between 1492 and 1600, European colonisation in the Americas contributed to the deaths of at least 55 million Indigenous people – around 90 per cent of the Native population, and 10 per cent of the world's total population – in an event some call the Great Dying.[16] It is worth noting that 'contributed to 55 million deaths' is a passive way of saying tens of millions of people experienced an apocalypse that destroyed millennia-old languages, ways of living and cultures – often violently.[17] These 55 million people suffered murder, war, enslavement, disease and famine: a series of atrocities on a scale monumental enough to change the temperature of the Earth. A study by scientists at University College London found that the deaths of those people by pandemic left behind vast tracts of cultivated land – about 56 million hectares – which grew back as forest and savannah before being captured by Euroamerican settlers. That regrowth drew carbon dioxide out of the atmosphere in quantities sufficient to cool the temperature of the Earth's surface, likely contributing to the 'Little Ice Age' of the sixteenth and seventeenth centuries. In addition to the humanitarian impact of this incursion of European economies in the Americas, the ecological impact resulted in the rapid, mass eradication of dozens of species of megafauna and the clear-cutting of millions of acres of forest.

It was not only the American host classes who were exploited by the Euroamerican parasitic classes. The host classes in Europe suffered from American colonisation as well. Though all this new raw material made 'Europe' wealthier, those resources mostly went to the parasitic classes of the continent and did little for Europe's host classes; the continent's peasants and workers saw little, if any, of it. One of the consequences of colonists amassing fortunes from Africa and the Americas was 'the catastrophe experienced by small farmers in Europe and England', who were undercut by cheaper labour and lived under the whims of ever-richer landowners with ever more tools for maintaining their supremacy.[18] In Ireland, the Great Famine killed a million people and displaced more than twice that many.[19] This was due to concentrated land ownership and the imperative to sell commodity crops on international markets rather than feeding those growing them.[20] In Scotland, lords and professional classes benefited from the British Empire and global slave networks, but tenant farmers and workers there still suffered. The Highland Clearances saw thousands of Scots killed and displaced. As author Chris Bambery notes, a new agricultural system was developed at this time in Scotland to dramatically increase yields, but it was based on mass eviction.[21]

The cool calculation of profit-making under the administrative regime of global capitalism and the tactical deployment of violence in an effort to seize land for the imperatives set by a parasitic human-ecology are necessary functions of this global system, but they entail actions that are difficult for most – though obviously not all – individual people to endure or justify without some sort of *reason*. It is under these circumstances that we see the wave of ideas that would make up a new intellectual age now called the Enlightenment. Those ideas served various functions – some emancipatory, others the opposite – and knitted together the progress narrative formula that would drive action during the five centuries of disparate parasitism between around 1400 and 1900 CE. It was necessary not only for the acquiescence of the host to hear about the grand narrative for whose facilitation they were being extracted from, but for all those involved, both those undertaking concrete energy capture and those organising it and undertaking abstract energy capture. That is, for the authority to believe they were working for a common good, for a future, for perfection, for the majority. The writings of philosophers and the diaries, letters and speeches of political leaders bear this out, as we will see.

Histories of Progress

Enlightenment thinkers were diverse. They frequently disagreed with one another, often vehemently, and many thinkers privately deplored and publicly decried the abuses of colonial and capitalist states, or what I am calling parasitic economies, during this period of European expansion.

But from the same period emerged arguments whose dominance came to depend on the extent to which this new intellectual machinery could accelerate disparate parasitic economies' conquest of the globe,

II NATION

transforming ecological systems into human systems, and transforming more diverse human systems into homogeneous imperial possessions. Often, the justifications for mass death and suffering were the ideals of Democracy, Equality and Liberty. But the tragic fact is that the new governments, economies and societies meant to represent the fulfilment of civilisational progress were often far less democratic, equal and free than the societies they destroyed and replaced. As anthropologist David Graeber and archaeologist David Wengrow document in great detail, many of the high-minded Enlightenment ideals debated in European salons and universities came directly *from* the peoples Euroamerican colonists were killing or displacing. In part, the Enlightenment debates taking place, even those steeped in progressive history-telling, recognised that Native societies already practised many of the virtues championed by Enlightenment European thinkers, like democracy, liberty, equality, fraternity and some form of free trade. As Graeber and Wengrow explain, the 'indigenous critique' shook many of these thinkers and the many Europeans reading and talking about them. The indigenous critique in this context refers to the reaction of many Native Americans to the Euroamerican society that arrived relatively suddenly, in which they 'developed their own, surprisingly consistent critique of European institutions' – critiques which 'came to be taken very seriously in Europe itself'.[22] Many cultures that already existed in North America put such values of liberty, democracy, and equality into practice effectively, honing and refining societies over millennia. Europe did not bring those values to the Americas; Native Americans brought them to Europe through reports brought back indirectly by missionaries and colonists, and in some cases directly by Native individuals. Wendat philosopher and statesman, Kandiaronk, from what is today northern Michigan, was by all accounts a preternaturally skilled orator and charismatic public intellectual. In addition to the political manoeuvring he accomplished in the Americas, engaging in complex diplomacy with European and Native Nations, ultimately with the aim of stemming settlement, there is evidence he may have gone to France himself. He 'took the position of rational skeptic' and, among other ideas, challenged Christianity in the heart of Christendom.[23]

Indigenous Americans saw Europeans as uncouth, poor conversationalists, greedy, and ultimately slaves to kings, lords and commanders in great need of liberation. Graeber and Wengrow quote the reports of missionaries who came into contact with Wendat and Mi'kmaq peoples. The Wendat were a confederation of tribes located around the eastern Great Lakes who constructed complex settlements as townsfolk, and the Mi'kmaq inhabited the region around Atlantic Canada and northern Maine. The major critiques made by members of these cultures focus on moral and practical issues. One prime source of critique was greed, with the Natives viewing European colonists as avaricious and selfish. Mi'kmaq natives reportedly said to the French, 'You are envious and are all the time slandering each other; you are thieves and deceivers; you are covetous and are neither generous nor kind.'[24] Even the European – in this case, French – manner of conversation was not safe from the indigenous critique:

> [French missionary Friary Brother Gabriel] Sagard was surprised and impressed by his [Wendat] hosts' eloquence and powers of reasoned argument, skills honed by near-daily public discussions of communal affairs; his hosts, in contrast, when they did get to see a group of Frenchmen gathered together, often remarked on the way they seemed to be constantly scrambling over each other and cutting each other off in conversation, employing weak arguments, and overall (or so the subtext seemed to be) not showing themselves to be particularly bright.[25]

Aside from scandalising Europeans with the apparent equality between men and women in these Native societies, the main difference and another major source of their critique were differences in attitudes towards individual liberty. 'That indigenous Americans lived in generally free societies, and that Europeans did not, was never really a matter of debate in these exchanges: both sides agreed this was the case.'[26] Indigenous Americans were aghast at the extreme hierarchy, the exertion of power over each other, and the imprisonment in cages that they saw in European settler society, viewing French settlers as

II NATION

'little better than slaves'.[27] It should come as no surprise that the intensely uneven means of energy capture, between individuals and between humans and nature, that was part of Euroamerican parasitism was strikingly alien to peoples who practised more mutualistic forms of extraction.

This period of intensive parasitism in the Americas accompanied a renewed and secular interest in progress during the Enlightenment. 'During the period 1750–1900 the idea of progress reached its zenith in the Western mind in popular as well as scholarly circles,' Robert Nisbet writes. 'From being *one* of the important ideas in the West it became the dominant idea.' But it was not just an isolated idea. Progress in this period 'becomes the developmental *context* in which other ideas important at the time, like equality and liberty, were framed'.[28] Even as Europeans debated such ideals, and even fought in revolutions on their behalf, ideas that came pure from the ink of the pen did not remain so when they moved from the page into the world. As in the Islamic Golden Age and classical and medieval European empires past, the progress narrative formula played a vital role in the reproduction of this parasitic economic and human-ecological form. Many of the period's most prominent thinkers published ideas that provided the foundation upon which secular progress narratives could be built. Some explicitly contributed to the project, while others did not intend for their words to be put toward the imperatives of a parasitic economy. As with thinkers of the so-called Axial Age – Buddha, Lao Tse, Confucius, Zoroaster, and others whose sermons of justice and liberty were bent towards Old Deal divisions of wealth and power – Enlightenment thinkers had little control over how their words would be used.[29]

In his brief thirty-seven years, Robert Burns became a central figure of the Scottish Enlightenment, itself a major contribution to the broader European Age of Enlightenment. The Enlightenment emerged during the 'long eighteenth century' (1685–1815), arising at the height of European colonisation, and formally ending during the transition towards fossil fuel-driven industrialisation. Scottish Enlightenment thinkers would directly inspire luminaries like Benjamin Franklin, Charles Darwin and John Stuart Mill. Scotland's capital, Edinburgh – a city that Thomas Jefferson considered peerless throughout the world[30] – became a global centre of learning, home to giants of the age like Adam Smith and David Hume, and, for a time, Burns himself.

Among Enlightenment thinkers, Burns is one of its better angels. An iconoclast from humble origins, a farmer-poet, a beloved fount of inspiration for liberals, socialists and freedom fighters alike, Burns represents the promise and nostalgia of an Enlightenment that could have been more than it was. Burns was a political radical, but he was careful to couch his views in historical retellings, and he published his more strident poems anonymously. He wrote lines imagining the battle speech of Robert the Bruce fighting for Scottish independence in 'Scots Wha Hae':

Lay the proud Usurpers low!
Tyrants fall in every foe!
Liberty's in every blow!
Let us Do – or Die!

Burns was also a radical environmentalist before that term existed. Having worked the fields, he was, Charlie McKinnon writes, 'acutely conscious of the environment and the delicate ecological balance between human activity and nature'.[31] He was a champion of the poor and the underdog. His famous verse 'To a Mouse' deplores contempt for even of the smallest creatures:

I'm truly sorry Man's dominion
Has broken Nature's social union,
An' justifies that ill opinion,
Which makes thee startle,
At me, thy poor, earth-born companion,
An' fellow-mortal!

Burns is for Scots as Thomas Paine – whose *Rights of Man* inspired Burns's poem 'A Man's A Man For A' That' and the American Revolution – is for Americans. Paine, too, came from humble origins, the son of a tenant farmer, and wrote important treatises on the value of broadly distributed liberty. He even articulated an early notion of a guaranteed basic income and wrote about the importance of workers' rights to the means of production about a century before Karl Marx. Paine characterised slavery as the 'height of outrage against Humanity and Justice . . . practised by pretended Christians' about a century before President Lincoln issued the Emancipation Proclamation, when so many were still performing mental and moral contortions to justify slavery.[32] The modern tale of the Enlightenment seen through the writing of Burns and Paine seems an untainted triumph of reason over bigotry, superstition and tyranny. But unfortunately, these were not the dominant voices, and the reality of the Enlightenment is not so clear-cut.

While Burns and Paine were busy writing revolutionary tracts, many other Enlightenment thinkers were writing high-level views of societal patterns and political philosophies. One thing most thinkers of the period agreed on was a particular understanding of history as progressive, recalling the historians of ancient Greece and Rome and the Islamic Golden Age. This has sometimes been called a 'Whig history' or Whig historiography after the British Whig party, who opposed the Tories and favoured Parliament.

The French thinker Anne Robert Jacques Turgot (1727–81) is sometimes credited as the first Enlightenment figure to propound a progressive historiography. For him, like most others of his time, history moves from primitive, savage stages through to classical civilisation and then modernity. Like other Enlightenment thinkers who followed, and like Hobbes before him, he deemed natives in the Americas 'barbarians' who represented earlier stages of human development. He weaves a brief early history from biblical myth, with a major flood bringing people together, but the abundance of languages sundering humans apart, 'strangers to one another, were almost all plunged into the same barbarism in which we still see the Americans'.[33] He continues his historiographic narrative: 'Thus the present state of the world,' which was the mid- to late eighteenth century, 'spreads out before us at once and the same time all the gradations from barbarism to refinement, thereby revealing to us at a single glance, as it were, the records and remains of all the steps taken by the human mind, a reflection of all the stages through which it has passed, and the history of all the ages'.[34]

Turgot illuminates the early connection between Christianity and secular progress narratives in his works, *On the Benefits which the Christian Religion has Conferred on Mankind*, and later, *On the Historical Progress of the Human Mind*. Soon after giving a lecture on these papers at the Sorbonne, Turgot pivoted away from a career in the Church and towards one in the French government. This personal shift coincides neatly with the secularisation of the idea of progress itself, one in which Turgot had a significant hand. The Christian monopoly on such myths in the European context gave way to one based more on reason, observation and natural philosophy (and incorrect history).[35] This secularisation of progress was important: it resulted in the new version of progress more familiar to contemporary readers, one that is rooted in material conditions rather than dependent on a post-death paradise. It was not simply that history unfolded according to the Christian God's plan, but that natural laws governed its development, and those laws were progressive, involving a 'succession of socioeconomic "stages"', as economist Ronald L. Meek, one of Turgot's few English translators, terms it: 'Through all its vicissitudes,' he writes, 'mankind in the long run advances toward greater perfection.'[36] While this secularisation is important, it can also be overstated. Historian John Baillie, in his book *The Belief in Progress*, has gone so far as to suggest that the sort of secular progress Turgot is writing on is simply a 'redisposition of the Christian ideas it seeks to replace',[37] and which a magazine review suggested is a 'by-product of the Jewish and Christian understanding of history'.[38] This is fair, given the reliance of Turgot's historiography on biblical history. The details of the progress narrative formula may change through cultures and religions, and even make the leap to secular ideas, as we saw in the Islamic Golden Age and can now see in the European Enlightenment, but the superstructure of the narrative remains the same, and it continues to serve the same purposes of extraction and expansion.

Turgot also sparks a theme common among those purporting a

progressive historical trajectory, claiming that 'self-interest, ambition, and vainglory' and other such selfish motivations are the primary drivers of progress. Or, in the words of Ronald L. Meek, Turgot believed 'the evil passions of ambitious princes promoted progress'.[39] It is from self-interested impulses rather than altruism that history and society move inevitably 'towards greater perfection'.[40] This is a convenient notion in a market economy lubricated by relationships based on self-interested transactions, and particularly a globally disparate one in which ties of community are less important. He connects the inequality of progress – that is, the unequal distribution of well-being – with a natural process whereby some are granted greater talent, and therefore greater reward, than others, a value system that will be familiar to readers in today's hypercompetitive neoliberal society. 'Circumstances either develop these talents or allow them to become buried in obscurity; and it is from the infinite variety of these circumstances that there springs the inequality in the progress of nations.'[41]

Perhaps unsurprisingly, Turgot is generally extolled by self-described libertarians today not only for his early championing of economic liberalism, but for connecting liberalism to ideas of progress and the 'naturalness' of the unequal distribution of such progress – a tradition that, as we will see, economic liberals today have adhered to with disciplined aplomb.[42] But while Turgot was integral to building an initial progressive historiography, he wasn't the only one.[43]

English historian and politician Edward Gibbon (1737–94) also contributed to this project. His six-volume *History of the Decline and Fall of the Roman Empire* brought him fame and elicited high praise from other Enlightenment luminaries like David Hume and Adam Smith. It is considered one of the foundational texts in the modern field of history, establishing many of the methodologies, practices and assumptions of historians that have persisted for two and a half centuries. Although the work is suffused with irony, one passage that can be read as more earnest perfectly captures the general attitude common among men of his class and time, and illustrates the Whig historian's belief in perpetual progress towards perfection:

Should these speculations be found doubtful or fallacious, there still remains a more humble source of comfort and hope. The discoveries of ancient and modern navigators, and the domestic history or tradition of the most enlightened nations, represent the *human savage* naked both in mind and body, and destitute of laws, of arts, of ideas, and almost of language. From this abject condition, perhaps the primitive and universal state of man, he has gradually arisen to command the animals, to fertilise the earth, to traverse the ocean, and to measure the heavens. His progress in the improvement and exercise of his mental and corporeal faculties has been irregular and various; infinitely slow in the beginning, and increasing by degrees with redoubled velocity: ages of laborious ascent have been followed by a moment of rapid downfall; and the several climates of the globe have felt the vicissitudes of light and darkness. Yet the experience of four thousand years should enlarge our hopes and diminish our apprehensions: we cannot determine to what height the human species may aspire in their advance towards perfection; but it may safely be presumed that no people, unless the face of nature is changed, will relapse into their original barbarism.[44]

This passage captures many of the concepts at play and the narrative formula: the belief that empires bring improvements rather than degradation in human well-being; the belief that progress compounds and is virtually permanent and irreversible; the belief that history moves towards something that could be considered universally perfect; and the belief that progress is measured in the extent to which a society 'commands the animals' or 'fertilises the earth'. Though Gibbon is contemptuous of Christianity (and Judaism and Islam) and argues its adoption was a major cause of the fall of Rome – contrary to the Christians who took credit for the Pax Romana – he is unmistakably echoing Abrahamic dominion in his value system.

The same chapter also includes the line: 'We may therefore acquiesce in the pleasing conclusion that every age of the world has increased, and still increases, the real wealth, the happiness, the knowledge, and perhaps the virtue of the human race.'[45] Though this

II NATION

passage drips with self-satisfied overconfidence, his work blends almost absurd-sounding statements of universality with insightful caveats and equivocations. After a reflection on the invention and adoption of gunpowder, for instance, Gibbon muses that, 'If we contrast the rapid progress of this mischievous discovery [of gunpowder] with the slow and laborious advances of reason, science, and the arts of peace, a philosopher, according to his temper, will laugh or weep at the folly of mankind.'[46] Even these very confident proponents of progress express anxiety about the collateral damage involved, or the risk of straying into the darker sides of technological development.

Moving from French to English back to French progressive historiography: Marie Jean Antoine Nicolas de Caritat, Marquis of Condorcet (1743–94) went even further in building a systematic, universal historiography of progress. He espoused a similar faith in the secular progress of humanity, rising through specific stages, even calling his theory a 'law of progress'. In his law, primitive man gradually develops into civilised man through ten stages, culminating in technocracy, and even anticipating early futurist notions of transhumanism – the capacity of technology to enhance or advance beyond the human mind, augment the body and bring about technologically derived immortality. Condorcet envisaged the course of history as culminating in a perfect future: 'The advantages that must result from the state of improvement, of which I have proved we may almost entertain the certain hope, can have no limit but the absolute perfection of the human species.'[47]

While these historiographies served many purposes, their most critical role was to justify and explain why it was appropriate for one nation to conquer another nation; and not only appropriate, but beneficial to both nations. To see this, it was not so important that these thinkers should explicitly advocate for conquest, but rather that they should build an intellectual foundation for including conquest as a natural element of progress in a grand narrative of taming or civilising or bringing the whole world into the dominion of domesticated, agrarian, market economies. Such an act of conquest, in this narrative, becomes both a necessary and a natural, therefore unstoppable, part of the process of a more developed nation – one closer on the path to perfection – civilising a less developed nation. But this dualism of savage/civilised, developed/less developed is not a benign, impartial observation about the world. In addition to being a factually erroneous view of history, it had real-world consequences. Embedded in the narrative of society advancing from barbarity towards civilisation was a racial hierarchy that parcelled out humans into types, often mediated by physical or geographic features, according to their supposed level of 'barbarism'. This made it easier to justify exploiting those deemed less civilised – purportedly for their own benefit – and to pursue the extraction of their labour and resources. Athenian expansionists had developed perhaps the first racial hierarchy in this lineage to justify enslavement, and the racial theory of Enlightenment narratives was used for a similar purpose: enslaving African workers for agricultural labour and killing or displacing Indigenous peoples to secure richly productive land. This racialism was similar to that used by the Islamic empires in their Golden Age scholarship, which possessed an essentialism: darker-skinned people frequently represented stand-ins for the 'savage inferior races', in contrast with paler Europeans (or in the Islamic case, Arabs). Enlightenment thinker David Hume considered 'negroes' to be 'naturally inferior to whites', while Voltaire agreed that non-whites' understanding of the world was 'greatly inferior'.[48] Even Europeans with the palest of skin were not always considered 'white'. Historian Roxanne Dubar-Ortiz argues that Scotland, Wales and Ireland served as proto-colonies for the British Empire. As with the implementation of capitalist techniques of statecraft and economy, the state honed practices of cultural assimilation, dispossession and genocide before deploying them against Indigenous peoples in colonies abroad. She writes of Ireland in the sixteenth century, just before the Enlightenment period:

> The English government paid bounties for the Irish heads. Later only the scalps or ears were required. A century later in North America, Indian heads and scalps were brought in for bounty in the same manner. Although the Irish were as 'white' as the

II NATION

English, transforming them into alien others to be exterminated previewed what came to be perceived as racialist when applied to Indigenous peoples of North America and to Africans.[49]

This practice of 'racialising' pale Europeans was not confined to the pre-Enlightenment and Enlightenment periods. It intensified into the mid-nineteenth century, when race science sought to prove Anglo-Saxons superior to the Irish through ethnographic studies and Irish skull measurements that associated the Irish with apes. Newspapers published images of the 'simianized Celt', claiming the Irish were at an earlier evolutionary stage than the English.[50] Victorian anthropologist John Beddoe created an 'Index of Nigrescence', based on hair and eye colour, in an attempt, as political scientist Sikata Banerjee writes, to 'confirm the impressions of many Victorians that the Celtic portions of the population in Wales, Cornwall, Scotland, and Ireland were considerably darker or more melanous than those descended from Saxon and Scandinavian forebears'.[51] This idea of 'dark Celts', which required straining at the limits of vision to see, served to position largely light-skinned colonised populations as less evolved and less civilised than 'Anglo-Saxon' agents of empire.

In the mid-eighteenth and mid-nineteenth centuries, the British monarchy and aristocracy colluded to evict thousands of Scots from their ancestral land, replacing them with sheep, and with deer for elites to hunt on massive estates. The process, called the Highland Clearances, significantly depopulated the rural Highlands and ended Scotland's clan society, with the British government demanding fealty to the Crown over clan chiefs.[52] This land-use pattern still shapes Scottish geography today: fewer than five hundred people own half the country's privately held land, making it a nation with among the most concentrated land ownership in Europe.[53] Those justifying the Clearances also drew on racial pseudoscience; one journalist referred to Highlanders as racially inferior to 'the Lowland Saxon', while Edinburgh surgeon Robert Knox wrote that they should be 'forced from the soil' by what he considered the superior 'Anglo-Saxon race'.[54]

In all of these cases, whether in Europe or the Americas, a parasitic relationship – of ecological systems and human classes – was established for the purposes of extraction and expansion. Enlightenment racial hierarchies were deployed as a means of rationally justifying and mediating that relationship. Racial hierarchies were not limited to gradients of skin colour – though skin was and continues to be significant – but instead focused on indicators like language, religion, bloodline and culture to mark a population as *coloniser* or *colonised*.[55] Identifying superior and inferior races was a fundamental component of the progressive historiography, and has been since at least the classical empires.[56] But while histories created a significant contextual framework, they alone did not suffice to build the value systems and intellectual arguments necessary for the secular progress narrative of disparate parasitism.

Philosophies of Progress

Progressive histories and theories of society were the most straightforward way in which Enlightenment thinkers influenced their contemporaries' social and economic policies, but other Enlightenment ideas were just as important in shaping the politics of the time. Those ideas provided either secularised veneers of Abrahamic ideas or other foundational beliefs that bolstered both a progressive view and the notion that the parasitic European colonialism was in the best interest of everyone, including those being colonised.

French thinker René Descartes (1596–1650) says little about progress itself, but from his thought comes Cartesian dualism – the fundamental separation between the material world and the immortal soul contained only within human beings – which provides the secular basis for a notion approximating Christian dominion and serving a similar practical purpose. Descartes casts all beings he believes to be unendowed with a self-conscious mind – every living creature but humans – as unfeeling and unthinking automata ready to be ethically exploited. There is ambiguity and complexity in Descartes' argument, but the portion of his thought that could be readily put towards practical ends was explicit: Animals are machines, are automata, do not think, have no language, have no self-consciousness, and have no consciousness.[57] They are no more victims of suffering than rocks that are quarried. This idea provides a secular rationale for parasitic economies to dispassionately denude ecological systems to the point of bringing about major extinctions. The idea also provided a rational-sounding justification to colonists seeking to destroy or displace the native peoples who occupied the land they wished to capture, as they could more readily dismiss such peoples as enemies of progress and more akin to unfeeling animals.

Such notions contradict the best research today, which finds that many species of non-human animals differ little from humans in their cognitive, emotive and sensory structures, and that Indigenous peoples are not inferior in any of the ways such theories can be twisted to suggest. The fact that animals are not automata was obvious at the time, as well, for anyone who had even a passing experience with them. But the empirical truth or falsehood of such claims was beside the point: believing animals to be automata was useful in practice to psychologically enable the mass exploitation of environments and animals, to reduce a population of tens of millions of bison to hundreds, for example, or to destroy billions of passenger pigeons to extinction in a single century.[58]

With many of these thinkers, one might reasonably ask whether their writing really can be proven to have led to these harmful actions. After all, where is the diary of a settler ruminating on the guilt he feels shooting hundreds of birds out of the sky, but then reads Descartes and suddenly feels better? Such diaries may exist, such internal struggles may be provable with documentary evidence, but making a demand of that kind misses the point. Philosophy does not typically infuse ideas into society through direct linear causation. You yourself may have a sense – or at least you may have had it at one time – that humans are intrinsically different from, maybe superior to, other animals. You may have sensed vaguely that you have a soul that is distinct from your body and that other animals may not have such souls. You did not have to read Descartes' *Meditations* to have such a feeling. Ideas like this have a way of melting into the collective cognition that constitutes a culture. They do not always integrate in a pure way, but instead fuse with other ideas to concoct a morass of notions that serve to either gum up brains or lubricate new thought. Furthermore, few of the ideas explored in this book necessarily serve as an origin for the behaviours or systems under scrutiny. Instead, they are often a retroactive rationalisation for behaviours conducted out of imperatives for parasitism, out of the expediency of political actors, out of impulses and needs realised in the course of settling land or venturing into unknown territory. 'Man is not a rational animal; he is a rationalizing animal,' wrote science fiction author Robert Heinlein. People often act first based on impulse or societal imperative, and come up with high-minded reasons later. Whether a given thinker or leader *truly believes* in the progress formula, or simply deploys it cynically as justification for some ulterior motive, or some mix of the two – or some other motivation altogether – may not really matter. If the outcome is the same either way, whether or not they believe does not necessarily change how one may intervene in the narrative, nor does it absolve them of culpability.

We have seen how English thinker Thomas Hobbes (1588–1679) contributed important philosophical arguments in service of a parasitic economy, most famously by building a memorable justification for a centralised state. Only a state with a competent bureaucracy and strong central authority, Hobbes argued, can save people from the 'state of nature' – an original condition of humankind, according to his theory – and the 'war of all against all' that those outside of such civilisations are doomed to suffer.[59] That is, the savage, primitive peoples live lives that are 'solitary, poor, nasty, brutish, and short', and only the intervention of a court, king and aristocracy can provide people with lives superior to such conditions. But Hobbes is often contrasted with French thinker Jean-Jacques Rousseau (1712–78) in their conceptions of a state of nature. Hobbes portrays humanity as corrupt and redeemed by the 'thin veneer' of civilisation, whereas Rousseau sees civilisation corrupting the inherent goodness of humanity.[60] While it is true that there are substantive differences in their understanding of humanity and history, in some areas Rousseau did not diverge as radically from Hobbes as is often suggested. For one,

II NATION

he was as invested in an Enlightenment progressive historiography as Hobbes: he contended that 'savage' man evolved through stages – agriculture, metallurgy, private property – towards civilised man. Unlike many other thinkers, though, Rousseau was less optimistic about this trajectory, and among the few to offer critiques of it. He argued that a more 'primitive' society would yield greater individual happiness. But rather than proposing a retreat from what he understood as civilisation, as his argument is often depicted, Rousseau argued instead for a kind of progress achieved through a state apparatus not unlike Hobbes's central authority. In place of a lone autocrat, Rousseau preferred a collectivist sovereign entity – a 'general will' – even if a monarch or 'elective aristocracy', he conceded, must be engaged as administrators.[61]

Scottish thinker Adam Smith (1723–90) is another Enlightenment figure whose philosophy, like that of many Enlightenment thinkers, investigated notions of human progress; Smith mentions them frequently in his works, arguing that society follows a linear progressive course through time. But his work is complex and his ideas have been instrumentally deployed to support notions that the man himself may not, and in some cases very definitively did not, support. Most famously today, and most frequently, his work provides a quasi-religious justification for markets. His 'invisible hand of the market' metaphor is still invoked in the twenty-first century. The original idea goes, basically, that markets are guided by divine will. Smith conflated the Christian God with the secular forces that move markets; the former still retained a supreme spiritual authority while the latter consisted of rational actors behaving in self-interested competition, though also using their own value systems to guide their actions in the market. This commonly invoked metaphor – of which Smith wrote only three times – is today taken as a primarily secular notion, with the Christian divinity removed. It is used to describe the natural power of a fictitious abstraction – the market – to efficiently and perfectly distribute wealth and resources according to the collective behaviour of selfish, rational actors. Smith's own meaning was more reliant on the Christian God and, like much of his work, was more complex than is given due today, or even by contemporaries who sought to use his writings for expedient political ends.[62] Whereas his most enduring work, *The Wealth of Nations*, is often used by defenders of the commercial system, Smith described it as a 'violent attack . . . upon the whole commercial system of Great Britain'.[63] Far from being an apologist for capitalism and the market, Smith was in his own time a sharp critic of European conquest and enslavement.[64] This, unfortunately, did not stop his work from being consistently appropriated by those less critical of conquest and enslavement.

Thomas Malthus (1766–1834), writing towards the end of the Enlightenment period, is often interpreted today as the quintessential pessimist, promoting a view of history, humanity and the future that could be interpreted as anti-progressive. But this is a misreading.[65] Malthus was just as invested in ideas of progress and parasitism as his contemporaries, writing, 'although we cannot expect that the virtue of happiness of mankind will keep pace with the brilliant career of physical discovery, yet if we are not wanting in ourselves we may confidently indulge the hope that . . . they will be influenced by its progress and will partake in its success.'[66] As economic historian Samuel M. Levin has shown, Malthus had a somewhat ambivalent approach to progress over his career and was sometimes sceptical of the most optimistic theorists mentioned earlier. But, Levin notes, he was friendly to the idea of progress and 'indicates the predilection of his thought' frequently through lines like 'the future improvement of society', 'the progress of mankind towards happiness', 'the further progress of wealth' and more. Although he is not limited to this optimism, Levin writes that, 'In most cases what he had in mind was a going forward, growth, change, or trend toward a better order of things, of life, or of society.'[67] And his trajectory was not one of youthful hope corrupted by hard experience. Later in his career, Malthus became *even more* optimistic about the future, contrary to his reputation today.

Malthus, like Turgot and Smith and other liberals of his time, echoes the idea that if everyone acts in a self-interested manner, society will gradually improve:

> On the whole . . . though our future prospects . . . may not be so bright as we could wish, yet they are far from being entirely

disheartening, and by no means preclude that gradual and progressive improvement in human satiety . . . to the apparently narrow principle of self-interest which prompts each individual to exert himself in bettering his condition, we are indebted for all the noblest exertions of human genius, for every thing that distinguishes the civilized from the savage state.[68]

Again, this notion that acting on self-interest yields broader societal progress is useful for those most invested in doing harm for their own personal enrichment: it is an obvious means of soothing the consciences of those in elite positions, or even common positions, who commit harm for personal gain. The parasitic class can take comfort from the idea that they are contributing to collective progress simply by continuing to extract from land and people. But such notions are also valuable to the recipients of harm, those who are most intensively being extracted from; they can take comfort in the notion that their suffering is still contributing to some collective improvement, and that their salvation, or their children's, resides in the future.

The Forging of Liberalism

If *disparate parasitism* is the human-ecological form that creates an imperative to expand and extract in a linear, hierarchical mode via distant oceanic voyages; and *capitalism* is an administrative form which systematises and implements that imperative in networks throughout the world; then *liberalism* is a specific ideology that enshrines those concepts – disparate parasitism and capitalism – into a moral framework that can appeal to broad political coalitions. Like anything that attempts to appeal to a broad swath of competing interests, liberalism contains irreconcilable contradictions.

Today, 'liberal' and 'liberalism' colloquially – and often pejoratively – refer to a broad coalition in anglophone politics that consists of 'centre-left' parties – like the Democrat, Labour or Liberal parties in Western English-speaking countries – which are often contrasted with 'conservative' parties in those same polities. This binary of liberal and conservative makes little sense from a standpoint of political theory and ideology, and is less common outside the anglophone world. Such a difference between liberals and conservatives is more one of aesthetic style than of substantive ideology. Most 'conservatives' adhere to liberal economics as much as many 'liberals' do – and many who are often included under the overly broad 'liberal' banner, like 'leftists', subscribe to very different economic and political programmes from either liberal or conservative parties.[69] In ideological matters, if not always practical ones, liberals and conservatives often share more in common than liberals and socialists, for example; the assumptions liberals and conservatives make, and the goals they have, tend to be more aligned than opposed with respect to important material concerns. This is because mainstream liberals and conservatives both tend to be invested in maintaining parasitism and capitalism, and these fundamentally override all the other issues on which they stake out distinct positions. At the time of writing, there has not yet been a politician from these parties to challenge the imperative for 'economic growth', the current term of art covering the many processes of parasitism. The solution to all problems, proclaimed by the majority parties of the legislative houses of Europe and North America, is simply *more* of everything that is at the root of those very problems.

The sort of 'liberalism' that emerged in the Enlightenment is still operational today, and it refers to two distinct strands of thought, what I will call *political* liberalism and *economic* liberalism. Political liberalism promotes civil rights, like freedom of speech, assembly and religion. Economic liberalism promotes economic entitlements, like the ownership of private property, profit, trade and markets. These two sets of principles are in tension. Economic liberal principles – through the normal operations of parasitism and capitalism – serve to consolidate property, wealth and, as a result, political power. This process leads to outcomes that directly undermine the goals and values of political liberalism – like representative government, civil rights and freedoms, and democracy – or even prevents them from being implemented in the first place.

Political liberalism, in a sense, has served as a moral compromise enabling the implementation of economic liberalism. In the Euroamerican context – the colonisation of the Americas by European states – in exchange for free speech, a partially and imperfectly representative government, and freedom of religion, populations would have to accept wealth concentration, the enclosure of land, rents, debts, and for a while, genocide and slavery. Enlightenment *political* liberalism helped usher in many positive outcomes that have undermined repressive governments, if sometimes only on behalf of already privileged populations; meanwhile the imperatives of *economic* liberalism have consistently denied those rights to people liberals consider racially inferior, or those belonging to a subordinate labour class, or to what economic liberals deem a primitive stage of development.

The tension intrinsic to this ideology means that liberalism is – or, at least, has been – ever doomed to fracture: for political liberalism to triumph, a society must curtail economic liberalism; if economic liberalism is to triumph, a society must abandon political liberalism.

Few personally embody this dynamic better than John Locke (1632–1704), the 'father of liberalism'. Locke inspired Enlightenment thinkers like Voltaire and Rousseau as well as American revolutionaries. He advocated powerfully for political liberalism, including a separation of powers, religious tolerance and natural civil rights, while he operated on economic liberal values in his own life. Locke was a major investor in the Royal Africa Company and the Bahama Adventurers Company, two of the biggest traders in enslaved people from Africa. In the history of the Atlantic slave trade, no single company captured and transported more people than the RAC.[70] In addition to his early support of slavery – he later rejected it except as a severe punishment – Locke's theories of labour helped to legitimise the seizure of land from Native Americans. Like other forms of Enlightenment thought, Locke's liberalism was racially hierarchical, drawing from a progressive historiography to rationalise the enslavement of Africans and the eradication of Native Americans. Here we have a clear microcosmic example of one of the founders of the ideology using the virtues of political liberalism to soften the vices of economic liberalism.

Liberalism has been an effective ideology for maintaining the legitimacy of multiple forms of parasitism and capitalism. As such, it will remain a running theme through the rest of this investigation. While Locke may have been among the first to light this torch – if we omit contributions by other thinkers like Khaldun – others have taken it and added their own colours to it, lighting, or burning, their trail through history.

Enlightenment thinkers can be read in many ways: as putting their powerful voices to the aims of revolution and democracy or the maintenance of oligarchy and monarchy; speaking eloquently on broad liberty and reason while sanctioning political hierarchies and expounding on the reasonableness of subjugating 'lesser races'. While the legacy of Enlightenment thinkers is complex, Enlightenment *political actors* – politicians, generals, profiteers – bear contradictions in both their thought and actions that are worth considering in depth to better understand how these narratives and processes have shaped the world we live in today, the consequences of which we will enjoy or endure for millennia.

A Nation of Destiny

Disparate parasitism sent ships from European nation-states, like Britain and France, Spain and Portugal, the Netherlands, Germany and Italy, across the world. But the colonies these countries established in Africa and Asia tended to be different from those in the Americas. African and Asian colonies more often involved a small population of colonial administrators who organised extractive operations and resource appropriation and transport back into their home economy. This is different from what is often referred to as 'settler colonialism', a phenomenon in which colonists establish permanent settlements and, in most cases, remove existing inhabitants from land that is then turned to the profit of settlers and investors.

The colonies of Britain, France, Spain and Portugal in the Americas yielded permanent nation-states that variably absorbed, eradicated or removed the nations of peoples already residing on the land. While

II NATION

Spanish and Portuguese colonies in South America were fragmented into multiple states, the original thirteen British colonies in northeast North America coalesced, after the American Revolution, into a single nation-state towards the end of the eighteenth century. From the end of that century and throughout the next, the new nation-state engaged in expansionism that perhaps more resembles the regional parasitism of the 'Old World', in that it was a land-based expansion of settlers absorbing contiguous territory. The key differences between this American expansion and those of the classical world were, first, a pre-existing basis of Enlightenment ideas and a global network of trade organised in a capitalist mode from which to build, and, second, the exploitation of fossilised biomass as a new form of energy capture that could increase the speed of expansion, production, communication and transportation.

These elements had a dual and contradictory effect. American expansionism was, like previous contiguous parasitic expansion, tinged with a mythic, religious form of progress. But, at the same time, it was developing more scientistic narratives of progress bequeathed from Enlightenment ideas rooted in rationalism and empiricism. As such, the United States offers a compelling case study in how progress narratives that utilised sometimes conflicting forms of propaganda were integral to the conquest of a continent, and how new technologies supercharged the parasitic economies of the nineteenth century; it also exemplifies the dynamic causal relationships between the two.

When Europeans first arrived in North America, the abundance of life on the continent was staggering, almost unbelievable to them. A few examples demonstrate how fast and thoroughly this new Euroamerican parasitism eradicated such abundance. The passenger pigeon, clad in vibrant fiery rust and iron-blue plumage, was once the most plentiful bird in all of North America, with a continental population of several billion. In 1813, the naturalist and painter John James Audubon recorded a migration of the pigeons. He estimated that over a billion birds made up a single flock, the sheer density of which blotted out the sky:

The air was literally filled with Pigeons; the light of noon-day was obscured as by an eclipse; the dung fell in spots, not unlike melting flakes of snow; and the continued buzz of wings had a tendency to lull my senses to repose. . . . Before sunset I reached Louisville, distant from Hardensburgh fifty-five miles. The Pigeons were still passing in undiminished numbers, and continued to do so for three days in succession.[71]

Such immense migrations were recorded throughout the nineteenth century. Observers claimed that the migrating birds resembled vast clouds and sounded like the crashing of distant waterfalls. When they nested, there was danger of branches falling from the collective weight of the pigeons' bodies and nests in the trees. Settlers hunted the birds for meat and to make feather beds, to prevent them from eating grain crops, and for sport. By 1914, the species was extinct.

In New England, cod were the passenger pigeons of the sea. Tales abounded that the fish were so numerous they would jump into boats, and you could walk across their backs to the shore. Historian Mark Kurlansky writes in his biography of the fish, 'the coast of North America was churning with codfish of a size never before seen and in schools of unprecedented density, at least in recorded European history.'[72] Kurlansky quotes a Puritan minister in Salem, Massachusetts writing in 1629, 'The abundance of fish are almost beyond believing.'[73] European markets soon made short work of this abundance; by the mid-sixteenth century, cod represented 60 per cent of fish eaten in Europe. It took a couple of short centuries for Euroamerican consumption to almost totally annihilate the fish stocks.

Nor was cod the only plentiful seafood. The New England shores were overflowing with sturgeon, oysters and clams, which the Naumkeag Natives taught English settlers to open and prepare. 'The waters were so rich in lobsters,' writes Kurlansky, 'that they were literally crawling out of the sea and piling up inhospitably on the beaches.'[74] But the settlers, 'being English,' he clarifies, 'did not want to eat unfamiliar food.' Until they learned how to monetise such creatures, the settlers shunned these delicacies and nearly starved.

At the end of the eighteenth century, there were between 30 million and 60 million wild bison on the continent of North America; by the 1890s, only 100 years later, there were fewer than five hundred individuals. This mass slaughter was partly driven by settler demand for hides and bones to turn into fertiliser, glue, ash or bone china for markets in European and North American cities, and was enabled by the use of new factory-produced guns.[75] But another reason settlers engaged in a deliberate process of exterminating the vast herds was to eradicate the Plains Nations peoples that depended on the bison for their survival. As one settler colonel put it, 'Kill every buffalo you can! Every buffalo dead is an Indian gone.'[76]

In the midst of this blood shower, settlers were eager to establish a heroic origin story for their new nation. Historian Colin Woodard traces one of the earliest projects of national mythmaking to George Bancroft, a descendant of Puritans and author of the ten-volume *History of the United States, from the Discovery of the American Continent* (1854–1878).[77] His *History*, Woodard writes, 'combined his Puritan intellectual birthright with his German mentors' notion that nations developed like organisms, following a plan that history had laid out for them. Americans, Bancroft argued, would implement the next stage of the progressive development of human liberty, equality, and freedom.'[78] Like other imperial frontiers of extraction throughout history, American westward expansion drew heavily on notions of progress. Whether in the private letters of Thomas Jefferson and George Washington, the public speeches of Abraham Lincoln and Andrew Jackson, or the country's newspapers, a particular American version of the progress narrative formula took hold. Combining secular ideas from the European Enlightenment with a persistent and variegated Christian ethic and frontier missionism, the US expanded with its own brand of progress. The result was a quasi-divine mandate to tame wild ecological spaces and subdue Native peoples. It was called 'manifest destiny'.

Although manifest destiny was not a coherent doctrine, and its virtues were debated even in its time, it represents a general ethic that motivated expansionism: because the founding of the United States, according to its creation story, was rooted in high ideals – liberty, equality, justice – it had a natural right to seize land in order to spread those values. American exceptionalism, the idea that the nation is exceptional among other nations throughout history, came from this same belief. Journalist John O'Sullivan coined the term 'manifest destiny' in 1845 in an essay arguing that the country's destiny was to 'overspread the continent allotted by Providence for the free development of our yearly multiplying millions'.[79] This notion resembles the contiguous parasitism of Rome and Virgil's Pax Romana-era mythology, in which Jupiter proclaims: 'To the Romans I assign limits neither to the extent nor to the duration of their empire; dominion have I given them without end.'

Four of the country's most important presidents to this process of expansion – Thomas Jefferson, George Washington, Andrew Jackson and Abraham Lincoln – illustrate how aspirations for progress were deployed publicly and internalised personally by the country's authorities in the process of westward parasitism that would come to represent the spirit of manifest destiny.[80]

The country's third president, Thomas Jefferson, drafted the most fundamental document to the nation's founding identity, the Declaration of Independence. The declaration was a radical document, calling for the end of monarchy and establishing natural rights and freedoms bequeathed from thinkers throughout the Enlightenment, many of whom, as mentioned, were themselves inspired by Indigenous American thinkers.[81]

But while it was indeed a radical document, much has been made of the hypocrisy of those who signed and composed it. While Jefferson wrote, 'All men are created equal,' he remained a lifelong slave owner. The equal men of his famous line were not the universal, rhetorical 'mankind', but specifically Euroamerican men, an idea often self-consciously reflecting the values of classical oligarchic republics like those of ancient Rome and Greece in which only landowning males were considered citizens and worthy of political representation. As such, Jefferson's attitude towards women, too, was far from generous: his respect for their intelligence was so meagre that he suggested women should be educated only in 'the amusements of life', like 'dancing, drawing, and music'.[82] But perhaps more insidious than these famous examples of hypocrisy, and more pertinent to

II NATION

understanding the parasitism of the early American state, is this biographical detail: he was also one of the major architects of the genocide of Native peoples.[83] Jefferson's belief in their inhumanity is evident in his writing: in the Declaration of Independence, he refers to Natives as 'merciless Indian savages'; in a 1807 letter to his friend Henry Dearborn, he writes, 'If ever we are constrained to lift the hatchet against any tribe, we will never lay it down till that tribe is exterminated, or driven beyond the Mississippi . . . in war, they will kill some of us; we shall destroy them all.'[84] In a letter of December 1813, Jefferson continued with this theme, writing of Natives: 'This unfortunate race, whom we had been taking so much pains to save and civilize have by their unexpected desertion and ferocious barbarities justified extermination and now await our decision on their fate.'[85]

This was not mere bloodthirsty bluster to impress a friend. He enacted policies that followed from this attitude. Jefferson made it a federal priority to pursue the permanent 'removal' of Creek and Cherokee peoples from Georgia.[86] Guiding his administration was a policy of outright extermination of Natives or, at best, forced migration and assimilation. Though assimilation seems more merciful than combat, historian Patrick Wolfe argues that assimilation projects like this often constitute a 'logic of elimination' or 'structural genocide' through the deliberate diluting of Native family lines, knowledge and ways of life: 'Though "softer" than the recourse to simple violence . . . these strategies are not necessarily less eliminatory.'[87]

It is clear that Jefferson's aim of parasitic expansionism across North America was rooted not only in the material imperative to steal land for voracious settlers, but also in notions of progress. His mind, according to Robert Nesbit, was 'steeped in faith in human progress'.[88] The Jefferson Foundation claims that the third president 'believed that if American Indians were made to adopt European-style agriculture and live in European-style towns and villages, then they would quickly "progress" from "savagery" to "civilisation" and eventually be equal, in his mind, to white men'.[89] Jefferson's administration forced natives to adopt Euroamerican agriculture and take on debt that could only be discharged by giving up more of their land. This tactic presaged structural adjustment policies like the loans made by global institutions such as the World Bank and IMF in the late twentieth century. Jefferson suggested that, through this process of giving up traditional ways of life and modes of energy capture, natives would inevitably be led 'to manufactures, and civilisation'. As historian Howard Zinn notes, this 'talk of "agriculture . . . manufactures . . . civilization" is crucial. Indian removal was necessary for the opening of the vast American lands to agriculture, to commerce, to markets, to money, to the development of the modern capitalist economy.'[90] While the official policy during Jefferson's administration was that of 'civilising' Natives by forcing them into debt and assimilating them, historian Jeffrey Ostler suggests that this was more propagandistic than genuine. Instead, he argues that 'U.S. officials were never seriously committed to a policy of civilisation. As early as Thomas Jefferson's presidency (1801–1809), U.S. actions made it clear that despite talk of civilisation and assimilation, the United States would ultimately pursue a third option for the elimination of Indians east of the Mississippi River.'[91]

Jefferson was putting into practice the chain of thought we have been investigating: he begins with the parasitic human-ecological imperative to intensify extraction, both concrete and abstract; he continues with the capitalist administrative system that encloses property for the accumulation of capital; and he uses the value framework of both political and economic liberalism to rationalise it – all of which is rooted in the metanarrative of the teleological development of history: progress from savage to civilised. He also demonstrates in his letters how such notions can ease the conscience of one whose individual actions had a monumental impact, in some cases an impact of the most horrific character imaginable. Yet Jefferson died content with his role in advancing this human progress, as his letter to William Ludlow in the introduction attests. The excerpt from that letter quoted earlier ends with this line: 'Barbarism has, in the meantime, been receding before the steady step of amelioration, and will in time, I trust, disappear from the earth.'

The passive language of 'disappearance' in this text transforms the tactical, deliberate acts of genocide, murder and theft into a mythic event, a natural course of history, devoid of human agency, culpability

II NATION

and responsibility, as its 'naturalness' was established in the work of Enlightenment historians like Gibbon. Jefferson could have added a disappearance from memory. Histories and political discourses today frequently attempt to erase, rationalise or justify the nation's founding genocide, which is harmful not simply as a betrayal of honest historical reckoning, but because echoes of this very same genocide still reverberate, not in memory but in fact. There are nearly three million Native Americans alive today who continue to suffer state violence and a hugely disproportionate rate of interpersonal violence by non-Natives, with an 'epidemic' of Native women suffering violent and sexual attacks, and who bear the brunt of environmental destruction and externally imposed economic hardship.[92] Jefferson's *progress* still rules.

And yet, even as he was executing this genocidal policy, Jefferson also expressed in a private letter the superiority of Native ways of life for increasing human happiness. He wrote the following to Edward Carrington in 1787:

> I am convinced that those societies (as the Indians) which live without government enjoy in their general mass an infinitely greater degree of happiness than those who live under European governments. Among the former, public opinion is in the place of law, and restrains morals as powerfully as laws ever did any where. Among the latter, under pretence of governing they have divided their nations into two classes, wolves and sheep. I do not exaggerate. This is a true picture of Europe.[93]

Even as he recognised that Native American lifestyles might be superior for the 'Pursuit of Happiness' to those bequeathed from Europe, he *still* pursued a policy of removing Natives, stealing their land for Euroamericans, and replacing their lifestyle with this other kind. When he became president fourteen years later, his office was involved in dividing *his own* nation into two classes: host and parasite, majority and authority, wolves and sheep.

* * *

When Rome sought to seize inhabited lands for extraction, they sent a general. Perhaps the most important general in US history was dedicated to the same work. The nation's first president and then-richest man in America was George Washington. He grew rich in part from land speculation, buying up huge tracts of Native land and cultivating it with the labour of over a hundred enslaved people. He was also a general in the American Revolutionary War and conducted some of the most important military campaigns aimed at opening land and pacifying Native peoples for American settlers – like those in the Ohio River Valley mentioned earlier. 'Indians and wolves,' he proclaimed, 'are both beasts of prey, tho' they differ in shape.'[94] Meanwhile, the Iroquois dubbed him 'Destroyer of towns' (an Algonquian translation: 'Devourer of villages'). Washington ordered more than forty Native villages destroyed in just one expedition, with more than ten thousand Natives dying in a single winter as a direct result.[95] Although many of the wars between Native and Euroamerican armies were brutal, with atrocities committed by all participants, historian Stuart Leibiger observes that, in that particular campaign, 'Whereas Indians did not kill prisoners (soldiers or civilians) and never raped female captives, Americans took the lives of combatant and noncombatant Indians alike, often scalping or skinning their victims. The women who were spared were often raped.'[96]

Yet, as with Jefferson, a strange dynamic is at play. Even while he is deploying ruthless tactics to terrorise and eradicate Native peoples, Washington knows that Natives are not simply savage, inhuman beasts, despite his rhetoric to the contrary. He grew up among Native people, and saw not only their humanity, but even their superiority. As historian Gillian Brockell notes, 'Washington started his military career alongside Native American politicians and warriors much more sophisticated than he was.'[97] The cognitive dissonance he must have felt intimately knowing Native Americans who were bilingual, gifted politicians and strategists, and who resided in established and well-run villages, while simultaneously organising scorched-earth campaigns to destroy those villages, must have been imperilling for a mind. This would be especially so if he were committing atrocities for largely selfish motivations: to further enrich himself. Such dissonance might

II NATION

be resolved with a grand narrative that could explain and morally justify such actions, that could convince people that self-interest was a vehicle for progress. There is good evidence that Washington believed he was contributing to human progress and that such atrocities could be understood as necessary preconditions for achieving this greater good. In a 1786 letter to his friend, French aristocrat and military officer Marquis de Lafeyette, Washington wrote:

> I endulge a fond, perhaps an enthusiastic idea, that as the world is evidently much less barbarous than it has been, its melioration must still be progressive – that nations are becoming more humanized in their policy – that the subjects of ambition & causes for hostility are daily diminishing – and in fine, that the period is not very remote when the benefits of a liberal & free commerce will, pretty generally, succeed to the devastations & horrors of war.[98]

Again, as with Jefferson, we see Washington connecting 'the devastations & horrors of war' as being necessary preconditions for establishing 'liberal & free commerce' – the moral and administrative frameworks for parasitism. Though we might see the tone Washington uses while confiding to his fellow aristocrat as strangely sanguine and self-satisfied for one who had engaged in war, weaponised rape, torture, slavery and genocide, for those in his parasitic class they would have fitted neatly into the framework of progress. In another letter to Lafayette three years later, he wrote explicitly tying the expansionist ambitions of his emerging country with progress:

> While you are quarreling among yourselves in Europe – while one King is running mad – and others acting as if they were already so, by cutting the throats of the subjects of their neighbours: I think you need not doubt, My Dear Marquis we shall continue in tranquility here – And that population will be progressive so long as there shall continue to be so many easy means for obtaining a subsistence, and so ample a field for the exertion of talents and industry.[99]

Before I provoke accusations of bias – or perhaps long after having done so – I want to recognise that, first, many settlers were not intrinsically violent, or any more so than the Natives they encountered, and in some cases less so. As mentioned earlier, many settlers were themselves victims of their own monarchies and empires from which they fled or were forced to serve. Further, many of those who have come to the Americas long after this process are not morally bound to the actions of these original settlers. It is also important to note that the term 'Native American' collapses into a monolith what was a richly complex continent full of different cultures – and to a lesser extent still is – with vastly varying attitudes towards violence, domination and extraction. Some warriors of various different Nations, in other conflicts throughout the country, did engage in ruthless forms of warfare and also committed dozens of massacres and atrocities. Some Native Nations, meanwhile, practised brutality against one another for generations before this period. The Iroquois and Nʉmʉnʉʉ (Comanche), for instance, would sometimes participate in the long-lasting ritual torture of captives. Some tribes, like many of those in the Eastern Woodlands, Great Lakes or Ohio River Valley, were more peace-loving than others. Many other Native Nations, meanwhile, used extreme violence for justifiable reasons, like as a deterrent against invasion by neighbours, or for putting down domestic upstarts with ambitions for tyranny, or, in some cases, joining together as a means of thwarting aspiring expansionism by other Nations. Meanwhile, some Native American Nations also engaged in forms of parasitism at various points in their potentially more than 20,000 years inhabiting the Americas, from the South American empires like the Aztec and Inca, to the 'capturing societies' across the continents that raided and kidnapped from other peoples.[100] Native foragers, horticulturalists and farmers throughout the Americas, from the Paraguay River basin to Amazonia to northern Florida, practised raiding and the taking of tribute and slaves from other peoples.[101]

However, an appraisal of historical facts suggests that Euroamerican settlers engaged in these things, violence, war, genocide, slavery, with far, far greater intensity, frequency, and wantonness than any Native nations did, either amongst each other or against settlers.

II NATION

Acknowledging these facts objectively is not 'bias', and is instead the *elimination of bias*. Like most others, I have been more sympathetic to my own culture and cultural ancestors throughout my life. As such, I have had to mitigate these sympathies in order to face the facts of the situation as impartially as possible, difficult as that may be. To confront this history without bias requires doing so.

Brutally violent as some Native peoples could be, there is no record of them totally annihilating the diversity of cultures and organisms across the whole continent, nor instituting systems for the mass enslavement and labour of multitudes of people. As David Graeber and David Wengrow note, 'Amerindian slavery had certain specific features that make it very different from ancient Greek or Roman household slavery, let alone European plantation slavery . . . while slavery of any sort was a fairly unusual institution among indigenous peoples of the Americas.'[102] Attacks and settlement by Euroamerican colonists practising a more intensive form of parasitism should not be seen as morally or practically equivalent to the violent actions committed by the societies they sought to eradicate, and certainly not as a reasonable means of justifying their eradication. Violence done in defence of oneself or one's loved ones is more justified than offensive violence done in the interest of theft, profit or ethnic annihilation. Native Americans practised a great diversity of political forms and cultures that cannot be easily generalised, but that persisted for millennia without causing catastrophic ecological harm or engaging in mass enslavement. This is a qualitatively different condition from an invading economy that seeks to eradicate millions of people and totally eliminate cultures, enslave millions of others, and make thousands of species extinct while eroding the ecological basis of all society, culture and life.

* * *

While Washington was undoubtedly among the greatest forces in bringing parasitism to North America, he was less bellicose – if more effective – than another general who followed in his smoking wake. Born thirty-five years after Washington, Andrew Jackson was the country's seventh president and shares other important features with the first. Like Washington, Jackson amassed a fortune as a land speculator and merchant of humans. He was among the country's most aggressive combatants in wars against Natives.[103] He was known to his followers by the nickname Old Hickory, but like Washington, he had another nickname among Natives. To the Creek, he was known simply as 'Sharp Knife' and 'Indian killer'.[104] In one notorious case, Jackson's forces burned a Creek village, killing men, women and children, and appropriated 23 million acres of land, which would later become cotton plantations worked by enslaved Africans. Of all the First Nations wars in which US forces seized Native land, the Seminole Wars in Florida lasted the longest, nearly forty years. By the end of the Seminole Wars, between three and four thousand people had been forced from their land. In their absence, the US government began draining the Everglades, clearing its biodiverse subtropical wilderness to make way for agricultural land and development. Jackson's general in these wars, Thomas S. Jesup, made the intentions of the US army and the campaigns explicit: 'The country can be rid of [the Seminoles] only by exterminating them.'[105] Jackson connected these campaigns in an address to Congress with the same progress rhetoric that had been deployed by previous presidents and elucidated by Enlightenment thinkers:

> What good man would prefer a country covered with forests and ranged by a few thousand savages to our extensive Republic, studded with cities, towns, and prosperous farms embellished with all the improvements which art can devise or industry execute, occupied by more than 12,000,000 happy people, and filled with all the blessings of liberty, civilisation and religion?[106]

Jackson's mention of '12,000,000 happy people' echoes the aforementioned tendency of progress propagandists to place greater weight on *quantity* of human life than on its *quality*. His description of these multitudes as 'happy', after all, was not based on any well-being indicators or surveys, which would have found far less happiness

II NATION

among settlers than among the Natives he was trying to eradicate. There were instead, simply, more of them.

However, we should treat Jackson's address to Congress for what it was: political rhetoric that did not necessarily bear much resemblance to reality, or even his personal beliefs. The Seminoles were one of the 'Five Civilized Tribes', in the American parlance. Along with the Creek, Choctaw, Chickasaw and Cherokee tribes, they adopted many of the practices of Euroamerican settlement. The Cherokee, for instance, had developed agriculture similar to that of the American settlers, even owning Black slaves, and wrote up a constitution to govern their territory. 'Why,' Patrick Wolfe asks, 'should genteel Georgians wish to rid themselves of such cultivated neighbours?' After all, he writes, the 'factor that most antagonized the Georgia state government . . . was not actually the recalcitrant savagery of which Indians were routinely accused, but the Cherokee's unmistakable aptitude for civilisation'.[107] One reason such things so antagonised the government is that they signalled permanence. In response, the government forced them to move out of Georgia and march 1,000 miles west into Indian territory. This became the Trail of Tears that killed 4,000 Natives. But why should this permanence be a problem, why removal necessary, if Jackson simply wished to see 'prosperous farms embellished with all the improvements which art can devise', and '12,000,000 happy people, and filled with all the blessings of liberty, civilisation and religion'? The answer, of course, is that the goal was not this definition of 'progress' – the so-called civilising of savage tribes – but rather the dominion of land under one authority – the US federal government – capable of monopolising extraction.

The fulfilment of the Euroamerican domination of the continent, and the transformation of the land and peoples from barbarism to civilisation, was implicitly an ideal of replacement and monopoly that rested on fragile myths of Euroamerican superiority. When Native Nations became 'civilised' in the way of settlers but outside their authority, they disrupted the anticipated Euroamerican course of history and conquest, posing a threat to the very foundations of belief motivating the project of manifest destiny. Just as Washington had to destroy the villages and towns of the 'civilised' Natives he grew up near – like those in the Ohio River Valley – in order to preserve the myth of Euroamerican superiority, so Jackson had to destroy the 'Five Civilized Tribes' to maintain the illusion of Euroamerican sophistication.

These many examples notwithstanding, politicians did not necessarily or always have to impose policies of expansion on a recalcitrant American public. They were politicians, actively responding to the reality that many settlers were independently seizing Native lands in a process of expansionism that was sometimes decentralised. Washington, for instance, once complained that a 'Chinese Wall, or a line of troops' would be necessary to prevent settlers from spreading into Native lands.[108] By the time Jefferson assumed the presidency in 1801, 700,000 settlers had already pushed beyond colonial borders. The progress narrative formula was not just an elite delusion of grandeur. It was a collective mass delusion, only effective if it could trickle down into the host class, who themselves become parasitic extractors at the point of concrete energy capture. Surely images of the savage Native pitted against the civilised, triumphant Euroamerican, who occupied a higher, more advanced stage of history on account of their language, race and geography, served as a moral salve that assuaged the guilt settlers may have felt, as much as it did politicians', at their contribution to acts of theft and murder.

It's also important to note that politicians swung back and forth in their rhetoric between that verging on bloodthirsty – like Jefferson's 'merciless Indian savages' and Jesup's inclination to 'exterminate' Natives – and a much softer posture towards Natives. Andrew Jackson, 'Indian killer' to the Creek, declared in his inaugural address in 1829, 'It will be my sincere and constant desire, to observe towards the Indian tribes within our limits, a just and liberal policy; and to give that humane and considerate attention to their rights and their wants, which are consistent with the habits of our government, and the feelings of our people.'[109] Just over a year after he said this, he urged Congress to pass the Removal Act, which forced Natives off their land and moved them west of the Mississippi River. But perhaps the most accomplished practitioner of this rhetorical manoeuvre – deploying the highest rhetoric while passing laws efficient to achieve the lowest ends – was also the country's most beloved and mythologised president.

* * *

Abraham Lincoln is often, justifiably, held up today as a paragon of liberal democratic virtue, and credited as one of those most responsible for abolishing slavery in the United States. This legacy has earned him nicknames like the 'Great Emancipator' and the 'Liberator'. Lincoln was an admirer of Karl Marx and once made the Marxian statement, 'Labor is prior to and independent of capital. Capital is only the fruit of labor, and could never have existed if labor had not first existed. Labor is the superior of capital, and deserves much the higher consideration.'[110] In his first inaugural address, Lincoln spoke like an anarchist: 'This country, with its institutions, belongs to the people who inhabit it. Whenever they shall grow weary of the existing Government, they can exercise their constitutional right of amending it or their revolutionary right to dismember or overthrow it.'[111] Lincoln sometimes even showed eloquent scepticism of progress, or that the United States was on an ever-improving path towards perfection. In a letter to a friend in 1855, he wrote that 'the country's growing degeneracy' appeared to him 'to be pretty rapid':

> As a nation we began by declaring that 'all men are created equal.' We now practically read it 'all men are created equal, except negroes.' When the Know-Nothings get control, it will read 'all men are created equal, except negroes and foreigners and Catholics.' When it comes to this, I shall prefer emigrating to some country where they make no pretense of loving liberty – to Russia, for instance, where despotism can be taken pure, and without the base alloy of hypocrisy.[112]

Lincoln was a heroic champion for equality; as such, he is often considered the country's greatest president, and he deserves adulation. He even spoke out publicly *against* manifest destiny. His early-career party, the Whigs, were publicly opposed to infinite expansionism.

There is a tale about him as a captain stopping his men from killing an apparently innocent Native man who carried papers of safe passage.[113]

But in his role as president, Lincoln, too, carried out the spirit of manifest destiny and promises of progress through conquest. In an annual speech to Congress, in 1861 during the Civil War, he proclaimed of the federal government, 'This is the just and generous and prosperous system which opens the way to all, gives hope to all, and consequent energy and progress and improvement of condition to all.' He went on to defend his government, saying, 'We thus have at one view what the popular principle, applied to government through the machinery of the States and the Union, has produced in a given time, and also what if firmly maintained it promises for the future. There are already among us those who, if the Union be preserved, will live to see it contain 200,000,000.' Here he echoes O'Sullivan's original essay coining the term 'manifest destiny', in which he celebrates 'the free development of our yearly multiplying millions', or Jackson's celebration of '12,000,000 happy people' in the unthinking fetish of quantity. Like other progress minded presidents, Lincoln was ever focused on a 'better' tomorrow: 'The struggle of today is not altogether for today; it is for a vast future also.'[114] In a call of 1861 for volunteers to join the US military, he appealed to them with the promise of 'practically restoring to the civilized world our great and good government'.[115]

Lincoln's role as an agent in bringing about emancipation for one ethnic group is complicated by his treatment of another. Many Union soldiers who had fought to free Black slaves in the South went on to commit atrocities like the massacre of Wounded Knee, in which they slaughtered not only Native warriors, but women, children, the elderly and infirm, in the US government's conquest of the west. Though he was personally less invested in conquest, and though his public addresses were less belligerent in their support of expansionism, like those of Washington, Jefferson and Jackson before him, Lincoln's administration sanctioned acts of settler and state violence against Native peoples for the sake of engaging in parasitic extraction, for opening and expanding territory. Less famous than his Emancipation

II NATION

Proclamation, he authored another proclamation in 1865 imposing martial law on Natives in the region of the frontier.[116] In an annual address to Congress in 1862, he urged that 'immense mineral resources of some of those [western Native] Territories ought to be developed as rapidly as possible'.[117] In the same speech, he warned that 'Indian tribes upon our frontiers have during the past year manifested a spirit of insubordination.' This led to the Dakota War of 1862, a campaign to remove and exterminate Sioux Natives, which resulted in a mass hanging of Sioux men and the largest single-day mass execution in the nation's history. The war started when the US government forced Dakota people in Minnesota onto small reservations where they were to practise agriculture even in areas where land was not arable. After treaties broken by the government, exploitation by settlers, mounting and unscrupulous debts, crop failures, a harsh winter and the huge influx of settlers, who overhunted game and deforested the region, the Dakotas suffered starvation and revolted against their resettlement. Minnesota Governor Ramsey called for them to be 'exterminated or driven forever' from the state.

Lincoln's administration passed the 1862 Homestead Act, legally empowering settlers to seize Native land. His administration carried out the Whitestone Hill massacre of 1863, in which between a hundred and three hundred Sioux people were killed, including women and children. In 1864, his government enforced the 'Long Walk of the Navajo' that forced 9,000 people to walk 300 miles in the arid southwestern United States, killing at least two hundred. In the same year, US soldiers carried out the Sand Creek massacre. Between two hundred and five hundred Natives – up to two-thirds of whom were women and children – were killed and mutilated. The massacre was apparently unprovoked: the Cheyenne and Arapaho people in the Sand Creek reservation complied with the Colorado governor's orders to stay within the reservation; nevertheless, an army of Colorado Volunteers attacked 'Without provocation or warning'.[118] Testimonies published by a Congressional investigation found that after the massacre, the soldiers 'decorated their weapons and caps with body parts – fetuses, penises, breasts, and vulvas'.[119]

The nation's presidents, enamoured with expansion and the myth of progress in manifest destiny, seemed eager to use the military to seize land and replace people and animals with farms and cities. But one commodity above all would enable them to stitch together a vast and wealthy country in the aftermath of the Civil War, and to make it a 'Great Power' on the global stage within a few decades.

The Civilisation of Density

The 'frontier' in the United States is an important part of the country's mythology and its sense of progress. We have been discussing 'frontiers' throughout this book. They are sites of extraction, where people move to farm, mine, ranch or fell trees. They have existed at the periphery and as the economic base of all parasitic economies. The frontier that opened with the first English settlers in North America in 1607 was not the first in the country, but it did maintain a long lineage whose legacy reached to 1890, when the frontier officially closed. When this happened, there was no more land left to absorb. When parasitic economies in the past have hit such resource limits, they have typically begun to fragment, gradually breaking into autonomous zones and conflict, the original polity shrinking to the size of a city, or cracking into dust. The United States itself nearly endured such a fracture, with the Civil War threatening to break it in two. The decisive element that saved it from disintegration was industrialisation, the regimenting of production into factory forms. But while various forms of 'industry' and 'factories' have existed for much longer than this industrialisation process, what really distinguished the period, often referred to as the Industrial Revolution, was the pairing of industrial forms with machines that could speed up extraction, production and distribution.

With virgin forests mostly harvested, game species made extinct or their abundance greatly reduced, with the ecological integrity that Indigenous peoples had cultivated for millennia dismantled and hauled into the increasingly global economy, with the over-exploitation of soil and the resultant creation of dustbowls, and the dwindling of previously abundant fish stocks, the last newly accessible lands

II NATION

of North America were approaching exhaustion. The Americas had been colonised and stripped of resources by organic machinery – livestock and hand tools – as Europe had long since been cleared by the same means, and with wood-and-cloth sailing vessels extracting fish and shipping goods at the speed of the wind and nets of natural fibres, limits to further expansion were beginning to show themselves. The parasitic economy was again bumping up against ecological barriers to expansion and extraction, threatening to dissipate the centralised energy gathered into monarchic and oligarchic power. The only place left to go was down.

Vast stores of buried fossilised energy could maintain the project of frontier conquest, and, importantly, the continued expansion of a parasitic economy whose only options were growth or collapse, boom or bust. If energy on land could not produce growth, energy beneath the land must do so. Fossil fuels are, after all, nothing more than the biomass of plants and animals, buried and compressed over millions of years. Digging up and burning fossil fuels is like harvesting the heavy ghosts of vast fields, forests and game, all those plants and animals had been pressed together by the steady passage of time, parallel dimensions collapsed upon each other; as if humans figured out how to send loggers, hunters and farmers back through time to carve up, shovel out and burn those sedimentary layers of life. The result was a massive injection of energy into an economy that would otherwise have confronted an energy- and resource-constrained future due to the hoarding of wealth by the rich and overpopulation by settlers. It was in Britain, at the heart of history's largest empire at the close of the eighteenth century, that coal was first put to industrial use. The organising principles of industry, an intensification of the Enlightenment capitalist administrative state, would be needed as well.

To understand the link between these new fossil fuel technologies and energy capture – and the societal consequences of that relationship – it helps to start with an understanding of energy *density*. All biological organisms, all their activities and the products of their interaction like societies and technology, are dependent on energy. When one hears the word 'energy', a number of images come to mind: light bulbs, electricity pylons, gas wells and oil drills, solar panels. But energy more broadly includes all the material things that living creatures consume. The calories in food, the carbon in wood, gravity pulling at water are all forms of energy. Energy from the world is transformed into heat or motion in the bodies of all those things that move by their minds. Material civilisation, which today includes aeroplanes, rockets, skyscrapers, nuclear bombs, cars, roads, hotels, theatres, farms, factories, and all the other inanimate physical objects that constitute the human world, is dependent on energy. The biological organisms – primarily humans and livestock – that inhabit that civilisation are also dependent on energy. Engines, an artificial metabolism, can harness energy for heat and motion as well as speeding up previously organic and biological metabolic processes.

Energy sources like wood, gravity and water (in the form of gravity-powered mills) tend to provide less 'density' than carbon that has been compressed by long spans of time and gravity. The energy density of dry wood, for example, is between 17 and 21 megajoules per kilogram; for bituminous coals, it's slightly more, between 18 and 25 MJ/kg; but crude oils have almost twice the energy density of wood or coal, between 40 and 44 MJ/kg.[120] A civilisation powered by wood and moving water will be less dense and populous than one powered by coal and oil because it controls less energy necessary to turn into biomass and physical infrastructure for *Homo sapiens*. A less energy-dense civilisation simply will not be able to build massive factories, fly aeroplanes or launch people into outer space; the latter depends not just on the copious hydrocarbons needed for jet propulsion, but also all the myriad material inputs – steel, glass, plastic, rare earth metals, and food (for workers) – and all their tangled production webs to achieve.

Human brains and bodies, like our material civilisations, are physical entities that need energy to function, every bit as much as a car needs fuel to move. Historical demographer E. A. Wrigley calculates that a man needs 2,500 kilocalories of food per day in order to work, of which the majority – 1,500 kilocalories – is needed just to stay alive; but '[w]ith a daily intake of 3,500 kilocalories a man could undertake double the amount of physical effort which he could

II NATION

perform if his intake was 2,500.'[121] A civilisation that captures more energy into the social, political and technological system that takes in, moves and processes all this energy – its socio-technical production metabolism – can afford to create more humans. It can also give those individuals more time than parasitic economies with lower energy inputs to pursue activities not directly related to food or tool production. We see explosions of invention in the nineteenth century because increased energy wealth could pay more wages to more people who could produce technologies that in some cases benefited the public, but also benefited their benefactors or employers, or supplied the military needs of states. Furthermore, industrialised societies could power more machines that could more efficiently harvest all the materials – timber, food, metals and minerals – used to build physical civilisation.

Energy density also relates to the amount of caloric energy that can be extracted from a given area of land or sea. If one civilisation can extract more energy per capita than another from a given square acre, we can say it is based on a higher density of energy. Using foraging as a way to extract calories from the land, for instance, a society can typically extract less energy than using subsistence farming, which, in turn, captures less than intensive industrial farming. This is because a region in which wild sources of food are foraged will be full of non-edible (to humans) biomass. Only a portion of that energy can be captured by an omnivorous primate with a particular digestive system that is only able to process certain forms of biomass. A farm, on the other hand, will usually push out all those non-edible or, to humans, non-productive sources of biomass – like many wild animals and plants – and replace them solely with things that humans, or their livestock, can consume. A society that augments its agricultural production with petroleum-powered machines like tree cutters, tractors and ocean trawlers, and petroleum-derived fertilisers, will be able to extract even more energy from a given area of land because they will be able to cultivate it more intensively and input more artificial nutrients at will instead of having to wait for natural nutrients to accumulate.

Archaeologist Ian Morris documents that in foraging societies, a person will capture around 4,000 to 8,000 kilocalories of energy per day from wild sources (depending on that person's distance from the equator – colder regions need more energy for heating, clothing and hunting).[122] Agriculturalists, with the help of domesticated crops and animal labour like ox-drawn ploughs, can capture up to about 30,000 kilocalories per day, which he estimates is the maximum energy capture possible for one person in a purely organic (non-fossil fuel-augmented) economy.[123] But with the introduction of fossil-fuelled machinery, energy capture jumped by an order of magnitude, enabling a person to capture, on average, 230,000 kilocalories in a single day by 1970.[124]

All this information about density may sound obvious; it is a simplification of complex processes, and it does not explain why civilisations suddenly began to consume so much more energy than they had during all of human existence. Coal was used for thousands of years in smelting metals and heating homes and palaces all over the world, from ancient China, to Greece and Rome, to Britain and Aztec Central America, without resulting in industrialisation, ecological collapse or global warming. That is not to say coal hasn't caused problems where it has been used, especially in urban areas. In medieval England, for instance, coal was used as a source of heat – for kitchen fires and heating – and caused a major air pollution crisis. London has been plagued by fossil fuel smog for over seven hundred years.[125] But societies only unlocked coal's full industrial potential when steam engines were invented and began producing electricity and motion rather than just heat. Some petroleum, too, has been readily available for millennia, but petroleum needs something like an internal combustion engine to maximise the delivery of its very dense energy into a civilisation's services and goods. The limits of technology in the capture of energy density are still very real: uranium, for instance, is much denser even than petroleum at a potential 3,900,000 MJ/kg to oil's 44–46 MJ/kg. But uranium is not equivalently useable.[126] It cannot be readily swapped into most practically useful technology, and it carries with it many considerable safety risks like radiation toxicity at the point of extraction, use and waste.[127] A civilisation's density is limited by these technological realities.

By the early nineteenth century, some states in the US were beginning to industrialise. It was this greater source of energy capture that gave the Northern states an advantage and enabled them to hold the Southern states under the federal government during the American Civil War. The burgeoning fossil fuel industrial economy of Northern states produced abundant material goods, accommodated more dense populations, and so produced better weapons carried by more soldiers than the agrarian Southern states. The Union army at its peak was double the size of the Confederate army, while Union states were over three times richer, more than twice as populous, and achieved an agricultural advantage (Confederate states led the nation in tobacco and rice yields, but the Union states produced more calorie-dense, strategically valuable commodities like wheat, corn, livestock, and horses). They also had ten times as many factory workers in five times as many factories (amounting to twice the density in factories).[128]

While industrialisation had complex effects, it is important to emphasise here that for well-being indicators for the majority of people, industrialisation was, on the whole, very bad. Life in industrialised cities was grim: air and water quickly grew polluted, people were packed together in overcrowded housing, and work was dominated by factories which themselves were often overcrowded, filthy, deafening, and caused devastating injuries, illness and death. Food passing through industrial systems often arrived at market unsafe: rotten, tainted, laced with poisons. Newly developed drugs also frequently contained toxins that maimed and killed many. New methods of social and political control arrived as well, with the advent of professional urban police and detective forces. This should not be surprising: industrialisation, and particularly the form augmented by fossil fuels, was not undertaken to improve human well-being, but to intensify energy capture, of both ecological systems and workers, and accumulate wealth for owners. Nor was the implementation of fossil fuels a 'natural' course of teleological history. It was imposed on its host by a parasitic class.

The decisions involved in abandoning water-powered factories for coal-powered factories had less to do with pure efficiency or the unstoppable logic of technological progress than with increasing the owners' ability to control production and workers. Historian and human-ecologist Andreas Malm describes the steam engine as 'an antagonist to labour *enabling* the capitalists to defeat a revolt, which threatened to drive the *infant* factory system into crisis'.[129] Using coal could allow owners to control the flow of labour, no longer relying on the rhythms of hydrology and seasonal light to run factories. 'Steam won,' in the battle against other sources of energy like water, 'because it augmented the [political] power of some over others.'[130] Coal and steam engines did not come to dominate because they were necessarily more efficient forces of production – early on, coal power was far more costly than water power – nor because they necessarily yielded better outcomes for most people. In *Capital*, Marx was already writing about how coal and steam could be useful for building elite political power and accumulating capital. Not until the development of the steam engine, he wrote in 1867, 'was a prime mover found which drew its own motive power from the consumption of coal and water, was entirely under man's control, was mobile and a means of locomotion, was urban and not – like the water-wheel – rural'.[131] With the primacy of the steam engine, workers could be crushed together in factories in the city, occupying less land and generating more energy where managers could keep a close eye on them.

Factories were already starting to emerge in industrial centres around 1800, at the end of the Enlightenment Age. In some ways, coal helped miners and factory employees to gain worker power, to collaborate and build the first labour movements. While factory owners used coal to control labour, labour used coal to control the flow of energy and demand better conditions. High-energy industrialisation also helped abolitionists make the case against slavery; by changing agrarian economies into industrial economies, coal helped to eliminate reliance on enslaved agricultural labour, replacing it with forced wage labour in factories and cities and with mechanised agricultural tools. The parasitic economy needed more affluent consumers to solve the overproduction crises that emerged, making wage labour a more effective way of growing it.

Even so, wealth inequality increased. In the final decade of the nineteenth century, 1 per cent of Americans controlled an equivalent

II NATION

income of 50 per cent, while holding more property than 99 per cent.[132] This was an increase from the richest 1 per cent holding just 8.5 per cent of the nation's wealth about a century earlier.[133] This growing inequality was partly thanks to fossil fuels, which could be easily concentrated, controlled, and transformed into liquid capital by a small management class. Because fossil fuels themselves are easy to concentrate, they often yield authoritarian outcomes. The idea of the 'resource curse' illustrates this well. A petrostate – a country whose economy is disproportionately dependent on oil and gas extraction and export – is 50 per cent more likely to be authoritarian and only a quarter as likely to transition to democratic government than a state without petroleum as a major economic base.[134] The oil curse does not exclusively plague countries of the Global South, as is often implied; it is simply the nature of oil, wherever it reigns.[135] But industrialists alone did not create this condition. In addition to granting them subsidies and state-sanctioned free labour, the federal government played its own role in suppressing workers, providing a repressive police and military presence to crush striking workers. The Great Railroad Strike of 1877, for instance, saw federal troops and militias murder dozens of workers.

While the frenetic production and urbanisation brought about by fossil-fuel industrialisation conjured a concrete illusion of progress – a steel-and-glass projection onto earth – what coal, oil, capitalism and industrialisation have never provided is stability. The 1800s saw one economic crisis after another, with recessions occurring in every single decade of the century. The Long Depression – originally called the Great Depression prior to the twentieth-century one – lasted twenty-three years beginning in 1873. The boom-and-bust of industrial America's financial capitalism – a rapid microcosmic reflection of the longer, larger boom-and-bust of all parasitic economies – began with the Panic of 1819. This event, the country's first financial crisis, was a direct result of the speculations of finance capitalism, and arguably of dependence on foreign commodities and the exhaustion of agricultural production in Europe.[136] It heralded a pattern of recession, depression and overconsumption that has been ingrained in financial capitalist societies ever since, with the US in particular suffering boom-and-bust economic bursts and crises in nearly every decade.[137] The more catastrophic sort of instability, linked with financial crises and fossil fuels both, was industrial-scale war. And fossil-fuelled progress, as the next section will cover, would deliver unto the world the greatest and most catastrophic wars ever to scar the earth.

This new form of energy capture supercharged the disparate parasitism of the European empires even as it fortified the continental territory of the United States. But this influx of energy had destabilising effects not just on the economies and societies in which it was infused, but on the ideas that made sense of these societies. The combination of fossil fuels, industrialisation, and the accompanying intensification of parasitic social and ecological relations, which often yielded poor working and living conditions, called into question the simplistic promises of progress espoused in liberalism, dreams of manifest destiny, or the optimistic scripts of Enlightenment *philosophes*. It contributed to a flurry of new debates around the direction societies were or should be going, and what the perfection of the course of history might really look like.

The Rise of Evolution

Survival of the fittest conjures an image of Charles Darwin's dense white beard hanging from a heavy-browed face as he describes natural selection: the process whereby species evolve or go extinct according to their performance in competition for scarce resources. While Darwin was happy to see the term associated with his theory, he didn't coin it. Credit for that goes to a contemporary, a philosopher who is less of a household name these days: Herbert Spencer.

Spencer was 'the most famous philosopher of his time', which was the latter nineteenth century.[138] He was a social theorist, not a naturalist, and did not coin his famous idiom to describe how speciation occurs based on impassive observation of wildlife. Instead, he applied the principle to his theory of societal progress. Spencer's musings on the evolution of human systems predate Darwin's seminal work *On the Origin of Species* and its application of evolution to

II NATION

biological systems. But whereas Darwin was generally content to observe and report,[139] Spencer felt compelled to imbue his observations with a more explicit moral and political claim. In a revealing passage from *The Man Versus the State*, he opines:

> All legislation which assists the people in the satisfaction of their natural wants – which provides a fund for their maintenance in illness and old age, educates their children, takes care of their religious instruction, looks after their bodily health . . . arises from a radically wrong understanding of human existence . . . Instead of diminishing suffering, it eventually increases it. It favours the multiplication of those worst fitted for existence, and, by consequence, hinders the multiplication of those best fitted for existence.[140]

In other words, just as a weak fish should be allowed to be removed from the school before it mates and passes on its genes or slows down the fitter fish, thus weakening the rest of the school, his programme could be understood as saying, if you're old or sick or poor or unable to access education, not only are you on your own, but you *should* be on your own for the good of the collective, of society or, indeed, all of humanity.

Spencer was one of the first theorists to develop what we might think of as an industrial-era story of societal progress, one that is often called 'social Darwinism' for its association with Darwinian evolution. Instead of favouring centralised states, monarchies or legislatures, he instead anoints centralised corporations as the prime movers of progress. In this view, natural law dictates that progress towards a perfect society is inevitable, but only if its evolution remains unhindered by state intervention. The nature of this progress will be so complete that the species and its societies will be perfect. Spencer's secular faith sounds strikingly similar to promises of Abrahamic progress. Spencer wrote eight years before Darwin's *Origin*:

> Progress, therefore, is not an accident, but a necessity. . . . The modifications mankind have undergone, and are still undergoing,

result from a law underlying the whole organic creation; and provided the human race continues . . . these modifications must end in completeness. . . . So surely must evil and immorality disappear; so surely must man become perfect.[141]

The human world of consumers, companies and captains of industry mirrors what Spencer imagined as the cut-throat competition between variably fit individuals in the natural world; and, if it works for nature, society ought to enforce it.[142] Darwin eventually followed Spencer's lead to some extent, declaring in his 1871 book *The Descent of Man* that 'the wonderful progress of the United States' was indeed the 'result of natural selection'.[143] Spencer's conflation of *is* and *ought* – evolution is this way, therefore society ought to be this way – may have stemmed from the fact that his theories predated moral philosopher G. E. Moore's idea of the 'naturalistic fallacy', a logical fallacy which holds that one should assume something is good *because* it is natural (which may sometimes or often be true, but should not be assumed to be true). Spencer's body of work rests a bit too heavily upon just such an assumption.

Spencer had one romantic infatuation in his life: the novelist George Eliot. The period was so suffused with the idea of progress that it came into Eliot's novels quite frequently, but in the form of a more ambivalent interrogation rather than as propaganda or polemic.[144] Characters in her magnum opus *Middlemarch* are tangled within a hope for progress that may improve their lives and the frustration of such improvement ever thwarted. Some characters engage in machine-breaking in the style of the mythical Ned Ludd and the Luddite movement of the time, attempting to fight off industrialisation and their own market obsolescence.[145] Other characters question the value of all railway technology.[146] Alongside Eliot's ambivalence, Spencer's definition and analysis of progress appears starkly simplistic, distilled to the utopian ideal that society should imitate what he understood as the ruthless natural selection of the wilderness, since doing so will result in a perfect human specimen and a perfect society.

Another famous author, Mary Wollstonecraft, also offered a more nuanced view of progress half a century before Spencer, writing in *Vindication of the Rights of Women*:

> Contending for the rights of women, my main argument is built on this simple principle, that if she be not prepared by education to become the companion of man, she will stop the progress of knowledge, for truth must be common to all, or it will be inefficacious with respect to its influence on general practice.[147]

Contrary to Spencer's view that universal education would assist those he deemed 'worst fitted for existence', and therefore harm the race and slow societal progress, Wollstonecraft considered universal education central to progress. Oscar Wilde also contributed to this debate in his dialogue 'The Decay of Lying':

> Life – poor, probable, uninteresting human life – tired of repeating herself for the benefit of Mr. Herbert Spencer, scientific historians, and the compilers of statistics in general, will follow meekly after him, and try to reproduce, in her own simple and untutored way, some of the marvels of which he talks.[148]

In this work, Wilde argues that progress myths function as self-fulfilling prophecies. If a society's most influential thinkers dream of a future of space travel, all efforts and funds will be put to the task of building rocket ships. If those thinkers say progress means leaving the poor and elderly to perish in squalor, then it becomes the rational, educated and compassionate thing to do, and so will happen.

Perhaps as important to Spencer's equation of progress with the laissez-faire operation of markets, societies and economies was the belief in self-interest as an engine for progress. This was central to Enlightenment notions of progress as well. Maintaining this older faith in self-interest, even as he sought to position his theory as novel and 'scientific', gained Spencer a loyal following among some of the wealthiest members of the burgeoning industrial class.

Scottish-American steel baron Andrew Carnegie was one such admirer.[149] This should come as no surprise: in the arena of companies competing to shape the industrialising United States, Carnegie was one of the 'fittest'. Through his steel company, he became one of the richest men of his time, or any time.[150] He achieved this in part by imposing extreme labour practices on workers, forcing employees to toil twelve hours per day, seven days per week for poverty-level wages. Injuries and deaths occurred frequently in his factories.[151] When workers organised the Homestead strike in 1892, the largest in US history, Carnegie responded by sending in armed mercenaries who shot many strikers, killing nine. Carnegie justified his actions by Spencer's reasoning, writing in his own essay 'Gospel of Wealth':[152]

> We accept and welcome, therefore, as conditions to which we must accommodate ourselves, great inequality of environment, the concentration of business, industrial and commercial, in the hands of a few, and the law of competition between these, as being not only beneficial, but essential for the future progress of the race.[153]

Carnegie was not alone. Rail, coal, oil and finance magnates of the time regularly engaged in actions that do not intuitively appear to demonstrate 'progress', like fatal strike-breaking, by which corporate mercenaries and state-sanctioned armed forces killed hundreds of workers; or creating unsafe working conditions, unregulated pollution of air and waterways, and toxic products that killed uncounted thousands of both workers and the general public; and – for more than half a century into the industrial era – slavery. Carnegie's invocation of competition as a motor of progress was disingenuous at best. These corporations, after all, colluded to build 'trusts', which consolidated the economy into industry-wide monopolies that fixed prices and controlled supply. And far from operating in opposition to the state, these trusts and corporations engaged in rampant political corruption, capturing the federal government and shaping policy through coercion and bribery. But theories like Spencer's could make such actions appear to both corporate elites and the public to be necessary evils in pushing the country forward to some imagined perfect state.

Besides Darwin, Spencer, Wollstonecraft and novelists like Eliot and Wilde, other thinkers contributed ideas to the scientistic version of progress important for the maintenance of this increasingly complicated industrial society. Russian anarchist and geographer Peter Kropotkin interpreted the relationship between evolution and societal progress differently from Spencer. Instead of emphasising competition, he suggested that evolution promotes cooperation between individuals within communities – and, by extension, competition *between* communities. Kropotkin called this 'mutual aid'.[154] The most cooperative communities will outcompete those that are less cooperative, and will thereby advance to a higher state. Kropotkin, too, drew from a naturalistic sense of evolution for his theory of progress: 'The animal species, in which individual struggle has been reduced to its narrowest limits, and the practice of mutual aid has attained the greatest development, are invariably the most numerous, the most prosperous, and the most open to further progress.'[155]

French philosopher Auguste Comte, one of the founders of modern social science, also had much to say on the matter of progress. In 1839, Comte expressed an optimistic view of history and inevitable human development: 'It is unquestionable that civilisation leads us on to a further and further development of our noblest dispositions and our most generous feelings, which are the only possible basis of human association, and which receive, by means of that association, a more and more special culture.'[156]

Societies inevitably go through three particular stages of progress: first, a theological stage in which all phenomena are believed to spring from tangible divine agency; second, a metaphysical stage in which abstract concepts become the driver of all phenomena, and in which even divinities are abstracted; and finally, a positivist phase, the *highest* stage of development, in which scientific explanations reveal physical laws of cause and effect as the primary source of all phenomena. Comte is often credited with developing positivism.[157]

The most comprehensive, internally cohesive, and probably most impactful narrative of human progress to emerge in the nineteenth century was that of German philosopher Karl Marx.[158] Instead of a wilderness of individuals or companies competing for supremacy, or competing communities of cooperative individuals, Marx saw historical progress as consisting of conflicting social-economic classes jostling for resources: history *is* class struggle. Marx argued that not only was the social development that led to industrial societies an inevitable form of progress-by-class-struggle that all civilisations would go through, but that we could predict their political course into the future, using historical materialist analysis. While Marx abandoned the strident determinism he proposed in his younger writing,[159] Marx*ism* has carried into even the twenty-first century a sense of inevitable progress towards a state in which class struggle will culminate in proletarian victory and a communist utopia that liberates people not just from oppressive human systems but ecological ones as well.

Darwin cast a long shadow over the age. Marx, in his own words, was a 'sincere admirer'. Marx and his collaborator Friedrich Engels saw their work as speaking to Darwin's scientific description of biological evolution. Marx wrote in a letter, 'Darwin's work is most important and suits my purpose in that it provides a basis in natural science for the historical class struggle.' Engels, meanwhile, suggested, 'Just as Darwin discovered the law of development of organic nature, so Marx discovered the law of development of human history.'[160] Marx saw history progressively unfolding from primitive communism, to slavery and feudalism, through capitalism to socialism, and culminating in proletarian communism. He wrote in *Capital*, 'The country that is more developed industrially only shows, to the less developed, the image of its own future.'[161] His adoption of progressive evolutionary history held a dark side. Even for Marx, who was a steadfast champion of workers, peasants and classes of marginalised or oppressed peoples around the world – the host classes of parasitic societies – the allure of societal evolution could lead him to offer an apologia for such ills as English colonisation in India:

> England, it is true, is causing a social revolution in Hindustan, was actuated only by the vilest interests, and was stupid in her manner of enforcing them. But that is not the question. The question is, can mankind fulfill its destiny without a fundamental

revolution in the social state of Asia. If not, whatever may have been the crimes of England she was the unconscious tool of history in bringing about that revolution.[162]

In other words, the British Raj might itself be immoral when viewed in isolation, but when viewed as a necessary step towards global communism, it is merely an unfortunate 'tool of history'. Such an argument looks little different from the liberal justification for colonisation as a means to achieving civilisation.[163] The race science that was to become fundamental to evolutionary progress tainted even revolutionary writing.

Friedrich Engels drew explicitly from Spencer, whose work, anthropologists Nancy Lindisfarne and Jonathan Neale contend, 'was saturated with white racism'. Engels imbued his own writing with racial hierarchies tied in with evolutionary progress:

> The plentiful supply of milk and meat and especially the beneficial effect of these foods on the growth of children account perhaps for the superior development of the Aryan and Semitic races. It is a fact that the Pueblo Indians of New Mexico, who are reduced to an almost entirely vegetarian diet have a smaller brain than the Indians at the lower stage of barbarism who eat more meat and fish.[164]

This talk of superior development and 'the lower stage of barbarism' is rife with racial binaries embedded in a progress narrative formula. But even as Marx furthered claims of historical progress, he also recognised the thoughtlessness that could accompany those claims; the more hysterical war cries in support of its triumphant march must have echoed loudly about his ears. Perhaps in that soot-stained era, with the poor packed tightly into cities and factories, it would have seemed as if too much carbonic energy dug up in coal mines and oil wells had accumulated like a black sand drift in the coils of the brains and bodies of the populace. That accumulation of energy had overheated the vibrating organic engine of nineteenth-century humanity, mass and elite alike, and the sputtering, rattling, pathological noise emanating from this calescent engine was the slurred moan of *Progress*.

Marx wrote in *Capital*:

> machinery, considered alone, shortens the hours of labour, but, when in the service of capital, lengthens them; since in itself it lightens labour, but when employed by capital, heightens the intensity of labour; . . . Exploitation of the workman by the machine is therefore, with him, identical with exploitation of the machine by the workman. Whoever, therefore, exposes the real state of things in the capitalistic employment of machinery, is against its employment in any way, and is an enemy of social progress![165]

Marx laments that in capitalist societies, technology is used to create more wealth for business owners rather than to simply produce what people need to live with less effort, thus giving workers more leisure time. It is not the development of more sophisticated technology that is at fault – after all, Indigenous societies have long invented tools and systems to make their work easier and the land more productive. Rather, it is technology pressed into the service of capitalist – or, more broadly, parasitic – logic that impoverishes those who should be liberated by it. Marx decries the still familiar claim that if you question the alleged benefits of technological and economic development, you will not only be accused of being the enemy of technology and a healthy economy, but an enemy of progress itself.

And yet, though the Marxist shape of history bending towards emancipation has been lastingly influential, there was a vital problem with its early theory of progressive stages, causing even Marx to denounce some of his early ideas. Later in life, Marx became keenly interested in anthropology; his huge, unfinished *Ethnological Notebooks* are filled with notes and commentaries on all manner of topics in world history. His notes demonstrate a particular fascination with the confederation of Iroquois peoples, whose Council of the Gens represented a 'democratic assembly where every adult male and female member had a voice upon all questions brought before it'.[166]

II NATION

How could history be the progression from 'primitive' to 'civilised' societies if the Iroquois, otherwise considered 'primitive', already had a participatory democracy and high gender equality long before – and well outside the cultural lineage of – capitalism or communism? Marx abandoned his previous ideas of pre-capitalist societies as rudimentary, or society and history as progressive, and found within them a new, multilinear hope for revolutionary forces. Now favouring a more cyclical view of history, in which future communism would be a version of existing and past Indigenous societies, he wrote:

> the historical trend of our epoch is the fatal crisis undergone by capitalist production in those European and American countries where it reached its highest peak. The crisis will come to an end with the elimination of capitalist production and the return of modern society to a higher form of the most archaic type – collective production and appropriation.[167]

Historian Franklin Rosemont observes that reading about the Iroquois and other First Nations gave Marx 'a vivid awareness of the *actuality of indigenous peoples*, and perhaps even a glimpse of the then-undreamed of possibility that such peoples could make *their own* contributions to the global struggle for human emancipation'.[168]

* * *

Before we move on, let's take a quick tally of the debate: Spencer declares that if we avoid regulation, competition between individuals will inevitably produce the perfect man through a process that looks like biological evolution. Novelists like Eliot and Wilde reply that the situation is more complicated than that and we ought to be more ambivalent about claims of progress. Perhaps appropriately or surprisingly, it is the professional storytellers who are more ambivalent than the scientists and theorists about such simplistic narratives. Kropotkin suggests that cooperation, rather than competition, between individuals achieves progress. Marx first agitated for progress through class struggle before abandoning this teleology for a view that Indigenous forms of society are most likely to deliver human well-being. As it happens, a ghost of this debate now quietly rages through eternity not just in their books but in the breath of wind and pollen and the droppings of wood pigeons: Eliot, Spencer and Marx are buried within mere metres of one another in London's Highgate Cemetery.

It's no surprise that the nineteenth century proved fecund in these new scientistic progress narratives: it was a time of monumental change. The age appeared to promise history's most rapid advancement in technology and social change, ushering in rapid urbanisation, social movements, and the invention of electric batteries, typewriters, cameras, telephones, telegraphs, elevators, escalators, automobiles, electric light, locomotives, matches, microphones, revolvers, antiseptics, pasteurisation, machine guns, plastic, movies, toilet paper, contact lenses, vacuum cleaners and zip fasteners, to name a few.[169] As steam engines, fuelled by coal and wood, and then internal combustion engines fuelled by petroleum, came to augment nearly every aspect of material society, from transportation and construction to agriculture and manufacturing, the *appearance* of progress suddenly accelerated. The abundance of theories that arose to explain it also aspired to present an ideal course that progress *ought* to follow. Theorists can be forgiven for getting carried away and letting a bit of destiny slip into their writing: such a newly frenetic pace of invention, spurred by the feedback loops of parasitism brought about by increasingly intensive use of fossil fuels and administrative tools like an ever more corporate capitalism, seemed to point to a future of endless, rapid growth and complexity in material civilisation. But was such frantic tinkering truly progress, truly moving the world in a superior direction? Was it blind wishful thinking, or deliberate political calculation, more than scientific observation that birthed these theories?

After all, industrialisation was a dark bargain that even those in the midst of it could see: aside from the eagerly erected soot-coated cities, the mass manufacture of ever deadlier weapons and the eradication of whole cultures, fossil-fuelled industrialisation rapidly increased the pollution of air and water even in rural areas, and continued the long project of eliminating biological diversity. The

II NATION

irony of the time was that even as nineteenth-century naturalists like Darwin were studying species variation and describing species previously unknown to Westerners in an explosion of scientific discovery, the mass extinction driven by disparate parasitism was well underway, eliminating life faster than anyone could document it. The rapid destruction of wild ecologies alarmed many, even some of the most enthusiastically bloodthirsty among them, like big game hunter and US president, Theodore Roosevelt.[170] In 1908, Roosevelt delivered a speech at the White House urging his fellow leaders to recognise the perils of environmental destruction:

> We have become great in a material sense because of the lavish use of our resources, and we have just reason to be proud of our growth. But the time has come to inquire seriously what will happen when our forests are gone, when the coal, the iron, the oil, and the gas are exhausted, when the soils shall have been still further impoverished and washed into the streams, polluting the rivers, denuding the fields, and obstructing navigation.[171]

Meanwhile, nineteenth-century scientists had discovered that burning fossil fuels could release carbon dioxide in quantities sufficient to warm the planet. In 1824, physicist Joseph Fourier hypothesised that atmospheric gases might create chemical conditions sufficient to trap the sun's heat; thirteen years later, he proposed that the amount of heat held in by the atmosphere could be changed by both natural fluctuations *and* human activity. In 1856, scientist and suffragette Eunice Newton Foote tested the heat-trapping abilities of 'carbonic acid' (what scientists then called CO_2) and observed that, 'An atmosphere of that gas would give to our earth a high temperature.' She was the first scientist to link carbon dioxide emissions specifically with global climate change.[172]

Fearing that fossil fuel use would cause a planetary 'greenhouse effect', Alexander Graham Bell – best known as the inventor of the telephone – advocated for the use of solar power instead.[173] And it was not just scientists and conservationists who were worried; people across Europe, North Africa and the Americas debated the 'climate question'. Science historian Deborah Coen writes that the field of climatology 'relied on people of all walks of life to report on the weather and its effects on their health and crops. Farmers and vintners supplied harvest dates to track the seasons from one year to the next, while sailors and fishermen informed early schemes for classifying clouds and winds.'[174] Climatology emerged as a 'citizen science' – it was, after all, in everyone's interest to know when the next big storm was coming.

Science fiction is often a reliable way to take the temperature of a society's fears and dreams about the inventions and discoveries of its time. Air pollution, resource shortages, deforestation and global warming haunted Europe's literary minds. In 1885, Richard Jefferies published *After London*, a Victorian post-apocalypse novel in which an environmental catastrophe turns London into a toxic swamp and England is reclaimed by forests, wild creatures and mutated humans. Mary Shelley, author of *Frankenstein*, mother of science fiction (and daughter of the aforementioned Mary Wollstonecraft), also published – in 1826 – a post-apocalypse novel called *The Last Man* set in the late twenty-first century, when humanity is almost entirely wiped out by a mysterious plague that spreads quickly in the overheated atmosphere. But such apocalypses still seemed far away. Rather than seeing this industrial growth as the first nails pounded into a coffin that would entrap not some distant, anonymous posterity but their own great-grandchildren, many understandably embraced the extra energy provided by fossil fuels and its attendant illusions of progress. But not all were so complacent.

The Fall of Revolution

The history of all hitherto existing society is the history of class struggles. Freeman and slave, . . . oppressor and oppressed, stood in constant opposition to one another, carried on an uninterrupted . . . fight, a fight that each time ended, either in a revolutionary reconstitution of society at large, or in the common ruin of the contending classes.[175]

II NATION

The *Manifesto of the Communist Party* by Karl Marx and Friedrich Engels was published in 1848, the year a wave of revolutions swept across Europe.

This rousing call to arms rings as true today as it did when it was published. Since the start of the lineage we are examining, 5,000 years ago in Mesopotamia, the parasitic human-ecology that has spread across the globe has remained resilient, passing to new cultures this heretic metabolism like the virus that it is. The individual cultures, empires and economies that have practised this form of human-ecology have not been so resilient. They have erupted into the world in fevered dreams of war and mercurial bureaucracy, and fallen, sometimes within decades, sometimes centuries, a few times even after a millennium or more. And mostly they have fallen in a long histrionic circus dance through puddles of blood and trash and viscera into a long night.

When parasitic societies have collapsed they have done so due to many compounding factors, including the breaching of ecological limits or through suffering sudden shocks, institutional decay, invasion by some other aspiring empire, or disease. It is strangely rare throughout history for such arrangements to be overthrown in a 'revolutionary reconstitution of society'. Such events likely number in the hundreds or the thousands, rather than the tens or hundreds of thousands that one might expect, accounting for the span of time involved and the numbers of individuals and societies that have been born and died in this span. Given the living conditions such parasitic economies have imposed, it is strange more people have not sought to upend them. The oldest documented labour strike occurred around 1157 BCE during the reign of Ramses III. Recorded on a scroll of 3,000-year-old papyrus is an account containing some details, though scant, of workers striking and conducting a sit-in to protest at not being paid their agreed wage rations. They were artisans working on intricate decoration of the mortuary temple for Thutmose III, a pharaoh who ruled for fifty-four years and expanded the Egyptian empire to new heights.[176] It is unclear from the scroll how successful their strike was.

What's more surprising than how few peasant, worker or slave rebellions there have been is how few of those rebellions actually succeeded: only a small proportion achieved their stated ends. (At least, this is a fair assumption based on a review of recorded revolts, but there are of course many reasons imperial record-keepers would not wish to maintain accurate records of revolts and unrest.) More surprising still is that even fewer rebels expressed the aim of fundamentally reconstituting society to start with. The Egyptian artisans were not looking to overthrow Ramses III; they just wanted their grain and beer rations.[177] But even when rebels took the more ambitious action of replacing a particularly corrupt or despotic regime, a singularly rotten dynasty, they rarely took aim at the *structures* oppressing them. They did not attempt to overthrow the imperial bureaucracy or monarchy itself, just its immediate agents.[178] Even the most famous slave revolt in Western history, the one in Rome led by Spartacus, did not set out to abolish the institution of slavery or overthrow the Roman government. They instead tried to institute their own independent group and escape Roman servitude, and failed.

When a group has organised to contest the rule of some oppressive structure, it has more often come from outside that structure, often on a frontier. People living outside the borders of a parasitic economy, with its myriad technologies of persuasion and entrapment, people enmeshed instead in the beliefs and structures of their own differentiated societies, *have* frequently resisted domination, at least for a time. One such warrior, the leader of resistance against European settlers in New England, was Abenaki war chief Gray Lock. Historian Colin G. Calloway records: 'For several years Gray Lock and his warriors conducted guerrilla raids against the Massachusetts frontier, harrying the settlements, carrying off captives, and draining the colonies of their resources and morale.'[179] Even when the Penobscots and other Eastern Abenaki groups made peace with the Massachusetts colony in 1725 and 1726, Gray Lock refused to treat with his enemies, remaining independent and undefeated, before disappearing from later records.

Pemulwuy, an Eora warrior who represented the original inhabitants of what is now Sydney, Australia, was another such figure to resist colonisation, also leading guerrilla attacks against British settlers.

II NATION

Pemulwuy posed such a threat to colonial authorities that large rewards were offered for his capture, including twenty gallons of spirits for free men, and for convicts a pardon and passage home.[180] After he escaped countless bullets and slipped free of his shackles in prison, two settlers shot him in 1802.

Although they won battles, on a national scale both men's efforts failed, and those who were conquered were killed, assimilated, imprisoned or pacified. Such opposition forces inhabiting the frontiers of expansive economies have won many battles throughout history, but theirs has been a slowly losing war as ever more land is ceded to parasitic states and empires.

Within such economies, the situation is just as bleak. Even though host classes – whether chattel slaves, wage slaves, peasants or subsistence serfs – have made up the greater bulk of parasitic economies, the condition in which a small group organises, administers and extracts in these societies remains steadfast. But even when more democratic incursions sprang up in such economies, as in Periclean Athens, in most cases political participation was still restricted to propertied male citizens who constituted a privileged minority; the Athenian political economy remained heavily dependent on slavery and oligarchy.[181]

There are material reasons for this, like the need for high degrees of coordination to undertake public works that, say, capture energy from riverine irrigation, and to organise complex defences in a region crowded with other militant economies, or to protect against the seizure by force of scarce, storable food and water by warlords and their mercenaries, or the protection rackets organised by generals who aspire to a more lofty imperial status. Despite these material realities, there have certainly been slave revolts and peasant rebellions within many parasitic societies, but they are exceptions to the rule. This does not make for a rousing manifesto, but if members of the majority are to succeed in subverting the dynamics of parasitism, that project must start with the truth. And the truth is, host classes have been remarkably compliant throughout history. Given the indignities and miseries foisted upon them, and given their relative capacity for strength in numbers and theoretical ability to coordinate collective actions, it is remarkable how few rebellions have overthrown these political dynamics, how rarely new sorts of societies have succeeded in replacing parasitic economies once they have taken hold.

This is not to lay any moral judgment on the docility of the oppressed classes – and certainly not to blame for their circumstances those violently forced into acquiescence – but rather to acknowledge the reasons for the success of parasitic relations so as to better understand them. In addition to the material reasons for the reproduction of these relations – like those already mentioned – equally important is the successful psychological engineering conducted by the parasitic classes on host classes – and on themselves and each other. To achieve this remarkable compliance has depended not just on force; the soldiers and police applying that force, after all, still need to be compelled to commit such acts. It has been necessary to collectively agree upon frameworks capable of dividing up classes to prevent their solidarity while simultaneously uniting them to toil in cooperation, either in the mine, the field or the battleground.

Recall Kropotkin's narrative of cooperative societies being more effective at outcompeting opponents. Societies, whatever their form, tend to yield extreme cooperation: far from being rational actors working in constant self-interest, humans, like other social predators, are almost incurably inclined to work together, even to their own individual detriment. In fact, humans are likely more cooperative than just about any other species of mammal. And it is precisely this willingness to cooperate that has made us too evolutionarily successful for our own good, and too willing to acquiesce to structures that are unfair or unfree or eliminationist. For all their important contributions, such grand narratives as the kind of constant war of all against all that Hobbes imagined, or even the perpetual war between classes that Marx illustrated, miss this vital reality: people want to get along. Of course there are conflicts undergirding every type of society, but, ultimately, people remain extremely cooperative even under nearly intolerable conditions, and groups that are able to achieve greater cooperation – whether in more regimented militaries or more collectivist economic projects – are able to outcompete others around them.

Those in charge of organising parasitic economies have put great

II NATION

effort into hijacking human beings' natural cooperativeness, their will-to-harmony, finding all sorts of psychological tools to compel them to work together and comply with whatever their place in the hierarchy demands. Debt morality – the ethic demanding repayment of debt whatever the cost – is one; others are nationalism and national duty, competitive entertainment like games, and religious tradition and membership. And progress is king of them all: progress collapses the conflicts that might more naturally arise from these other strategies and from the frictions of hierarchy. Progress gives people a collective goal, and it works.

In the nineteenth century, this took the shape of corporate management and the use by their intellectual vanguard, Spencer most prominently, of newly developed evolutionary theory to provide (1) a moral justification for the existence of an industrial oligarchy: they're the fittest, smartest, most capable, so, by the very laws of nature, they deserve to accumulate the most wealth and power; (2) a scientific-sounding justification: this sort of parcelling out of success is simply how nature, race and the world work – just ask Darwin; and (3) a practical justification: the captains of industry are the evolutionary agents leading us to a glittering future of plenty and ultimate perfection – just look at these railroads, steam ships and factories. These narratives of progress served as the glue holding together the moral, scientific and practical reasoning that compelled people to cooperate even under extreme conditions.

Evolutionary theory, appropriately, is not static and has been refined since this period. As more recent scholars note, evolution is not a linear or hierarchical process, the supplanting of inferior beings with superior ones; indeed, we can just as easily imagine evolution to be a *co-constructive* project:

> Every new species inherits parts of its body plan from earlier organisms. For those who want to admire the diversity of life, the trick is not to imagine this inheritance as teleological progress, the climbing of a ladder toward the sun. Instead, we might appreciate the ghostly presence of ancestors inside us, which makes it possible for us to do whatever we do.[182]

Science today casts evolution as a more chaotic process of mutations and adaptations, significantly shaped by environmental chance, and certainly not as motion towards 'perfect' organisms. Not only does society not reflect the 'survival of the fittest', even nature doesn't reflect such a model, at least not so simplistically or consistently. Natural life is governed by the random luck and the tyranny of chance that govern many human destinies as well. Industrialists rose to positions in which they could extract and organise society in part because they made the right decisions at the right times, but also because they were in the right place at the right time, thanks to the larger systemic circumstances that arose just at that particular time: the confluence of hydrocarbons being readily available, the westward expansion over the North American continent in chaotic zones of extraction, and their being the right age, class and disposition to be able to take advantage of these conditions.

Even if we can understand that this grand progress narrative was not scientific, not based on empirical facts, and not really even *meant* to describe reality as it is so much as craft a credible delusion capable of enchanting both host and parasitic classes, I do not wish to imply that nothing has ever improved, even in the rigid and narrow opportunities allowed by parasitic economies. Nor do I wish to imply that making improvements through organised struggle is impossible or should be abandoned. Instead, what I am suggesting is that when conditions are made as bad as they can be made, then anything can look like progress. If conditions have worsened over millennia, years or even decades in which they improve are not evidence of a grand arc towards perfection. Furthermore, when there have been victories in improving conditions for the public, these themselves have been used as evidence of progress, in some cases in order to deter faster or further improvements. As such, it is worth looking at examples where some positive changes did occur and why, in order to see how they fit into the progress myths of the period.

Not all people naturally adapt to harsh conditions, and not all people will cooperate with those who exploit them. The moral and practical justifications for industrial parasitism, though powerful, were not universally convincing. As through other periods of history,

II NATION

during industrialisation there were movements of people who sought to subvert conditions that forced children into hard labour, workers into dangerous mines, the urban poor into squalid conditions, and the non-productive, destitute or criminal into harsh new prisons.[183] A small proportion of the population was willing to take risks to achieve better conditions for everyone. It is in those populations that we can find some genuine improvements, if not a grand moral arc.

While trains, planes and cars are glittering manifestations of the technological boom, perhaps the most important technological innovation for improving human life and health, particularly in urban areas, was sanitation. Although public health researchers had advocated the health benefits of public sanitation since the mid-nineteenth century, basic measures like separating sewage from drinking water were long opposed by the parasitic classes, who considered them too expensive and disruptive.[184] Anthropologist Jason Hickel documents that in Britain, with commoners winning the right to vote and mounting political pressure from workers' unions, agitators eventually succeeded in persuading the state to implement public sanitation, and later 'public healthcare, vaccination coverage, public education, public housing, better wages and safer working conditions'.[185] The improvements in health and well-being made during this era were not ushered in by captains of industry or the telephone, but by people demanding such simple measures, which caused life expectancy to increase into the twentieth century, largely due to decreasing infant mortality rates. As Professor Hickel attests, 'Recent data shows that water sanitation measures alone explain 75% of the decline in infant mortality in the United States between 1900 and 1936, and half the total decline in mortality rates.'[186] Their implementation directly contradicted theorists of progress who conjectured that a better society would only arise from ruthless natural selection and rational self-interest. Instead, they were demanded by common people for the benefit of everybody. Oscar Wilde noted, 'Disobedience, in the eyes of anyone who has read history, is man's original virtue. It is through disobedience that progress has been made, through disobedience and through rebellion.'[187]

The mid-nineteenth century saw the greatest wave of revolutionary action in European history, known as the Revolutions of 1848, or the 'Springtime of the Peoples'. This series of near-simultaneous political uprisings in over fifty countries sought to overthrow monarchies and empires, and replace them with independent democracies or representative government. All over Europe, the working classes briefly united against monarchs and oligarchs to demand constitutions, regular elections, and equal voting rights for all men (women's suffrage became more common only early in the twentieth century), freedom of assembly, expression and of the press, and restrictions on the powers of the crown and church. Why did all of these revolutions occur so suddenly at the same time?

One explanation is that the disruption brought about by new fossil energy capture opened new possibilities for social and political organisation. Parasitic economies have almost always depended on slavery, and while some form of abolitionist values has sometimes coalesced at the fringes of these economies, they rarely gained power sufficient to abolish forced labour for extended periods. The caloric limits of complex agriculture and the resources it can yield per person simply favour stratified, unwaged labour and create broad poverty, particularly when there are no limits to human reproduction and the necessity of expansion. Petroleum-synthesised and mechanised agricultural production helped undermine this ancient lineage of subjugation. The work of abolitionists like Sojourner Truth, Harriet Tubman, Frederick Douglass and William Still began the cause that industrial economies would end.

But the disruption that came from fossil fuel expansion had complex effects.[188] Historian Timothy Mitchell argues that coal helped empower mass movements: 'Fossil fuels helped create both the possibility of modern democracy and its limits.'[189] Even as coal improved the ability of states and corporations to subjugate land and people, it also helped workers organise. Mines fostered intimacy among employees and offered privacy from bosses, two conditions necessary for building strong relationships between workers, something labour solidarity depends on. As a result, miners and factory workers built some of the first collective action labour movements in the modern world. By the latter quarter of the nineteenth century, US coal miners were going on strike at a rate twice that of workers in the next-highest industry.[190] Given coal's importance in generating the electricity and heat used by cities throughout the country, a miners' strike wielded the power to halt production across the country. As in the United States, coal miners in the United Kingdom played an important role in forcing concessions from British rulers to improve working conditions and implement more policies for the common good. They were so effective that, in response, the British government turned to foreign oil as a means of undercutting the power of coal miners and other trade unions. Winston Churchill led the charge, starting with the costly project of transforming the Royal Navy's steam engines from coal to oil.[191]

Factories played a similar role to coal mines. The first major factory strike of the Industrial Revolution in the United States took place in 1824 in Pawtucket, Rhode Island.[192] Some of the country's first textile factories were located in the town, and their owners had recently begun employing young women rather than child labour to run the newly mechanised 'power-looms'; they imagined that young female workers would be just as easily controlled as children, while being stronger and more reliable. But managers made the mistake of cutting pay by a quarter and expanding the working day by an hour when working days already stretched to twelve hours in winter and sixteen in summer. In response, 102 workers closed off the entrances to the factories. The community had been waging a thirty-year war against industrialisation and supported the striking women. On the last day of the strike, the women burned down one of the factories.[193] The next day, the owners agreed to negotiate.

Although early successes were quickly suppressed, the revolutions had lasting effects: dissolving the absolutist King's Law in Denmark, birthing the second French Republic and delivering suffrage for men there, bringing peasants new rights throughout Germany with a ban on manorialism (a kind of serfdom), and establishing a new parliament in Prussia, for a few examples. Maybe the most enduring impact, positive and negative, were the new identities in which political fervour found expression. As historian Christopher Clark observes, 'Political movements and ideas, from socialism and democratic radicalism to liberalism, nationalism, corporatism, syndicalism and conservatism, were tested in this chamber; all were transformed, with profound consequences for the modern history of Europe.'[194]

In addition to technological advances, political revolutions, independence struggles and abolitionism, the first stirrings of modern mass politics that built the labour and suffrage movements were beginning to gain momentum in the nineteenth century. Though studded with frequent setbacks and defeats, they also carried what seemed to be an unstoppable force of history. Suddenly these movements and the ideas that guided them began breaking apart the power structures that had dominated for millennia prior, to prise open the possibilities of egalitarianism and promise a parallel sort of progress towards an increasingly just world of prosperity and liberty. It was in the midst of this century that abolitionist preacher Theodore Parker made his famous statement in an 1863 sermon:

> I do not pretend to understand the moral universe; the arc is a long one, my eye reaches but little ways; I cannot calculate the curve and complete the figure by the experience of sight; I can divine it by conscience. And from what I see I am sure it bends toward justice.[195]

This sentiment has echoed through the decades since in paraphrased proclamations from political leaders like Martin Luther King, Jr. Such nineteenth-century notions of progress maintain a tremendous grasp on contemporary beliefs about what progress means and what it should mean. It's a lineage of thought that we've all been bequeathed, whether we're liberals, capitalists, communists or anarchists, raised in China or India, Chile or Canada, England or Kenya.

Historians often refer to the period between around 1890 and 1920 in the US as the 'Progressive Era', due to some of the major social changes that journalists, unions and activists forced through the federal government. President Theodore Roosevelt, who was in office between 1901 and 1908, was one of those rare politicians willing to make unorthodox decisions that verged on courage. When combined with the actions of agitators like, on one end of the social change spectrum, anarchists assassinating political leaders, investigative journalists making public the conditions caused by industry, workers organising history's largest strikes, environmentalists recording the destruction of wilderness, and activists like the suffragettes marching and destroying property in the fight to gain women the right to vote, Roosevelt's willingness to act against his own party and economic class yielded significant changes in the federal government and, as a result, in the rest of American society. Within this period, around 230 million acres of land were legally protected by a new National Park system, federal laws were passed to regulate the health and safety of food and pharmaceuticals, women won the right to vote, the Senate was made an electoral branch (before this, Senators had been appointed), some of the most important laws prohibiting trusts and monopolies were passed, and unions won important victories. [196] With these interventions by the federal government, this was a direct repudiation of Herbert Spencer's belief that government should not interfere in protecting vulnerable citizens. It undoubtedly expanded the freedoms of much of the public and improved public health and well-being. But inequality remained high and would continue to grow into the 1920s until the crash and the Great Depression, which would test new revolutionary ideologies. Ideologies that would be the engine of the next phase of parasitism and progress.

Coda

The great innovation of the early modern period, around 1400 CE, was the expansion of parasitism from a primarily regional, contiguous form to a global, disparate form. This change was made possible first by the large oceanic vessels that could move goods, weapons, messages and colonists rapidly across large waterways, and then again by the deployment of fossilised energy in machines that could speed up extraction, communications, transportation and manufacturing even more.

The result of this new mode of extraction was the invention of a new administrative form governing global exchange – what came to be called capitalism – and a flurry of ideas based on progress, like liberalism, Marxism, republicanism, manifest destiny, and scientistic societal evolution based in part on biological evolution. The world that these material forces and ideas built was one of great misery: the genocide of the Americas represented potentially the greatest loss of human life and cultural diversity in any single event up to that point, eradicating tens of millions of lives and dozens or hundreds of cultures. It represented what was likely the fastest extinction event to occur to that point, killing billions of animals in the process. It entailed a newly systematised form of plantation slavery that captured, transported and killed millions. It spurred rapid urbanisation in which great numbers of people were pushed together in excessively dense cities, full of air and water pollution and other filth. Urban workers were frequently employed in equally dense and dangerous factories, toiling in dark, cold, monotonous environments.

Nevertheless, the extraction of great wealth from previously verdant lands in the Americas, combined with the technological invention made possible by fossil energy, helped to create a veneer of improvement that made some of those philosophies of progress seem plausible. Conditions were intolerable enough that many rose up in revolt, leading in some cases to better conditions and in others, to a futile bloodletting. Even so, all of these conflicts, forces and ideas set the course for the twentieth century, in which a new form of parasitism would emerge and, where once nation-states had been the agents of geopolitics, now ideological systems would represent the primary demarcation point between conflicting agents. The narrative formula of progress would be adapted to ideological systems whose primary interest was in governing a new form of global parasitism and distributing resources throughout networks of extraction and consumption. The result was a burst of blood and fire.

III System

Overview

Date: 1900 to Present (2020s)
Progress: economistic and ideological
Parasitism: networked and fossil fuel
Agents: nation-states, corporations, ideological coalitions

Context

By the end of the nineteenth century, there was another change in the form that parasitic economies began to take, the biggest development in nearly five hundred years since maritime economies began to spread wood, steel and cloth tentacles across huge expanses of sea and land. The combination of intensively industrialised production in the form of the mass factory, fossil fuel technologies that expanded and sped up extraction – like harvesting wood, clearing land, planting and synthetically fertilising fields – distribution, communications, and a global system of trade and finance capitalism, led to the emergence of a new form of parasitism. I call this form 'networked' parasitism because it is dependent on a dense, interdependent interaction between networks of individuals and institutions. In place of the simpler linear extraction of value, time and material resources that characterised past forms of parasitism, now with the sheer increase in energy, these processes were carried out by and between more complicatedly related actors.

The nineteenth century saw the rise of a new kind of entity: the multinational corporation, a large, centrally governed bureaucracy engaged in extracting profit from the processing of resources and labour. With it came other new institutions, like non-profit organisations, research laboratories and an expanded government bureaucracy. Electrified forms of communication, like the telegram, radio and telephone, enabled instant coordination across many miles, in turn facilitating the coordination of business and government across vast geographic areas. Contiguous parasitism had captured energy from Indigenous societies and native wild and domestic species. Disparate parasitism, meanwhile, had captured energy from Indigenous societies, imperial subjects, and both exotic wild and intensively domesticated species abroad. The new form, networked parasitism, captured energy from all these as well, but with the addition of ancient species of plant and animal in the form of fossils. This enabled concrete energy capture through increasing electrification and then the digitisation of extraction and production, and abstract energy capture from extremely large, dense populations of urban subjects. Though the foundations for this system were built in the nineteenth century, it was only in the twentieth that it came to dominate.

The term that emerged to encompass the totality of this interdependent network of networks was 'the economy'. Economist Giorgos Kallis notes that the idea that we have today about the economy as a large, interconnected system emerged at this time.[1] As parasitic relations became an interconnected global process, it made sense to understand that process, measure it and attempt to direct it with a concept like *the economy*. The narratives that arose in attempting to shape the present and future of human-ecological parasitism were no longer just theological, no longer just rational or secular, nor even just scientistic, but were now economistic. They focused entirely on frameworks and models measuring the exchange of resources through the systems of extraction, production, distribution and waste that crisscrossed the world. Everything – values, morals, priorities, aspirations – became about 'the economy' and how it was to be ordered, its spoils distributed, its mechanics governed. The ideologies that sprang up at the turn of the twentieth century are all about who gets to steer the great steamship of a new kind of parasitism, rather than, say, whether there ought to be a steamship at all.

In previous sections, I mentioned ideologies a few times in passing. Since this section is mostly focused on ideologies and their dynamics, it's a good time to interrogate the term 'ideology'. Critical theorist Terry Eagleton, author of a critical text on the concept, contends that 'Nobody has yet come up with a single adequate definition of ideology,' but I think it's worth putting out a working definition.[2] In the present analysis of progress and parasitism, ideology refers to a system of moral values making normative claims about how the world *should* work, and of empirical beliefs about how the world *does* work, based on interpretative claims – true or false – that primarily refer to social, political and economic phenomena. Ideologies are fluid and somewhat amorphous, developing over time, coalescing after brokered compromises or alliances of necessity – like enemy-of-my-enemy friendships – fracturing into antagonistic pieces, or being refined through debate and research. The twentieth century was a world stage for the dramas of ideologies. Orthodoxies that had their basis in nineteenth-century ideas coalesced into more fully developed systems in the twentieth. Their theoretical debates were taken up by political parties and organisations who acted upon the world with great effect. Ideologies and the institutions they governed made up these networks of extraction, production, consumption, distribution and excretion. Let's take a brief example.

Capitalism, in this understanding, is not an ideology. That sort of claim might produce some fruitless semantic debate, but for our purposes, 'capitalism' refers to an administrative system that governs the flow of value, and a capitalist is someone who controls capital within this system. Your average capitalist is more likely to subscribe to an ideology like *economic liberalism*, a belief in the moral primacy of markets and private property, because it is congruous with his role as an owner and mover of capital. But he could just as well adopt an ideology aligned more with Marxism, even if he suffers some cognitive dissonance in his life as a result, or has to rationalise to himself the conflict between his life and his beliefs. Ideology, after all, is not just a pure reflection of one's circumstances or stated beliefs, but an identity, often a chosen one, and membership of an in-group hostile to some out-group. The same could be said of communism. 'Communism' is an administrative system governing the exchange of wealth, labour and resources in a given economy, rather than an ideology. A communist ideology would include Marxism, Leninism, Maoism, Trotskyism and so on. People may work as administrators of a communist system while pining for, and identifying themselves with, economic liberalism, for instance.

The importance of an ideological identity can hardly be overstated. For an illustration let's consider the twentieth century's world wars. The First World War was *not* demarcated by sharp divisions between ideologies. Instead, it was a conflict between nation-states, primarily those engaging in disparate parasitic energy capture, mostly comprising nineteenth-century empires that generally shared similar forms of government and attitudes towards political economy, even if their cultures and histories differed. The reasons for the war were paper thin; the monarchs involved were often related by blood or marriage, the economic interests could all be worked out in trade deals and indeed made more prosperous by the maintenance of peace. In seeking a material reason for the war, you could look at the foreign policy of empires: their competing over control of distant resources and supply chains in colonies throughout Asia and Africa. Or point to these empires' domestic policies: like their perceived need to eliminate what amounted to excess populations of workers both created and made superfluous by the surplus-generating forces of high-energy industrialisation. At this time, millions of people of working age and fitness suffered starvation, disease and indignity in overcrowded cities and buildings. Those 'lucky' enough to be in employment experienced harsh treatment in factories and mines. And on top, they all witnessed the loss of the rural and wild. Leaders saw that such conditions threatened revolution if, as they saw it, these workers were not soon disposed of in the trenches. Looking for cultural reasons for the war, you might suggest that the peace maintained for generations in Europe to that point had left a population ignorant of war and eager for the chance of glory. Or you might blame the pettiness of elites who found themselves at the calamitous end of long-incestuous dynasties, for whom something like continental war was a matter of noble sport. In any case, few can look back at the First World War and think of it fondly as a just or glorious or even meaningful war.

The Second World War is different. For many in the Allied countries, it *was* a just and meaningful war. Not only was it just, it was noble and great and glorious, all the things war is sold as but rarely is. It is a war that even now descendants in those countries are eager to give a romantic flourish, to even feel nostalgia for. But why? It was conducted with the same senseless brutality as the First, it was even more destructive, killing even more civilians and soldiers, and it brought even more terrible firepower into the world. It broke men with no less certainty, doled out indignity with no less largesse. So why are these twin calamities remembered so differently? The most important answer is ideology.

III SYSTEM

The second war was just because, for the victors, it was a battle of good against evil. They fought to subdue forces that bothered little with ambiguity in the darkness embedded in their ideology, the harsh tenets in which they threatened to engulf the world. Imperial Japan invaded Korea and China with such ruthless depravity that their atrocities and human-subject experiments rival even their German National Socialist allies. The Italian Fascists espoused a cruel authoritarianism based on hysterical machismo and a racially mediated colonialism. The Nazi ideology is the most potent emblem of undiluted evil. With their industrial mass murder, their blatant, self-aware deployment of propaganda, their merciless social Darwinism, their efficiency at social control, their attempt to cultivate a fear-inducing aesthetic, and their offensive belligerence that started the war in the first place, they are a perfect enemy against which nations could wage a war that might be looked back on with pride and meaning long into the future.

Like the examples in previous modes of parasitism and progress, the system of networks that made up this new global form of parasitism was not stable. In addition to the aforementioned economic crises embedded in the system – the decadal recessions, resource crises, and collapses – there were more catastrophic disruptions: global wars, reaching new levels of intensity, scale and sophistication. In past modes of parasitism, the agents that went to war had tended to be divided along lines of culture and geography. Empires from one or a small confederation of peoples united to battle the disparate tribes of other peoples; nation-states battled one another for territory or colonial possessions. Now, the agents that waged war were ideologies. These ideologies did not at first align with cultural or national boundaries, and many countries went through civil wars or other intranational political violence before one particular ideology gained a position of hegemony within the nation-state. That state itself then engaged in war, offensive or defensive, with other states whose ruling party was based on some opposing ideology.

This new form of parasitism relied on dense and complex networks of individuals and institutions for energy capture, and so required an intellectual mechanism capable of ordering this more convoluted form of economy. Ideologies are that mechanism, designed explicitly to decide how and why resources are distributed in particular ways, how and why certain groups engage in parasitism of other groups, and developing ideas about the world to justify and reproduce those relationships. They did what religious belief systems had done before, but they were secular, and they had a veneer of intellectual legitimacy from their scientistic and rationalistic basis; still, they affected human minds in ways similar to religions, and so yielded similar results.

What all of these ideologies have in common is a teleological sense of historical movement towards utopian futures. We can understand both the dynamics of these ideologies in the twentieth century, and their role in the world today and into the future, only by understanding both their relationship to human-ecological parasitism and their deployment of progress narratives as a tactical tool and an epistemological foundation alike.

In the fifth century BCE, Athenian historian Thucydides – one of the Greek thinkers who saw Athens as the height of civilisation – sketched out a principle that would become known as the 'Thucydides trap'. It goes like this: when a major, hegemonic power is challenged by another rising power, the challenge usually culminates in war. Another way of looking at such a situation is to think that, when one parasitic state's expanding zone of extraction bumps against that of another parasitic state, the conflict often results in organised violence. There are of course exceptions: trade agreements, treaties and alliances, or some form of devolved power in which the challenger becomes a vassal or some other high-status subordinate among the hegemonic power's other subordinate states. But mostly the outcome has been war.

The first half of the twentieth century could be seen as a series of Thucydides traps, a drama of states seeking to usurp one another as they came to grips with oil-fuelled industrialisation, while having to negotiate colonial boundaries and the need for their spoils to pass through a more complex global network of states and corporations, all vying for competitive advantage. Alliances and antagonisms often formed along the lines of the in-groups and out-groups defined by ideologies. The anglophone world was torn between its homegrown

III SYSTEM

economic liberalism and democratic socialism, but eventually compromised on social democracy; continental European countries were riven by civil war between socialists, communists, anarchists, fascists, liberals, monarchists, imperialists and others. After a series of left-right conflicts, Germany, Spain and Italy turned decisively to fascism, Russia and those in its sphere of influence adopted communism, while much of the rest of Western Europe dithered around various shades of liberalism and social democracy.

The interwar period, between roughly 1920 and 1940, reveals a fascinating moment of global uncertainty in which these various ideologies and, as we will see, the progress narratives that propelled them strove for supremacy. George Orwell, a major voice in these debates, sneered at the 'smelly little orthodoxies which are now contending for our souls'.[3]

'World war', war that pulls countries from nearly every continent into a single conflict, can only make any sense, only come about when economies are so knitted together by instant communication, by rapid transit, by, ultimately, the dense energy that makes these possible, as they were in the twentieth century. The Second World War was the first war, but not the last, to pit ideologies against one another on a physical battlefield.[4]

This century was a time of competing utopias romping in the abstract potential of the dream, awakened by the harsh glare of radioactive firestorms. After world wars, famines, genocides, dozens of regional conflicts, overthrows and realignments, by the end of the century, there was one primary ideology left standing – economic liberalism – and it still rules today. This section will examine each of the major ideologies and how they fell, one by one, until only today's remains. And this one is as dependent on progress narratives as any other has ever been.

The Future of Fascism

Fascism was born in Italy in the early twentieth century. The word 'fascism' is from the Latin *fascis*, the word for a ceremonial axe wielded by ancient Roman kings before the Republic. Taking this ancient weapon for its name and symbol suggests the romanticising of an imagined past characteristic of a conservative viewpoint. But fascism was more substantively forward-looking than backward. It was fundamentally progressive. It emerged out of an art movement called Italian futurism.[5]

The Italian futurists were enamoured with modernity, and with the power and domination promised by petroleum machines. They despised nature and women, and craved the metallic and macho. Italian writer F. T. Marinetti's *Manifesto of Futurism*, published as a preface to his poems in 1909, proclaims: 'We affirm that the world's magnificence has been enriched by a new beauty: the beauty of speed.'[6] It wails, 'No more contact with the vile earth!' The *Manifesto* explicitly invokes progress by positioning modernity as the culmination of historical evolution: 'Why should we look back, when what we want is to break down the mysterious doors of the Impossible? Time and Space died yesterday. We already live in the absolute, because we have created eternal, omnipresent speed.'[7] Marinetti's manifesto expresses the prototypical value system of the fascist movement, its tenets including war, a death and purity fetish, nationalism, authoritarianism and anti-intellectualism: 'We will glorify war – the world's only hygiene – militarism, patriotism, the destructive gesture of freedom-bringers, beautiful ideas worth dying for, and scorn for woman.'[8] The inclination to call the movement 'futurism' suggests an orientation to the future and faith in progress. Yet its proponents craved a future severed from history and all those who valued its lessons, the 'professors, archaeologists, ciceroni and antiquarians'. Recall the Marduk cult and the Babylonians breaking from their past to pursue a bright new future with a bright new emperor. Lust for futurity yields a neurotic fixation on youth: 'When we are forty, other younger and stronger men will probably throw us in the wastebasket like useless manuscripts – we want it to happen!' This masochistic urge to be dominated by the 'strong' recalls the social Darwinism espoused by Spencer, a tendency that further develops as fascism spreads through Europe. Even the destruction of knowledge intrinsic to fascism has roots in this hysterical little essay, which advocates the burning of books. And finally, undergirding everything, the futurists worshipped one thing above all – violence: 'the racer's stride, the mortal leap, the punch and the slap'.

Reading these lines, one might be tempted to laugh off such histrionic excess as performative comic-book villainy. The writing is genuinely funny.[9] The author's villain origin story is even funnier: Marinetti, a wealthy Italian poet, was originally inspired to write his manifesto after flipping his new Fiat to avoid cyclists.[10] Tongue-in-cheek hyperbole is no doubt deployed to leave its writer the escape hatch of irony, a tactic still used by fascists today: an ejector seat of comic disavowal should they feel the need to flee from their convictions. The British eventually succeeded in laughing off their own homegrown British Union of Fascists by the start of the Second World War – despite its support among many aristocrats, newspapers like the *Daily Mail*, and members of the British royal family – but other nations did not find such luck in ridicule. Comical as the futurists may sound, the fascists that followed were mortally serious.[11]

Central to Italian fascism was the mission of expanding and restoring Italian colonies in Africa and across the Mediterranean. The ideology incorporated a state collectivism they called 'corporativism' (now better known as 'corporatism'), a 'third-way' (*terza via*) economic programme purporting to offer an alternative to economic liberalism and socialism. They sought to centralise management of both economy and government while maintaining the traditional labour hierarchies characteristic of parasitic economies, consolidating state and corporate bureaucracies into one in order to increase both abstract and concrete energy capture. One of the main aims of Fascist Italy was the restoration of – or restoration of the appearance of – imperial Rome, a central aspiration being *spazio vitale*, or 'living space', whereby certain territories would be staked out for occupation by Italians only. Italy had been considered the 'least of the great powers' during the nineteenth-century rush to colonise the world. The fossil-fuel industrialism of the early twentieth century, combined with an intensively bellicose, authoritarian and collectivist ideology, could give it the opportunity to stake out a higher status among the world's colonial powers and catch up with the exclusionary colonialism indulged in by other Western empires throughout the nineteenth century.

Benito Mussolini founded the National Fascist Party in 1921 and led the fascist movement from its inception. He co-wrote the movement's manifesto, *The Doctrine of Fascism*, with Giovanni Gentile. 'If the 19th century was the century of the individual (liberalism implies individualism),' they claimed, 'we are free to believe that this is the "collective" century, and therefore the century of the State.'[12] For the fascists, the State – the centralised bureaucracy ruled by one man, Mussolini – married to the corporate sector is the primary vehicle for expansion and extraction. It derives its legitimacy from the extent to which it generates progress:

> A nation, as expressed in the State, is a living, ethical entity only insofar as it is progressive. . . . For Fascism the State is absolute, individuals and groups relative . . . Instead of directing the game and guiding the material and moral progress of the community, the liberal State restricts its activities to recording results. The Fascist State is wide awake and has a will of its own.[13]

Though the essay claims to reject liberalism and its dream of 'indefinite progress', the fascists only rejected *political* liberalism, eliminating opposition parties, suppressing trade unions, and curtailing civil liberties like free speech. The corporatist economic policy of fascism resembled in many important ways the economic liberalism that had spread European economies across the globe through the previous centuries. Influential Italian scholars like fascist Giuseppe di Michelis openly acknowledged their intellectual debts to liberal thinkers. Di Michelis cites liberal politician Richard Coudenhove-Kalergi, for instance, as an inspiration for ideas of 'Eurafrican' integration, which became central to fascist ideology with its colonial hierarchy and resource extraction in Africa.[14] Thinkers in favour of the corporatist vision of the early fascists justified this intensified expansionism as the 'unavoidable condition for the economic (as well as intellectual and moral) progress of the nations'.[15] The Italian Fascists were very clear that their colonisation was intended as a means of economic development. Their technocratic language often sounds indistinguishable from modern discourse on the burden of 'developed' countries to help 'less developed countries' achieve economic growth: 'They conceived of some sort of "white man's burden" falling to the Europeans who have the task of helping the native populations on their path to development.'[16]

Fascist imperialism not only sought the technological trappings of progress like combustion engines and aeronautics; audaciously, fascists even portrayed their form of imperialism as a vehicle for *social* and *moral* progress. Like the narratives embedded in futurism, the third way economic ideology of the Italian fascists pitched this 'new' kind of imperialism as morally superior to others – kinder, gentler, less exploitative and crass than empires past. In reality, it was no less extractive or expansionist or exclusionary. Within a few years of Mussolini gaining power through a 1922 coup d'état and appointment to prime minister by the king of Italy, tens of thousands of Italians moved to colonies in Italian East Africa, with territories in modern states like Libya, Somalia, Eritrea and Ethiopia, killing as many as 20,000 people in Ethiopia alone.[17] Historian Giulia Barrera writes that Mussolini 'intended to forge the empire following a strict hierarchy of racial relations' and sought to legally segregate Italians from Africans.[18]

The fantasy of building an empire in the image of Classical Rome was core to the Fascist doctrine. This impulse was, to Mussolini, 'not only territorial, or military, or commercial; it is also spiritual and ethical . . . Fascism sees in the imperialistic spirit – i.e., in the tendency of nations to expand – a manifestation of their vitality.'[19] There have been few more explicit expressions of the marriage of parasitic expansionism and a progress mythos. This impulse did indeed yield an Italian Empire and it looked like other fossil fuel empires in practice.

But the Italian Empire built by fascism did not achieve the 500-year longevity of the Roman Empire. It lasted only five years, from 1936 to 1941. Italy was a kingdom and Mussolini took power and governed at the consent of King Victor Emmanuel III. The monarchy and aristocracy supported fascism – as they did in Nazi Germany, and as many in Britain supported British fascist aristocrat Oswald Mosley

– because fascism is more compatible with monarchism and aristocracy than the egalitarian alternatives like socialism. Socialism and communism were growing in popularity in Italy and threatened to redistribute the country's consolidated wealth and power. The fascists were a wall against that threat, successfully co-opting the desire for worker justice through promises of progress. Indeed, fascism's rise to power was the result of Italy's property-owning classes seeking to restore an 'old order'.[20] The fascists themselves used populist rhetoric promising prosperity to the many and resolving class struggle through nationalist state collectivism. It was an attempt to placate the host class in a parasitic dynamic with the binding ethic of nationalism – an identity collapsed into association with the nation rather than a class or creed – and the collective goal of progress. If there was a difference between monarchy and fascism, it was more in style than substance. Fascism was monarchy with a modern, populist veneer. Despite their mutual affinity, the fall of Mussolini's regime came not from political dissidents alone, but through plots orchestrated by aristocrat Count Dino Grandi and the king. Seeing Mussolini's failure to govern and win the war, and aware of a likely Allied victory, Italy's elite intervened again to protect their class position.[21] As the world war he had eagerly joined was about to be lost, Mussolini attempted to flee to Switzerland, but was intercepted by Italian communists and executed.

* * *

Fascism failed in Italy, while it succeeded in Spain only after a civil war between the fascist Nationalist party and a coalition of international fighters from the spectrum of egalitarian ideologies like anarchism, political liberalism, socialism and communism supporting the Spanish Republic. English socialist writer George Orwell joined the international coalition with the explicit purpose to 'fight against fascism', as he put it in *Homage to Catalonia*, his book reporting on his time in Spain. The Civil War lasted only three years. It was a microcosm of the ideological conflicts occurring elsewhere at the time within other countries: the Soviet Union funded communist fighters in Spain, while Nazi Germany and Fascist Italy funded the Nationalists. Western

III SYSTEM

Europeans flocked to fight with the Republicans against the Nationalists even as their governments, virtually all the major Western powers, remained neutral, giving no aid to the Republican fighters, fearing a Russian foothold in Spain should the Communist Party take control. Again, fascism serves as a means of preventing wealth redistribution.

The infighting between the different factions in the resistance that were meant to be unified against the Nationalists limited their effectiveness. Soviet-backed communists arrested the leadership of the militia Orwell was a part of before torturing and executing them. Orwell and his wife Eileen O'Shaughnessy barely escaped the Stalinist purges in Barcelona with their lives.[22] Soviet foreign policy sought alliances with neutral neighbouring France and Britain, but, meanwhile, participated in fomenting conflict between the various factions in Spain.[23] Orwell notes how Russian communists in Spain used promises of revolution to get men to fight the Nationalists – *'The war first and the revolution afterwards'* – but had no intention of following through on those promises:

> The thing for which the Communists were working was not to postpone the Spanish revolution till a more suitable time, but to make sure that it never happened. This became more and more obvious as time went on, as power was twisted more and more out of working-class hands, and as more and more revolutionaries of every shade were flung into jail . . . the effect was to drive the workers back from an advantageous position and into a position in which, when the war was over, they would find it impossible to resist the reintroduction of capitalism.[24]

By Orwell's account, anarchists wished to forge ahead with the revolution; the communists fought them while the fascists defeated them both and took power. It may have been this experience that inspired some of Orwell's other writing. The author displays a sensitivity to the role progress myths can play in subjugating populations in his most famous work, *Nineteen Eighty-Four*. The book includes several passages meditating on the role of progress narratives in the exercise of power:

> . . . the Party member, like the proletarian, tolerates present-day conditions partly because he has no standards of comparison. He must be cut off from the past, just as he must be cut off from foreign countries, because it is necessary for him to believe that he is better off than his ancestors and that the average level of material comfort is constantly rising.[25]

The Party, the omnipotent force governing Nineteen Eighty-Four's dystopia, gains at least part of its power from convincing everyone – rulers and ruled – that there is a progressive course to history and that today is better than yesterday, and tomorrow will be better than today. Not only is the present better than the past, and material comfort constantly increasing, but the Party is responsible for this progress, and deviation from the Party would necessitate deviation from progress itself. It also reveals the importance of cutting people off from history to convince them that the present is better than the past, just as Marduk worshippers and Italian futurists wished to break from the past and, in the latter case, purge society of its historians and archaeologists. A citizenry blind to history or, worse, convinced of a mythical past, is easily led into parasitic excesses of conquest and ultimate self-destruction.

The general leading the Nationalists, Francisco Franco, won the Civil War in 1939 and dominated the country as dictator until 1975. In the chaos during and immediately following the war, the Nationalists unleashed rape and execution on the population and killed hundreds of thousands, while conducting medical experiments on prisoners in Franco's concentration camps.[26] By the end of the war, his regime began to implement economic reforms that resembled technocratic *economic* liberalism, while the state political repression remained, denying *political* liberalism. This is a common theme: liberal capitalism combined with illiberal political repression. We'll see more examples arise later in the twentieth century, and in the twenty-first. The economic elites and companies that thrived during Franco's regime, benefiting from the enslaved labour of defeated foes, remain among the largest and wealthiest companies in Spain today.[27] The cultural impact on Franco's victory was also long-lasting. As Spanish journalist Antonio Maestre portrays it, 'Still today, we are living amid the moral rubble of that defeat . . . For four decades, national-Catholic culture subjugated freedom, literature, theater, critical thought, and educated thinking.'[28] It's important to remember for our own time that this era of repression was immediately preceded by 'the incipient ideas, consciousness, and morality that had provided an illusion of progress to the popular classes, to women, for culture and the arts'.[29]

The Ambivalence of Modernism

Since the First World War, economic liberalism had faced an existential crisis.[30] All through the latter nineteenth century, as markets and wealth were consolidated and the progress promised by voices like Herbert Spencer's failed to play out in reality, public support for economic liberalism waned. The Progressive Era, as mentioned earlier, saw interventions between 1890 and 1920, led by an active populace – or at least by active segments within the public – that did yield some improvements, like women's suffrage, the regulation of corporations, elections for Senators, public sanitation, and basic labour protections. But the Roaring Twenties produced a spectacle of industrialist wealth funding opulent leisure and conspicuous consumption, even as inequality and destitution spiked and economic crises threatened to collapse the economy. It was in the vacuum of waning liberal hegemony that communism, socialism and fascism fought for supremacy.

Like Italian futurism, modernism was an aesthetic movement that arose in the early twentieth century to contend with the rapid industrialisation, urbanisation, and changes in lifestyles that accompanied this new mode of networked parasitism – and to confront anxieties about what appeared to be a rush into an uncertain future. But unlike futurism, many of those leading the movement were authors sceptical of progress narratives, who rejected the authoritarianism of fascism and the power hierarchies of capitalism. Much of this book so far has focused on the thinkers who have promoted narratives of progress in service to expansion and extraction. It has touched only lightly on agents in society who were not sufficiently convinced by progress narratives to carry out the actions necessary for expanding extraction, or those who suffered from the process. There is also value in looking at some of those who questioned such narratives and how their ambivalence was deployed through their own arts of creating narrative.

Old literary forms were not up to the task of integrating the rapid changes of this new kind of society. Experimental literary forms emerged, from free verse poetry to stream-of-conscious prose. Modernism's characteristic interest in interiority – turning 'inward' to follow characters' thoughts – came in part from new psychological insights made by the founders of a new science of psychology, most popularly Sigmund Freud and Carl Jung, and in part from growing fear about the effect such tumultuous changes – like the traumas of trench warfare and the decline in wilderness – were having on individual minds. Narratives were collaged from foreign languages, invented words, folk songs and obscure vocabularies; poetry borrowed rhythms from American jazz and blues music. Modernism's rallying cry was 'Make It New'.

That phrase was popularised by Ezra Pound, a founding modernist figure not only famous for his poems, but for his role in discovering, editing and publishing giants of the movement like T. S. Eliot, Ernest Hemingway and James Joyce. Pound was also an anti-Semitic fascist who was indicted for treason against the United States for a series of radio broadcasts he made while living in Italy during the Second World War, speaking in support of Mussolini and against US President Roosevelt and the Allies. Helped by the pleas of his friends, Pound avoided prison – and possible execution – and was instead institutionalised for twelve years at St Elizabeth's in Washington, DC, the United States' oldest federally funded mental hospital.[31]

Pound's protégés rarely shared his politics. T. S. Eliot was accused of reactionary conservatism and anti-Semitism, but never directly aligned himself with Italian or German fascism. Fearing his friend would be hanged for treason, Hemingway emphasised Pound's possible insanity, writing, 'It is impossible to believe that anyone in his right mind could utter the vile, absolutely idiotic drivel he has broadcast.'[32] Joyce favoured decentralised governance and the empowerment of the proletariat, while refusing to attach himself long term to any particular organisation.[33] He read broadly in anarchist theory and was inspired by the revolutionary socialism of fellow Irishman Oscar Wilde, who once stated, 'The form of government that is most suitable to the artist is no government at all.'[34] Joyce was suspicious of restrictions on freedom whether implemented by church or state, and was a champion of the common man. These ideas play out in works like *Ulysses*, Joyce's reinterpretation of Homer's epic *Odyssey* as a day in the lives of ordinary Dubliners.

In *Homage*, Orwell wrote of Barcelona that the barbers were mostly anarchists, and evidence suggests the modernists were, too, or at least some shade of anti-fascist. A pamphlet of questionnaire responses entitled 'Authors Take Sides on the Spanish War' published in 1937 gives some insight into the intelligentsia's political attitudes between the world wars. Of 147 leading writers, 126 voiced support for the Spanish Republican government, sixteen were neutral, and only five wrote in support of Franco and fascism. Supporters of the Republic included W. H. Auden, Samuel Beckett, Aldous Huxley, Hugh Macdiarmid and Leonard Woolf, many writing explicitly as anarchists, communists, pacifists or anti-capitalists. Those refusing to take sides included T. S. Eliot, Ezra Pound, H. G. Wells and Virginia Woolf's lover Vita Sackville-West. Evelyn Waugh, famous for *Brideshead Revisited*, wrote in support of Franco, though he claimed, 'I am not a Fascist nor shall I become one unless it were the only alternative to Marxism.' Orwell famously declined to vote, writing instead: 'Will you please stop sending me this bloody rubbish . . . I was six months in Spain, most of the time fighting, I have a bullet-hole in me at present and I am not going to write blah about defending democracy or gallant little anybody.'[35]

The survey was sent out by Nancy Cunard, a poet, publisher, activist and anarchist, heiress to her father's Cunard Line shipping business – one of the first steamship companies – and later disinherited by her aristocratic mother (a close friend of British fascist Oswald Mosley) for her radical politics.[36] Cunard railed against Franco and Mussolini, and later worked for the French Resistance fighting Nazi occupation from London. Her relationship with the Black American jazz musician Henry Crowder caused a public scandal; Cunard went on to advocate vocally for civil rights. Together with Crowder, she published *Negro*, an anthology of poems, fiction and essays by Black American writers that highlighted prominent figures of the Harlem Renaissance, including the anthropologist and novelist Zora Neale Hurston, communist jazz poet Langston Hughes and pioneering sociologist W. E. B. Du Bois. As a cultural phenomenon, the Harlem Renaissance was a predecessor to the more explicitly political Black Pride and Civil Rights movements in the United States.

The ideological mix of the modernists and the narratives they created reflected the intellectual diversity and conflict of the interwar period. Modernism grappled with the contradictions of liberal modernity, with its complicated and unfulfilled promises of progress and prosperity. 'Modernity's big story is progress,' wrote anthropologist Deborah Bird Rose,

> It works with a concept of time that is . . . one-directional, [a] future-facing orientation of history. . . . It is teleological in that it is aimed towards a known outcome – always more progress, always more of the good life. It is a secularized version of Messianic time. And it is elastic in that the end point is always just at the edge of the horizon – the Messiah never comes.[37]

So what did the modernists think about progress? Literary scholar Gerald Graff writes that, 'Paradoxically, much literary "modernism" arose in revolt against "modernisation", against the menacing, yet apparently irreversible forward progress of the steam engine of history.'[38] This was a time when the old institutions of politics, science, medicine, religion and literary tradition were being interrogated, and visions for the future fought over. Meanwhile, wilderness and pastoral havens were being rapidly paved over by roads, railroads and cities without any improvement in human or animal well-being to show for it. Archivist Thomas Putnam notes, 'In the early 1920s, in reaction to their experience of world war, Hemingway and other modernists lost faith in the central institutions of Western civilisation.'[39] How could modernity represent the giddy heights of progress when across the globe people were being drawn out of disintegrating rural communities into increasingly alienated cities, grim factories, and devastating modes of warfare?

Other modernists like F. Scott Fitzgerald, Gertrude Stein and Virginia Woolf would complicate the progress narratives with their fiction and essays. Woolf's *Three Guineas*, for instance, is written from an anti-fascist perspective, invested in women's rights. Woolf was raised among the cultural elite and came from a liberal family where notable guests like Henry James visited for dinner. Her work is pacifistic and tinged with environmental thinking, but – while influenced by her husband's socialism – Woolf never committed explicitly to the ideology. Fitzgerald's works incisively capture the mood of the Jazz Age, a period he described in his debut novel, *This Side of Paradise*, as 'a new generation . . . grown up to find all Gods dead, all wars fought, all faiths in man shaken'.[40] Fitzgerald's character Jay Gatsby is among the more eloquent fictitious depictions of the uncertainty of the time, of the promise of progress, and the delivery of progress always just deferred. For Gatsby, the green light is ever out of reach, representing a goal for the future already lost in the past. Gatsby is an adventurous, conquest-oriented man, Herbert Spencer's 'perfect man', the fascists' Übermensch who outcompetes others in Manhattan's blurred illicit and legitimate markets. Gatsby was a conduit for the unrestrained, electric energy of this highly unequal period, and held within him the era's own self-destructive destiny.

Gatsby is a vivid vehicle to deliver a certain pessimism about progress that must have been simmering beneath the golden decade of the Roaring Twenties, of which Fitzgerald and his wife Zelda were central figures. Jay Gatsby starts his life penniless but hopeful. Hope is a frequent invocation of those who wish to instil faith in a progress narrative. Hope is the faith that the kingdom of heaven will be built, the promised land reached – but not now. Gatsby pursues an American Dream that dangles wealth before those clever and lucky enough to navigate competitive markets and hierarchical institutions. And his ambition does elevate him, but economic mobility is only available to Gatsby through the mentorship of a drunk aristocrat and the patronage of a rich criminal. His rise is enabled by illicit activity, which is also true of typical aristocrats and industrialists; but they, benefiting from periods of chaos like industrialisation, manage to hold onto their wealth and status anyway. Gatsby is too late. He is killed by another member of the underclass, Wilson, who is tricked into the murder by the aristocrat Tom Buchanan. Tom has only just denied Gatsby membership to his rarefied class when he whispers to Wilson that it was Gatsby who slept with and murdered his wife. In fact, it was Tom who had done the first crime and Tom's wife who did the second. And so the generational elite maintains the purity of his class by pitting two lower-class rivals against one another while getting away with the crime.

Fitzgerald had always been concerned with the seductive but corrupting hope apparent in progress. His first novel catapulted him to literary fame and contains this meditation: 'Progress was a labyrinth . . . people plunging blindly in and then rushing wildly back, shouting that they had found it . . . the invisible king – the *elan vital* – the principle of evolution.'[41] He often showed scepticism about the idea, as his biographers attest: 'it was indeed conventional to think that prosperity incarnated the Idea of Progress. F. Scott Fitzgerald was among those who took the notion with a grain of salt.'[42] Literature professor Ronald Berman continues in *Modernity and Progress*:

> Towns, buildings, and markets are ephemeral. They inevitably become reminders of material limits. The same images used by advertisers to celebrate growth were used by writers of the twenties to reverse the common judgment about it . . . Fitzgerald wrote about the entropic ruins of the American landscape. . . . We know that writers resisted the confusion of progress with prosperity. They were not satisfied by industrial democracy and resented its commercialism. More than anything, they resented its claims.[43]

The inequality of the age mingled with the vast wealth encapsulates the contradiction of the American Dream: a dream of progress denied, with few exceptions, to those who do not already reside among its heights.

Ernest Hemingway, a dear and long-time friend of Fitzgerald's, was an important figure of the era as well. Hemingway travelled to Spain to report on the Civil War. He claimed not to take sides, but, like Orwell, expressed leftist sympathies, even supporting Fidel Castro's revolution in Cuba. Hemingway's childhood trips to northern Michigan inspired a lifelong devotion to the outdoors and the wild world,[44] but that devotion was tainted by his lust for hunting big game, including elephants, tigers and lions. John Walsh, dissecting the author's death, suggests Hemingway sought to portray himself as 'the perfect man, the perfect synthesis of brain and brawn'.[45] Such striving for 'perfection' suggests a mind seeking to fashion itself into an interchangeable ideal of those confused times: standing at the apex of Herbert Spencer's evolutionary hierarchy, or burgeoning fascism's pinnacle of fast cars and guns, or even communism's working man of action, a rough warrior of the blue-collar intelligentsia. If Gatsby embodies the contradictions of the age in fiction, Hemingway did so in fact. Like the other modernists, Hemingway's work, too, was ambivalent about modernity and sceptical of its tidy narratives.

Yet for all its promise of progress, modernity was not kind to its modernist authors: Hemingway took his own life aged sixty-one, the decline in his mental health in part the result of a congenital disease and the sporadic head trauma the sportsman suffered throughout his life.[46] The despair gained from an abusive father must have contributed, on top of legitimate paranoia in an increasingly carceral and complex world. For years, Hemingway was afraid he was being stalked by government agents. Though long dismissed as mere paranoia, his fears were vindicated: J. Edgar Hoover's FBI did indeed monitor and stalk Hemingway closely for years as a result of his leftist sympathies and ties to Cuba.

George Orwell died young, just forty-six, of tuberculosis. Despite being given 'excessive doses' of an experimental new 'wonder drug' to treat TB – with painful side effects – his long-time illness and gruelling work schedule did him in.[47] Virginia Woolf took her own life at the age of fifty-nine. Having struggled throughout her life with mental illness from which she feared she would not recover, she filled her coat pockets with stones and walked into the river near her home. Nancy Cunard died alone in a hospital aged sixty-nine after she was found by police on the streets of Paris weighing less than sixty pounds, afflicted by alcoholism, poverty and mental illness. F. Scott Fitzgerald, meanwhile, drank himself to death aged forty-five after descending into obscurity – a few years before a revival of his work – with the darkness surrounding one who believed he'd fallen far short of his potential.

What the modernists' relationship with progress narratives suggests is that, as with the nineteenth-century debates about progress as evolution, novelists remained some of the most ambivalent about ideas of progress. Modernism presented a level-headed counterpart to the fascists' futurism. Unlike the political leaders of the time, and even well-meaning ideologues, modernists maintained a thoughtful scepticism that even many scientists and philosophers could only aspire to.

The Destiny of Nazism

By the 1930s, the Great Depression sparked in the US financial system had burned through the rest of the world of financial capitalism. In Germany, the depression combined with the post-war conditions lingering from 1918 to create a situation in which the ideologies vying for power across the globally networked parasitic economy found a fertile battleground there. There were frequent murders, and political assassinations occurred nearly every week.[48] The situation also created a desperate imperative to enrich the German economy by any means, and fostered a generation of men broken by the senselessness of the First World War and now idling in another meaningless quagmire, seeking their way out. After years of political conflict between communist, socialist, liberal, fascist and social democratic parties, the National Socialist German Workers Party emerged triumphant, led by Adolf Hitler. Germany soon embarked on the conquest of Europe, attempting to open frontiers of extraction both east and west. The Nazi Party was motivated by the intensification of both concrete and

III SYSTEM

abstract energy capture. Fascist Italy pursued such economic growth by attempting to re-establish colonies in Africa, while Germany pursued growth by attempting to colonise in an older mode: the contiguous parasitism of past European empires and the American manifest destiny. Germany also intensified abstract energy capture by engaging in forced labour and mobilising an extremely collectivised populace for war.

Like Mussolini's notion of *spazio vitale*, Hitler accelerated an older imperial imperative, *Lebensraum*, also meaning 'living space', a notion based on conquering land for habitation and cultivation by the conqueror. But though deploying some similar rhetorics and allying in shared projects of colonisation, in one important way Germany's example is an anomaly of the time. By engaging in a more contiguous form of parasitism, the Nazis relied on an older narrative of progress, one more akin to those deployed by settlers in the nineteenth-century westward expansion of America (and those of the classical world, like that deployed along with the Pax Romana of Rome's destiny to colonise and rule indefinitely). Hitler was deeply inspired by the American example and drew from many of the same tropes, including an evolutionary sense of progress in which cultures and peoples are ranked in a hierarchy of their development. The notion held that it was the destiny of the dominant race – in this case Aryan or Nordic instead of Anglo-Saxon – to populate land cleared through genocide or the removal of Indigenous peoples, and not only the destiny, but the duty of such a race, given that it would lead to the evolution and perfection of humanity. Hitler explicitly voiced his inspiration for Nazi concentration camps as deriving from Native American reservations and explicitly deployed these tropes in their support.[49] Historian Norman Rich has claimed, 'The United States policy of westward expansion . . . served as the model for Hitler's entire conception of *Lebensraum*.'[50] Professor Robert J. Miller writes, 'Hitler envisioned the "German East" of Poland, Ukraine, and Russia as an analogy to the American frontier of the "West," and he expected to expand the German empire eastwards in the same fashion as the United States had westwards. Hitler and other Nazi leaders often referred to Jews, Poles and Ukrainians as "Indians".'[51] 'There is only one task,' Hitler himself proclaimed in 1941. 'To set about the Germanisation of the land by bringing in Germans and to regard the indigenous inhabitants as Indians.'[52] His meaning was clear: treat Jews and Eastern Europeans as Euroamericans treated Native Americans.

Though writing mostly in the late nineteenth century, Karl May remains one of Germany's best-selling authors today, with around 200 million worldwide book sales. His most famous stories were set on the American western frontier, with Native American characters central to his narratives.[53] May's Westerns entertained and shaped the worldviews of millions of Germans, including famous ones like Albert Einstein. But also Adolf Hitler, who read May's books voraciously as a child, and into adulthood, keeping the author's entire collection in his bedroom.[54] Hitler told Albert Speer that they did for him what 'works of philosophy [did] for others or the Bible for elderly people'.[55] May himself died in 1912 and so never saw his admirer's career, but his centring of Native protagonists in his stories rather than exclusively settlers, along with his pacifism, suggests that he likely would have abhorred this association with Hitler.[56]

Central to the manifest destiny of Nazism was faith in progress, as it was in the United States. Depictions of Nazi Germany in retrospect have cast it as a dark dystopia suffused with fear and anger. Many imagine Hitler seething and gesticulating at a podium in a tantrum of barely controlled rage, and a society something like the English dictatorship depicted in *Nineteen Eighty-Four*, with its dreary grey tedium, drab and dirty clothes and cities, and blatant tools of social control. For those in Nazi Germany who were gradually pushed to the margins and increasingly surveilled and controlled – leftists, Jews, Roma, the openly non-heterosexual and the disabled, for instance – Germany did indeed creep toward that sort of dystopia. And something like the crass subversion of, even contempt for, language and truth, along with the Nazis' manic public exercise regimens resembling those depicted in Orwell's England, hit close to dystopian fictions. But for most German citizens, those considered part of the Aryan demographic, there was a pervasive sense of hope and optimism in Nazi Germany. In the lead-up to the war, and even in its early years, a sense of collective struggle, camaraderie and faith

III SYSTEM

in a glorious future pervaded society. Hitler's primary appeal was his ability to channel this optimism and faith in the future among multitudes of Germans. Novelist Karl Ove Knausgård, who embedded a rigorous contemplative biography of Hitler in his six-volume *My Struggle* series, wrote eloquently on the source of Hitler's power and magnetism:

> What made Hitler so different [from other aspiring leaders], however, was the flame he ignited in all who listened to him speak . . . in which the entire register of his inner being, his reservoir of pent-up emotions and suppressed desire, could find an outlet and pervade his words with such intensity and conviction that people wanted to be there, in the hatred on one side, the hope and utopia on the other, the gleaming, almost divine future that was theirs for the taking if only they would follow him and obey his words.[57]

Widespread hate, fear and anger were perfectly compatible with widespread hope, optimism and grandiosity. Knausgård notes that, 'reading accounts from early 1930s Germany' when the Nazi Party was just taking power, 'there is a striking optimism radiating from everywhere'.[58] Nazi propaganda contained many explicit visions of utopia and promises of a positive future, provoking in people great jubilation and exhilaration during the Nazis' rapid rise. Hitler's autobiographical manifesto, *Mein Kampf*, and his speeches offer examples of the kind of faith in progress that would characterise the Nazi worldview. The book mentions *progress* as a momentum of history explicitly a dozen times, and implicitly many more:

> The progress of mankind resembles the ascent on an endless ladder; one cannot arrive at the top without first having taken the lower steps. . . . It is no accident that the first cultures originated in those places where the Aryan, by meeting lower peoples, subdued them and made them subject to his will. They, then, were the first technical instrument in the service of a growing culture.[59]

Here he justifies a mythical conception of the Aryan 'race' and its conquest as a tool for subsequent progress. Hitler and some of his closest collaborators were famously interested in the occult and Germanic mythology. Their rallies engaged in ceremony and symbolism and deployed mystical rhetoric. In *Mein Kampf*, Hitler writes, 'I believe today that I am acting in the sense of the Almighty Creator: By warding off the Jews I am fighting for the Lord's work.'[60] He claimed to be doing so explicitly to prevent Jews from instituting 'the Marxian creed'[61] or, in other words, to prevent them undermining traditional hierarchies of extraction. The Nazis were greatly motivated by an idea of progress rooted in building a perfect utopia that would prevail far into the future. The language of the 'Third Reich', or third realm or empire, promised a stable future in eternal victory. In a speech he gave in the midst of the war, just two years before he would kill himself, Hitler said,

> A new life will begin to bloom on the sacrifices of the dead and the ruins of our cities and villages. We will then continue to fashion that state in which we believe, for which we fight and work: the Germanic state of the German nation as the eternal and identical homeland of all men and women of our Volk, the National Socialist Greater German Reich. . . . Beyond this, the Greater German Reich and the allied nations will have to secure jointly those Lebensraume that are indispensable to securing the material existence of these people.[62]

Hitler's speeches also contain telling elements that reveal his attempt to directly transgress the networked parasitism characteristic of the Allied nations. In a speech in Munich in 1922 in which he lays out the major tenets of his ideology, he proclaims, 'It is only the international Stock Exchange and loan-capital, the so-called "supra-state capital," which has profited from the collapse of our economic life, the capital which receives its character from the single supra-state nation which is itself national to the core.'[63] He speaks of Germany as a colony to the networked parasitism of global financial capital, though not in those words. His anti-Semitism was deeply rooted in anti-capitalism – or perhaps the reverse; it is debatable which came first. But there is little doubt that his popularity came from an attempt to supplant one type of parasitism – networked, global, capitalist – with another, older form: contiguous, mythic, imperialist. The Holocaust was directly inspired by this mission. Hitler and the Nazis saw the Jews as a supranational and trans-ideological force, attempting to constrain Germany through finance capitalism in the West and Bolshevism in the East. Jews were the greatest force preventing Aryan progress. The Nazis believed that in order to secure a prosperous future for Germany and Aryans – and therefore the rest of humanity, who would benefit from their genetic superiority – they had not only to expel or eradicate German Jews, who made up a relatively small portion of the total number of European Jews killed in the Holocaust, but to defeat the countries that housed Jews and seize the land for Aryans, while eradicating the Jews who lived there.[64]

To accomplish this, the Nazis took many lessons from the United States' conquest of the west of the continent that went beyond a vague sense of manifest destiny and instead included specific frameworks for law and policies. Historian Alexandra Minna Stern reflects, 'The Third Reich's 1933 "Law for the Prevention of Offspring with Hereditary Diseases" was modeled on laws in Indiana and California. Under this law, the Nazis sterilized approximately 400,000 children and adults, mostly Jews and other "undesirables," labeled "defective."'[65] In the United States, forced sterilisation continued as standard practice for most of the twentieth century, with 'procedures thought to have been performed on one out of every four Native American women [in the 1960s and 1970s], against their knowledge or consent'.[66] In the United States until 1981, more than 60,000 people were forcibly sterilised across thirty-two states. Primarily they were immigrants, people of colour, Native people, people with mental illnesses and physical disabilities, people who scored low on IQ tests, poor people, prisoners and unmarried mothers.[67]

One of the most substantive inspirations the Nazis took from the United States was how to legally and culturally build a racially segregationist society. Racism in the US has always served material ends. It is not some phantom that springs fully formed from the brains of ignorant Euroamericans, nor a malevolent ghost haunting dark corners of the human mind. It is a political tool used globally to serve in creating hierarchies of abstract energy capture. Yale law professor James Q. Whitman quotes the famous segregationist John C. Calhoun – a man admired greatly by the Nazis – justifying race-based slavery because it served as 'the best guarantee to equality among the whites'.[68] In other words, by giving poor and rich White people alike a singular identity – Whiteness – the wealthy White people who benefited from slavery could maintain the support of poorer White people by casting non-Whites as a villainous outgroup. Poor whites would compete with poor Black and Indigenous people instead of competing against rich Whites, leaving the multiracial host class divided and making it easier for – eventually, multiracial – parasitic classes to extract energy from them.

The Nazis took this lesson well. Race science was one way they segregated the country and cast non-Aryan populations as the outgroups against whom 'Aryan' Germans, both rich and poor, should unite in competition. Whereas socialism engages in class conflict in order that the working classes may gain more agency in determining their own destinies, *national* socialism solved the problem of economic class not by empowering workers, but by eliminating class conflict through the supremacy of identity both with the 'nation' – the blood of the people and soil of the land – and with the (fictional) race tied to it: Aryans. Hitler and the Nazis could divert the redistribution of wealth along class lines by redistributing wealth along racial lines, from Jews to Aryans, thus protecting Germany's (non-Jewish) wealthy classes from expropriation by workers. European anti-Semitism has often carried this material aspect – the taking of wealth from Jews – as other forms of racism have elsewhere, to protect wealth hierarchies. The Nazi period was an intensification of this process, forcing Jews to register their property before it was simply stolen. As historian Götz Aly has written, 'Aryanisation was essentially a gigantic, trans-European trafficking operation in stolen goods.'[69] In the same 1922 Munich speech mentioned earlier, Hitler outlined this principle explicitly:

III SYSTEM

There are no such thing as classes: they cannot be. Class means caste and caste means race. If there are castes in India, well and good; there it is possible, for there there were formerly Aryans and dark aborigines. So it was in Egypt and in Rome. But with us in Germany where everyone who is a German at all has the same blood, has the same eyes, and speaks the same language, here there can be no class, here there can be only a single people and beyond that nothing else.[70]

Of course, scientifically and theoretically, this passage is far from factually accurate. But as political rhetoric, removing economics from class and replacing it with race and identity proved an effective trick. Many today, including some progress propagandists, continue this tradition of attempting to erase material class differences and focus instead on racial or national differences. And they do it for the same reason: to prevent host classes from pulling out the parasitic classes embedded in their skin.

The Nuremberg Laws, which codified the repression of non-Aryans and others deemed genetically unfit, were substantially and explicitly modelled on American legal precedents. 'It is particularly startling to discover,' writes James Q. Whitman, 'that the most radical Nazis . . . were the most ardent champions of the lessons that American approaches held for Germany.'[71] He goes on to show that 'when Nazis rejected the American example, it was sometimes because they thought that American practices were overly harsh: for Nazis of the early 1930s, even radical ones, American race law sometimes looked *too* racist.'[72] The role of these laws and rhetoric was, like the example of Calhoun's tactic, a nationalism with the promise that 'Society would no longer be divided into noble Germans and commoner Germans, master Germans and servant Germans. Now every German would count as a coequal member of the ruling class by simple virtue of membership in the Master Race.'[73]

In Italy, the aristocracy and monarchy were sympathetic to fascism. In Germany, though the monarchy had been abolished, generational elites generally supported Hitler and Hitler generally supported them. National socialism held a certain mystique for the elites of other countries as well. Historian Adam J. Sacks points out that, in Britain, 'leading lords were captivated by fascism, and realised that it served their interests quite nicely.'[74] King Edward VIII was a major Nazi sympathiser. After abdicating the throne, he took a tour of Nazi Germany, indulging in Nazi salutes and reminding Hitler of their shared racial brotherhood. This was perhaps more ostentatious but less egregious than his actions as king: for instance, he was happy to ignore the way British fascists terrorised Jews and immigrants throughout Britain, while he played a passive role in allowing the remilitarisation of the Rhineland, which had been a bulwark protecting against Germany's rearmament and a second world war. Charles Edward, grandson of Queen Victoria and former British prince supported Hitler, even joining his SA paramilitary. Oswald Mosley, who founded and led the British Union of Fascists (BUF), was a baronet. Though Mosley is remembered more for his bluster, his acolytes actively sought to sabotage British defences during Nazi air raids on southern England, guiding Luftwaffe bombers to their targets by burning homes.[75] Despite actively undermining Britain's national security during the most vulnerable period in its modern history, Mosley was never charged with treason or any other crime and, aside from a brief stint in custody, was never held accountable. His notoriety faded and moderated with time, and he died comfortably in 1980 in France at the ripe age of eighty-four. Harold Harmsworth,[76] 1st Viscount Rothermere, founder of the *Daily Mail* and *Daily Mirror*, supported Hitler and Mussolini, meeting the former on several occasions. He and his newspapers supported the BUF, which helped to popularise the party in Britain. More obscure nobles joined the movement as well: Dutch noblewoman Baroness Ella van Heemstra and her British husband Joseph Hepburn-Ruston collected money for the BUF. Joseph eventually left his family to spend more time with the Nazis. Their much more famous daughter, Audrey Hepburn, would later speak out against the Nazis during the war. In both Italy and Germany, aristocrats and royals supported fascism for both the practical reason of seeking to protect their elite positions from expropriation by left-wing ideologies, and the more sentimental one

III SYSTEM

that both groups were enamoured with blood purity, traditional hierarchies, and hereditary rank and lineage.[77]

The United States was not free from such elite sympathies towards Nazism, largely in the 1930s before entering the war. America's own aristocracy was enchanted with Nazi and Nazi-adjacent ideas, with some actively aiding Germany in the lead-up to the war. Henry Ford openly supported the Nazis, and the Ford Motor Company had ties to the Auschwitz death camp, while the company's German subsidiary used slave labour.[78] General Motors also profited from Nazi ties.[79] Thomas Watson, founder of IBM, retained intimate business links with Nazis well into the war.[80] George W. Bush's grandfather, US Senator and Wall Street banker Prescott Bush, profited from ties with some of Hitler's most important financiers.[81]

All this is not to suggest that the United States and Nazi Germany were morally equivalent, nor to say that Nazism only existed because of the American example of manifest destiny. Given the other examples of expansionism Germany's allies pursued – colonial fascism in Italy and imperialism in Japan (which was driven by a concept similar to *Lebensraum*: *Hakkō ichiu*, meaning 'the world under one roof') – it likely would have found some other example to follow, or another way to comprehend its own version of fascism. Nevertheless, American settlers and Nazis were often motivated by the same thing: building parasitic relations between Indigenous and settler populations and within the settler population. And they used similar justifications: both claimed to represent a more civilised society, to repopulate conquered land with a higher race, and to achieve progress towards the perfection of humanity.

Just as Enlightenment thinkers developed theories to rationalise the depopulation of the Americas of Indigenous people and the continent's repopulation with Europeans, and just as great American politicians like Washington, Jefferson, Lincoln and Jackson carried out these theories, Hitler drew from the same theories, pursued the same policies, for the same reasons. The German people, he argued, simply needed the land and the space that other peoples occupied in order to progress, and to help the whole gene pool of humanity progress. While these are striking similarities, at the time, the United

States had already completed its conquest of the continent. It had departed from the mode of manifest destiny and contiguous parasitism, was thoroughly embedded in global networked parasitism, and was now grappling with its own internal conflicts between ideologies. The big difference in the anglophone world, as the modernists attest, was that they decisively turned away from fascism and towards a truce between economic liberalism and socialism that resulted in its own sort of ideology.

Stability of Keynesianism

The short lives of the Italian Empire under Mussolini and the German Empire under Hitler reveal that the older forms of parasitism they deployed in an attempt to grow their economies could not last long in a world now composed of interdependent networks of institutions performing intensive fossil energy capture. Franco's fascism in Spain accommodated itself to this new order by focusing on domestic extraction instead of foreign conquest, by remaining generally neutral during the Second World War, and then by fully accommodating economic liberalism in the 1970s after Franco's death.[82]

Fascism, with its bellicose nationalism and closed-border collectivism, was simply ill-suited to this new form of globalised energy capture, dependent now upon networks of diverse states, intergovernmental institutions and multinational corporations. European fascism and Nazism were defeated by two distinct ideological forces: state communism to their east and social democracy to their west. In Germany, the Social Democrats had been an obstacle to the rise of the Nazi Party and were among those parties banned by Hitler's regime after he gained power. Stalin had convinced many German leftists to abandon the social democrats, which had the consequence of dividing the left in the face of Nazi opposition. This, combined with Soviet intervention in the Spanish Civil War, which also fatally divided leftist opposition and led to Franco's victory, makes a strong case that Stalin was the greatest ally of fascism in Europe. But he was also its greatest enemy: the casualties borne by the Russian Red

Army in defeating Germany were much greater than those of the West; their industrial war machine was essential in breaking Germany's eastern front, which weakened its western front and hastened the end of the war.

The depression that helped turn Germany, Italy and Japan towards their respective versions of authoritarianism had a different effect in the English-speaking world. Whereas economic liberals in the Axis countries ultimately allied with fascists to stop leftists, they ended up doing the opposite in the US and UK in the interwar period and during the war. As in the Progressive Era of the early twentieth century, the early 1930s and the 1940s saw a Roosevelt presidency responding to pressure from below and leftist agitation by brokering a compromise between capitalists and labour. The rapid degradation in well-being sparked by the Great Depression, combined with the collectivist urgency of wartime, created both the material conditions and the political justification for the passage of a broad range of policy packages that represented the emerging ideology that has come to be called social democracy. Through the war and for two decades after, the US and UK built the foundations for many of the policies that enabled strong union representation, built social safety nets and welfare states, and created the mid-twentieth century model of five days per week, eight hours per day waged labour. Franklin Roosevelt's New Deal in the United States and Labour Prime Minister Clement Attlee's Welfare State in Britain redistributed much wealth by creating public goods. In the United States, these included programmes like Medicare, Medicaid, Social Security, and the sixty others in the New Deal legislative packages. In Britain, the National Health Service was created to provide free healthcare alongside the introduction of unemployment benefits and sick pay, pensions for the elderly, free education until the age of fifteen, and the nationalisation of British manufacturing industries like steel, gas, coal, electricity and rail.

The First World War saw a clear redistribution of wealth, but only from the super-rich – the top 1 per cent of earners – to the rich, the next 9 per cent of earners. The Second World War, on the other hand, marked the beginning of a greater equalisation of wealth that would continue for three decades. By the end of the 1970s, the richest 1 per cent owned 20 per cent, falling from 50 per cent in the 1930s and two-thirds prior to the First World War.[83] Meanwhile, the wealth held by the poorest four-fifths of adults increased from 8 per cent in the 1930s to 35 per cent in the 1970s. Wealth taxes prevented the immediate restoration of large fortunes among the upper classes; in the United States, union membership peaked in 1945, with many other countries' peaks following soon after. Voting rights were extended, and a general atmosphere of national solidarity helped engineer policies geared towards broader distributions of wealth and opportunities among the working – or in the parlance of this book, host – classes. In historian Walter Scheidel's words, 'WWII created both the political will and the fiscal and organisational capacities required for ambitious redistributive programmes.'[84]

The New Deal had a dark side: it played a major role in accelerating carbon emissions, funding the creation of high-polluting industries like the chemical and personal vehicle industries, and initiated the modern culture of overconsumption.[85] In effect, it built the modern world as well as the torch lighting the fires that are burning it down. This process had a predictably harmful impact on Indigenous Americans who were further dispossessed and dislocated by expanding development driven by the federal government.[86] But even as the New Deal reflected some of the country's racial exclusion, failing to extend benefits equally to Black or Indigenous Americans, all this new wealth created the largest 'middle' class – a precarious position between the upper reaches of the wealthiest parasitic classes and those most intensively extracted from – in the country's history.[87] This injection of wealth into the economy, and the activity of production and consumption that resulted from redistribution, is sometimes referred to as the Great Acceleration, which saw growth in populations and economies all over the world.[88] The combination of huge efforts to rebuild infrastructure and a flood of coal- and oil-based technology, like automobile manufacturing and air travel, spurred what may be the largest middle-class expansion not just in US history, but in all of recorded history (or at least as long as something like a 'middle class' has been a meaningful category). Whereas the coal economy of the nineteenth century accelerated industrialisation and yielded all kinds of new technological inventions, the mid-twentieth century acceleration utilised increasing floods of oil and gas to generate another shift in technological development. Consumer electronics like televisions, refrigerators and washing machines, commercial air travel, personal vehicles, car-dependent infrastructure and sprawling suburban homes rapidly spread.

British economist John Maynard Keynes is the primary theorist most closely tied to the economics and policies that formed the basis of this social democratic ideology. Keynesianism encompasses a broad anglophone social democratic programme that fuses corporate market policy with social-good regulations. It was a form of liberalism, but one that gave priority to more centralised industrial policy, regulated finance, and favoured checks on inequality in order to achieve a more stable system; it was these interventions, after all, that helped to stabilise the economy after the Great Depression and Second World War. Keynesianism was an attempt to put out the fires started by classical economic liberalism by restraining its impulses in favour of greater political liberalism. In continental Europe, the parallel to this project was ordoliberalism, which helped to guide the establishment of welfare states during post-war reconstruction, and was also aimed at building a stable sort of 'social' capitalism. Whereas economic liberalism engages in what journalist Naomi Klein has popularised as 'disaster capitalism' that deliberately destabilises polities, ordoliberalism contains a tendency to stability.[89] Liberal capitalism and democracy cannot long co-exist unless the former is restrained sufficiently to ensure that the latter remains viable. The economic crises that yielded fascism, Nazism, communism, and the many conflicts of the twentieth century, were ultimately crises of capitalism and liberalism failing to adequately deliver democracy, equality and broad prosperity.

Keynes rejected a purely teleological view of history, seeing instead the 2,000 years of the common era as for the most part economically stagnant rather than ascendant. Still, he was a believer in progress, or at least in the potential of the state and economy to work together to continue growing and expanding the middle class. Perhaps more so, he held a faith in technology to develop an automated economy that would liberate people from work. He predicted that, soon, there would be enough wealth in the world that people would simply stop coveting it. 'When the accumulation of wealth is no longer of high social importance,' he wrote in 1930, 'there will be great changes in the code of morals.'[90] The real problem would become, he predicted, what to do with all this wealth and leisure time. Such a view today, when automation has only yielded *more* work and exploitation – as Marx perceived and predicted – and covetousness of money has reached new pathological extremes, looks tragically naïve.

Even so, Keynes had a profound, often positive impact on the mid-twentieth century social democratic economies. But these economies were still parasitic, and they still depended on ever-expanding extraction to avoid collapse. The economistic idea that emerged to put that dynamic into operation was a new concept called *growth*.

* * *

In 1957, the United Kingdom's Conservative Prime Minister Harold Macmillan gave a speech to the nation in which he declared 'most of our people have never had it so good.'[91] In the speech, Macmillan called for increases in production to ensure economic growth while attacking central planning and nationalisation, suggesting that the Labour Party would yield 'rationing, shortages, inflation, and one crisis after another' – a strange and ahistorical assertion given the role of social democratic interventions in *preventing* these very crises of unrestrained finance capitalism. In a different speech, also in 1957, Macmillan declared, 'We've been in the lead of industrial and scientific progress at least since James Watt invented the steam engine and we are still in the lead in these days of atoms and aeroplanes, so we have no reason to be frightened of temporary difficulties.'[92] All the progress that Britons should be so grateful for was thanks to the vigilance of the Conservative Party, he said, wrapping up with the assertion that 'Expansion and freedom, trade and progress, remain our main theme and objective.' Macmillan used this speech to promote 'Free Trade Area' proposals, which were attempts to avoid British economic integration with Europe and to allow Britain, at least nominally, to introduce greater market competition, presaging the

Brexit debates led by the Conservative Party sixty years later.[93] His meaning was clear: keep us in power and we will continue progress via economic growth.

There was indeed some cause for Britons to be optimistic. The war that had claimed at least 60 million lives and birthed novel atrocities, from the Holocaust to the detonation of the atom bomb, was over. In its wake, a new mechanised era promised to enrich Western economies almost beyond measure. But not quite.

Economic growth, to have the political and rhetorical value that it seemed to promise, needed to be measured. The burgeoning discipline of economics sought to quantify growth through a new framework: gross domestic product, or GDP, which, to use energy scientist Vaclav Smil's succinct definition, 'expresses the monetary value of all final goods and services that are produced or provided within the borders of a country during a specified period of time'.[94] This single number became the measure of nations' success and well-being. But, as economist Giorgos Kallis notes, measuring all of a country's 'economic activity' – which includes phenomena that range from difficult to impossible to measure with great precision – is not a straightforward technical equation. Many economists refused to even take the notion seriously as an academic exercise. Instead of being a purely scholarly endeavour, 'It was politicized to its core.' From the beginning, politicians and those economists who sought to engage directly with policymaking deployed GDP as a political tool.[95] The GDP framework was first used widely to measure the growth of post-Second World War economies. Smil calculates that though per-capita GDP growth was slow between 1913 and 1950, after 1950 growth rates burst up to 4.05 per cent in western Europe and 2.45 per cent in the United States.[96] This was due to a unique historical circumstance: the confluence of abundant fossil energy, the rise of consumer industries, Keynesian public investment projects, and the imperative to rebuild Europe after a devastating war, yielded massive economic activity. But instead of being seen as a historical aberration, this accelerated growth was taken for granted as a new normal that would persist indefinitely. Soon, an expectation of perpetual compound annual growth of around 3 per cent became the standard for a 'healthy' economy.

This was also when economics as a scholarly discipline began to position itself at the centre of political power. Economists and political actors alike began to take the growth of the post-war economic boom for granted. This expectation of endless growth transformed both the economics field and the political landscape. Historian Christopher F. Jones notes: 'The founding fathers of economics – luminaries including Adam Smith, David Ricardo, and John Stuart Mill – shared a belief that growth was finite, and that the reason for limits lay in the natural world.'[97] They understood that if economics is to be a science that measures real things in the world and predicts how they will behave through time, they have to take natural laws and physical facts about the world into account. And in the real, physical world, there are real physical limits to resources and, therefore, growth.

But in 1957, the same year Macmillan demanded Britons' gratitude, Keynesian economist Robert Solow published a pair of articles that departed from the intellectual foundations of the field. Solow's models simply removed natural resources like land, fossil fuels, forests, rare metals and minerals from his modelling as factors affecting economic growth. When mentioned at all, they were abstracted as capital, an intangible and theoretically inexhaustible fiction. By removing natural resources from his equations, on paper, Solow convincingly argued that growth could continue eternally. In place of evolutionary science as the basis for understanding progress, economic science – or a politically warped version of it – became ascendent. Whereas evolution claimed to discover laws of nature in biology, economists claimed to discover laws in how economies functioned – as Engels had previously – but in this case by removing laws of nature themselves from the equations.

The narrative of perpetual growth became an attractive façade for endowing networked parasitism with economistic and moral legitimacy. Many accepted the argument uncritically not just because it sounded scientific, but because it sounded nice: it served material interests across classes – more stuff for all! – and it provided a scientific-sounding basis for the appealing idea of infinite progress via the perpetual introduction of new wealth into this relatively new

III SYSTEM

abstraction, the 'economy'. Like nationalism in fascist countries, 'growth' in multiethnic liberal ones could resolve the perennial problem of class conflict. Instead of the hard work of redistributing wealth from a powerful and violent financial oligarchy to the public, the economy could simply *grow*, and all the new wealth would reach the public, eventually.

Solow's papers and the idea of infinite growth became a foundational tenet for much subsequent modelling, but were even more important for those economists positioning themselves at the nexus of their profession and political power. And it took deep hold in the field – or at least those in the field most enamoured with power. Thirty-four years later, in 1991, Lawrence Summers – then Chief Economist of the World Bank, later US Treasury Secretary and President of Harvard University – stated, 'There are no . . . limits to the carrying capacity of the earth that are likely to bind any time in the foreseeable future. There isn't a risk of an apocalypse due to global warming or anything else. The idea that we should put limits on growth because of some natural limit, is a profound error.'[98] Essentially, he is saying that there is practically no limit to planetary resources, and by consequence, to how many people can live on the planet, which opens the door to the fanciful conclusion that tens of trillions of humans could reside on a planet where limits are already being passed far beyond safe amounts at 8 billion.

Growth is a sanitising word. Growth sounds natural and life-affirming. Living things are the only things that grow. Mountains may become taller by the forces of the earth's crust pushing vast plates together, scraping land into the sky. Stars may swell towards the end of their lives to an enormous feverish inferno. And by the winds, dunes may accrue in slow waves. None of this is 'growth', it is accumulation. *Growth* implies a biological process. In biology growth is also accumulation of physical, material properties. Cells push together, replicate and expand, transforming biomass into biomass. The important difference is that the growth of an organism is bound up with agency, mind, motion and death. That is, with the rare and precious flame that is living. In economics, the term growth refers to a process that is neither natural nor life affirming.

Even setting aside the clear deficiencies in a theory that ignores the limited nature of obviously limited resources, GDP growth alone has not been intrinsically or causally related to net improvements in general human welfare. In fact, often the opposite is true: increases in GDP have frequently caused decreases in general well-being. As Jason Hickel has noted,

> If you cut down a forest for timber, GDP goes up. If you extend the work day and push back the retirement age, GDP goes up. If pollution causes hospital visits to rise, GDP goes up. But GDP includes no cost accounting. It says nothing about the loss of the forest as habitat for wildlife, or as a sink for emissions. It says nothing about the toll that too much work and pollution takes on people's bodies and minds.[99]

The economistic narrative of growth-as-progress is detached from empirical reality, and from what makes human life enjoyable, sustainable and meaningful. This detachment from reality is vividly captured in Solow's 1974 pronouncement: 'The world can, in effect, get along without natural resources.'[100] Consider this statement for a moment: the world can get along without air, water, minerals, food, textiles, or all the other things on which life and society depend. In defending why he took this position, he simply said it was the 'natural assumption' to make.[101]

Perhaps the notion of infinite economic growth was so easily accepted at the time, despite its clear contradiction with observed reality – people in the 1950s still needed to breathe air, drink water, eat food, wear textiles – because the 1950s presented a sliver of history in which fossil fuel-augmented production could superficially appear to yield something close to an endless supply of raw, artificially synthesised resources developed in a lab from nowhere. But Solow's assumptions are not based on empirical data, they are based on a particular faith – one he shared with Keynes – that technological progress would continue to develop infinitely, a faith that has continued into the twenty-first century.

The United States and Western Europe enjoyed great post-war

levels of GDP growth, and though ideologically motivated actors – politicians, corporate heads and some economists – sought to credit liberalism and capitalism with this growth, they weren't the only economies to grow in this period, or to draw on growth as a central narrative. Ideologically anti-capitalist countries to the East, too, fabricated growth-as-progress narratives and managed, at some points, to eclipse their liberal counterparts in achieving rapid growth.

Cults of Stalinism-Maoism

The socialist pamphlet *Proceedings of the 1932 National Convention of the Socialist Party* (of America) proclaimed:

> Above all things we need new and revolutionarily different philosophy and loyalty to remake our world into a federation of cooperative commonwealths instead of a chaos of quarreling nations, suspicious races and contending classes. The more complete is the acceptance of this revolutionary philosophy the sooner the class struggle will end in victorious establishment of a classless society . . . The hope of the world is Socialism.[102]

The United Kingdom's Labour Party manifesto of 1924 made similar promises. Using a self-consciously religious heading, *The spirit that giveth life*, the manifesto states:

> We appeal to the People to support us in our steadfast march . . . using each success as the beginning of further achievements towards a really Socialist Commonwealth, in which there shall at least be opportunity for Good Will to conquer Hate and Strife, and for Brotherhood, if not to supersede Greed, at least to set due bounds to that competition which leads only to loss and death.[103]

Lines between twentieth-century progress ideologies – particularly fascism, communism, liberalism and socialism – were often unclear, with ideas, rhetorics and allegiances overlapping, flipping and cross-pollinating. Liberal anglophone economists collaborated with Nazi and fascist counterparts in the interwar period. Mussolini was a socialist politician before founding the National Fascist Party; Oswald Mosley joined the democratic socialist Independent Labour Party before founding the British Union of Fascists; Adolf Hitler and his associates co-opted working-class populism and socialist language in often successful efforts to siphon members from leftist parties, even going as far as naming their party the 'National Socialist German Workers' Party' ('Nazi' is a portmanteau of '*National-Sozialistische*').[104] Meanwhile, many thinkers and writers who are frequently considered liberals today, like Leonard Woolf, Norman Angell and George Orwell, at the time self-identified as socialists.[105]

As mentioned, in the anglophone world, liberalism and socialism found a tenuous compromise in social democratic Keynesianism. In Europe, fascism mostly won the ideological battles of the interwar period. In Russia, the communist revolution was victorious. Socialists and communists were interested in the more equitable distribution of the fruits of growth, if not critical of growth itself. Marxian progress myths were based on a more equitable and fair distribution of value, but they were not without risk of authoritarian co-opting and orientation towards parasitic imperatives. After the revolution, communism as practised in Russia almost immediately descended into assassinations, executions, purges and authoritarianism while espousing its own version of a growth-as-progress mythos. What Marxian progressives did when they seized power was ultimately similar to what other progress ideologies in other parasitic economies did: continued to run their society according to an imperative to expand and extract from both human and non-human populations. And they were very effective.

Prior to 1970, the Soviet Union's per capita GDP growth rate was one of the world's highest, coming in second only to Japan.[106] Between 1928 and 1970, the USSR's annual gross national product grew at a rate of nearly 5 per cent, higher than the average US growth rate at the time and at any time in the twenty-first century.[107]

Joseph Stalin and the Bolsheviks established their regime in 1924. His administration took control of the economy using Five-Year Plans. One such plan involved the centralised control of settlement patterns, industry and agriculture, evicting millions of peasants and putting them to work on state-owned farms. The collectivisation and mechanisation of Soviet agriculture resulted in millions of deaths in the 1932–3 famine.[108] Despite the supposed abundance delivered by growth and progress, the twentieth century was, for many millions, uniquely defined by hunger. More people died of famine in the twentieth century than in any century prior, and of the estimated 70.1–80.4 million deaths, 80 per cent of the victims died in just two countries: the Soviet Union and China.[109]

Such policies were achieved in communist states so rapidly and uniformly with the autocratic power of dictatorship. Despite purported ideological dissimilarities and the frequent antagonism between communists and fascists, fascism in Europe granted one man the role of head of state with extensive powers. Soviet communism did the same. Stalin became the physical embodiment of the proletarian dictatorship. Vladimir Lenin, leader of the Bolsheviks and Stalin's mentor and predecessor, had expressed regret just before his death over Stalin's rise to power in the Party, recognising that his autocratic ambitions did not reflect the spirit of the revolution. Anarchists in Moscow publishing anonymously early in the Soviet takeover, in 1922, predicted such an outcome, writing:

> State Communism, the contemporary Soviet government, is not and can never become the threshold of a free, voluntary, non-authoritarian Communist society, because the very essence and nature of governmental, compulsory Communism excludes such an evolution. Its consistent economic and political centralisation, its governmentalisation and bureaucratisation of every sphere of human activity and effort, its inevitable militarisation and degradation of the human spirit mechanically destroy every germ of new life and extinguish the stimuli of creative, constructive work.[110]

Given that this turn was a betrayal of the revolution and its ideological basis – Marxism and Leninism – Stalin needed a strong narrative to justify his position. He deployed economistic growth-as-progress narratives to great effect.

In a speech to voters in 1946, Stalin gave his own 'never had it so good' address eleven years before Macmillan.[111] And like Macmillan, Stalin begins with the triumphs of his state and party, proclaiming that 'the Soviet social system is a better form of organisation of society than any non-Soviet social system.' Despite the massive loss of life suffered by the USSR during the Second World War, Stalin frames the event as a momentous victory proving the superiority of state communism: 'our victory signifies that our Soviet *state* system was victorious, that our multinational Soviet state passed all the tests of the war and proved its viability'. Like Macmillan, Stalin touts his party's success in the prior six years of war, praising the three successive Five-Year Plans that preceded it as producing sufficient pig iron, steel, grain, fuel and other goods necessary to win the war, going into great detail about how far the country had come since the First World War and revolution: 'This unprecedented growth of production cannot be regarded as the simple and ordinary development of a country from backwardness to progress.' It was 'a leap', transforming the Motherland 'from a backward country into an advanced country, from an agrarian into an industrial country'.

The catastrophic collectivisation and mechanisation of agriculture that had claimed between 6 and 8 million lives were framed as necessary to 'put an end' to the 'backwardness in agriculture'. Speaking about 'utilizing all the achievements of agricultural science and of providing the country with the largest possible quantity of market produce', Stalin sounds more like a corporate executive giving a presentation to his board of directors than a revolutionary leader addressing the Soviet Communist Party. Stalin briefly brushes off the popular resistance to his policies, much of which was led by women and peasants, saying, 'the Party yielded neither to the threats of some nor to the howling of others and confidently marched forward in spite of everything.'[112] This language of putting an end to backwardness and marching forward are rooted in a progressive, teleological sense of historical movement, with not the proletariat, but Stalin himself as its primary agent. Such an attitude proved harmful both for Russian workers and for the country's Indigenous peoples who, combined, formed the Soviet host class to the parasitic class represented by party elites.

Siberia's Indigenous cultures and Russia's environment suffered from this rapid economic growth. Reindeer herders and walrus hunters were seen as 'backwards' and in need of modernisation via integration into the industrial state farms. A local Party leader tasked with imposing the collectivisation policy in Anadyr, just below the Arctic Circle, stated that treating 'kulaks and shamans with kindness, will make them spawn like flies. We must seize all of them so they will not be hostile to our work.'[113]

The USSR attempted to implement their own version of the US reservation system by forcibly resettling Indigenous groups like the Yupik into more densely populated villages.[114] Soviets justified this the same way earlier Western liberals did, by seeing the Indigenous groups as representing a primitive past in need of being violently pulled into civilised modernity. 'In Marxist terms,' historian Bathsheba Demuth writes, '[Indigenous peoples in Russia] lived on the first rung of history's ladder, before the rise of agriculture and industry, let alone socialism . . .' Instead of considering Marx's scholarly maturation, his embrace of Indigenous modes of government, and eventual rejection of a teleological view of history, 'the Bolsheviks preached liberation to people who did not see themselves as oppressed.'[115] Scholars Dennis and Alice Bartels write that, 'With the assistance from the new Soviet state,' Soviet planners argued, 'Northern Peoples could, theoretically, "progress from vestigial clan-based social organisation to socialist nations, skipping the socio-cultural evolutionary stages of feudalism and capitalism."'[116] Even on the basis of heavily propagandised Soviet sources, scholars have 'characterized Soviet resource extraction in the north as a form of colonialism that excluded Northerners'.[117]

In addition to this attempt to impose material parasitism upon foraging peoples' economies, the Soviet state practised cultural imperialism as well, promoting a process of 'Sovietisation' which required

Indigenous peoples to adopt the Russian language, education, and settled lifestyles. As part of this process, they forced Indigenous children into schools with sub-par educational standards.[118] Scholars Igor Krupnik and Michael Chlenov write of the impact on the Russian Yupik in particular: 'The psychological shock of the relocations, of bidding farewell to one's homeland, and the shutting down of old villages curtailed people's ability to resist further governmental actions.'[119] State forces took over at all levels of governance, banned native subsistence whaling in the 1960s, and authorised the end of Yupik language teaching in the 1970s.

The process of Sovietisation involved coercing Indigenous students – along with the rest of the Soviet population of men – to abandon traditional roles and remodel themselves in the image of the 'New Soviet Man', an ideal recalling Spencer's perfect man, Hitler's invocation of an Übermensch, and the Italian fascists' futuristic masculine ideal. The New Soviet Man, the 'man of the future' in Russian revolutionary Leon Trotsky's terms, would reflect the progress-oriented values of communism: a muscular evangelist of Marxism who practised exertion, austerity and manual labour, possessing an iron will that enabled him to sacrifice himself for the benefit of the collective.

There was a New Soviet Woman as well, purportedly emancipated and equal to her male comrades. Much publicity was devoted to recruiting women into traditionally male-dominated professions in medicine and the sciences, aviation and factories.[120] But as the USSR adopted more conservative values through time, the New Soviet Woman came to be idealised as a full-time worker *and* a homemaker – maximising extraction from her labour as in the West – finding her destiny as a loyal wife and mother who contributed to both the economic productivity of the state and the biological reproduction of its workers and armies, the hosts on which the parasitic classes fed.[121] Chivalry and 'family values' dominated; abortion was banned. The Soviet state used women as bludgeons of progress. Cultural imperialism imposed on Muslims and Indigenous peoples could be justified by progressive Soviets claiming to oppose what they deemed the gender 'backwardness' of these peoples, even as they expected

III SYSTEM

the nominally liberated Soviet woman to languish under the Soviet version of patriarchy.[122]

As became the norm with all growth-as-progress economies – and as is intrinsic to all forms of parasitic human-ecologies – the Soviet Union harmed the environment. The USSR's industrial economy caused numerous environmental disasters, including oil spills, chemical and heavy metals pollution, radioactive contamination, desertification, deforestation and more. Like their Western analogs, the USSR practised the systematic repression of environmentalists while privileging polluting industries.[123] At times in the twentieth century, the rapid economic growth and unrestrained industry of the USSR, combined with their even more successful suppression of environmental dissent, led to worse environmental records than some of their capitalist counterparts. Although the United States was the greatest emitter of harmful air pollution, the Soviet Union was a close second and contributed to a 'severely degraded environment' throughout the Soviet bloc countries.[124] But the Soviet Union was not a lone communist state engaging in growth-as-progress parasitism. The story of Communist China bears many important parallels with, and a major difference from, the Soviet story.

* * *

After nearly thirty years of intermittent civil wars, China consolidated the regions of the mainland – which excluded Taiwan – in the 1950s under the administration of Mao Zedong, and swiftly began its own rapid process of industrialisation, urbanisation and agricultural mechanisation. As in the USSR – and like the earlier industrialising of the West – this process led to famines and displacements. Like Stalin, Mao drew heavily on progress narratives. The 'Great Leap Forward' – the name for Mao's Second Five-Year Plan, which focused on mechanising and collectivising China's agricultural system following the Soviet example – contained a promise of progress in its very name. The policy resulted in potentially 30 million deaths, due to the famine that resulted from agricultural and ecological mismanagement.[125]

Quotations from Chairman Mao Zedong, more colourfully called 'The Little Red Book', is one of the most prolifically printed books in history. First published in 1964, the book helped solidify Mao's cult of personality while laying the foundation of his theoretical contribution to Marxism-Leninism through more than two hundred short aphorisms. During his regime, the Ministry of Culture attempted to distribute a copy of the book to every Chinese citizen, mandating that everyone read it.[126] The book frequently invokes 'progress' to sanction potentially unpopular positions, like, for instance, waging war: 'History shows that wars are divided into two kinds, just and unjust. All wars that are progressive are just, and all wars that impede progress are unjust,' a claim familiar from Augustine's just war theory. 'We Communists oppose all unjust wars that impede progress, but we do not oppose progressive, just wars.'[127] But progress is also used to assure readers that war will soon be obsolete: 'War, this monster of mutual slaughter among men, will be finally eliminated by the progress of human society, and in the not too distant future too.'

'Progress' is repeated through Mao's homilies like a religious mantra. Another passage reads: 'The world is progressing, the future is bright and no one can change this general trend of history. We should carry on constant propaganda among the people on the facts of world progress and the bright future ahead so that they will build their confidence in victory.' But, like all such predictions or promises, this one's fulfilment may be indefinitely deferred: 'At the same time, we must tell the people and tell our comrades that there will be twists and turns in our road. There are still many obstacles and difficulties along the road of revolution.'

Though Mao defines progress as working towards a classless society, the vehicles he valorises are the same ones that many technocrats used in the capitalist states then, and still do today: industrial production and faith in inevitable technological advancement. Economic growth is the explicit goal of all this progress: 'the social conditions are being created for a tremendous expansion of industrial and agricultural production.' Mao, like many proponents of progress narratives today, has no room for pessimism about such goals: 'Ideas of stagnation, pessimism, inertia and complacency are all wrong. They are wrong because they agree neither with the historical facts of social development over the past million years, nor with the historical facts of nature so far known to us.'

These sorts of hopeful adages, similar to Western liberals' benedictions of 'progress' uttered both long before Mao and long after, were vehicles likely helping to propel the dictator to power. His image can frequently be found looking to a bright future, presenting himself as the embodiment of hope.[128] Although first published in the 1960s, the quotations stretch back to the 1920s and were central to Chinese Communist Party rhetoric in the lead-up to the revolution. Political science professor David Shambaugh notes: 'The Chinese Communists learned much from their study of the Soviet, Nazi and other totalitarian states' propaganda methods'.[129]

The environment again suffered under the industrialisation and growth-as-progress economy of Maoist China. Mao himself disdained nature. One of the first programmes in the Great Leap Forward was the Four Pests Campaign from 1958 to 1962, which sought to eradicate sparrows, mosquitoes, rats and flies. The result was an ecological disaster.[130] The widespread elimination of sparrows led to a boom in locust populations, which decimated crops. The mechanisation and intensification of agriculture and the increase in industrial intensity were accompanied by all the familiar effects of soil degradation, habitat destruction, and air and water pollution. The Little Red Book contains many quotations casting society as fundamentally in conflict with nature. Mao's attitude can be captured in one passage: 'man must use natural science to understand, conquer and change nature and thus attain freedom from nature.'

The Soviet Union and Communist China can trace their ideological lineage to Marxism-Leninism. While early in Marx's career, he and his co-author Engels emphasised a progressive story in which history develops according to stages of linear progress – from simple to complex, through stages of feudalism, capitalism and socialism – Marx eventually disavowed this. His later writing, as mentioned earlier, equivocates far more with the neat narrative of progress, particularly as Marx recognises the long success of Indigenous cultures like the Iroquois in cultivating something like large-scale communism. But Lenin saw the rhetorical potential of such a progress narrative and implemented Marx's earlier historiography in his own writing. In the *Historical Destiny of the Doctrine of Karl Marx*, Lenin writes: 'Since the appearance of Marxism, each of the three great periods of world history has brought Marxism new confirmation and new triumphs. But a still greater triumph awaits Marxism, as the doctrine of the proletariat, in the coming period of history.'[131] While the Soviet Union stuck more or less to this story until its dissolution, China took a different path. The Great Leap Forward promised growth and progress, but failed to deliver it. Growth stagnated during Mao's rule and Five-Year Plans. His death in 1976 was well-timed, allowing China to break from its Marxist-Leninist ideological basis and adopt tenets from an ideology more equipped to pursue economic growth. One born in the mountains of Switzerland.

The Theocracy of Neoliberalism

In 1947, two years after the end of the Second World War, ten years before Solow's infinite growth papers, and two years before Mao founded the People's Republic of China, Austrian economist Friedrich von Hayek convened a meeting of thirty-nine scholars in Mont Pèlerin, Switzerland. This was a time when the USSR seemed at its strongest, the Chinese Communist Party was winning its civil war, and Keynesian policies in the West were redistributing wealth from the richest members of society via public-good programmes. The share of income held by the top 10 per cent in the United States dropped from 45 per cent in the 1930s to its lowest point of just under 35 per cent in 1970.[132] When Hayek had convened his meeting in 1947, that share was already nearing its historic low and showed no sign of stopping.[133] Similar inequality trend lines were visible in Europe as well where these social democratic interventions were taking place. Greater equality in wealth had positive real-world effects for a majority of Americans, Britons and Europeans, from increased life expectancy to wider access to healthcare and nutrition, to more leisure time.

But Hayek was an aristocrat with generational wealth and status – he also happened to be a friend of Keynes when they were students at the University of Cambridge together – and he worried about threats to his own status and to what he saw as the most important foundation of a free society. He termed that foundation 'classical liberalism', which primarily referred to what this book calls *economic liberalism*: the primacy of markets, property, corporations as agents, and capital.

The purpose of the meeting in Switzerland was to determine how the balance of power could be tipped towards economic liberalism and away from political liberalism by stopping the spread of Keynesian social welfare policies like those contained in the New Deal. Such policies were contested by those who, like Hayek, would be targeted by efforts at redistribution – the generationally wealthy, corporate management heads and the new rich – and so they sought a way to work together to undermine them. The group also acted in anticipation of the Labour government's looming 'Five Giants' reforms in the United Kingdom that would create the National Health Service and many other social welfare programmes a year later. If Keynesianism attempted to reconcile the fundamental conflict within liberalism between capitalism and democracy by restraining the former in the interest of the latter, Hayek's project could be seen as a direct response to that, and an attempt to do the opposite, restraining democracy in the interest of capital. While Keynesianism was a modification of liberal capitalism rather than a deviation from it, Hayek feared that these relatively modest social welfare programmes represented not merely a slightly different form of liberalism, but instead, to use his words – the title of his most famous polemic – a 'road to serfdom'.

The Great Depression and the recessions that had occurred in each decade throughout the nineteenth century made it clear that the economic liberalism of Herbert Spencer and the Gilded Age – a figure and period Hayek and his acolytes vocally approved of – was incapable of guiding the state or the economy in a way that ensured long-term stability, even for its own stated ends of peaceful, open commerce. Recall that it was born in the midst of disparate parasitism when the violence of opening new lands and resources to marketisation and extraction might achieve efficiency through laissez-faire, hands-off economic chaos. In the densely networked world of the twentieth century, more deliberate state interventions suggested that planning was a path to greater growth and stability. But state planning and intervention, in certain hands, also threatened to somewhat erode classical wealth hierarchies and take some of the decision-making power out of the hands of economic elites, which included business leaders and aristocrats like Hayek. What this meeting of Hayek and his scholars would have to achieve was a 'new' ideology that could effectively integrate into a tightly connected world of networked parasitism, which meant retaining a significant role in the economy for the state, while also keeping the balance of power in the hands of economic elites rather than state planners or a democratically represented public.

The meeting in Mont Pèlerin set off an intellectual and political movement to restore the supremacy of economic liberalism over social democracy's political liberalism. Seeing themselves as building a 'new' kind of liberalism, they began to call themselves 'neoliberals' and their ideology 'neoliberalism'. From this meeting, a transatlantic network grew over the course of the next twenty years, consisting of think tanks, media organisations, research institutes at universities, and industry trade groups. The central tenets of neoliberalism spread so effectively in part because they had big money behind them, but also due to their stripped-down simplicity. Adherents evangelised these tenets through their media networks and, eventually, through global, intergovernmental institutions like the International Monetary Fund, the World Bank and the World Trade Organisation, where such values and beliefs came to dominate.[134] By the late 1970s, neoliberalism would achieve a new political consensus in the anglophone world, shaping the administrations of eight or more US presidents and at least as many UK prime ministers (at time of writing), and spread far beyond their borders. Although, in an echo of Herbert Spencer,

III SYSTEM

neoliberal messaging often focused on 'freedom' from and antagonism towards the state, in practice these values included belief in the primacy of the market backed up and enforced by heavily armed governments. Their goal was to capture governments so that they intervened on behalf of business entities against the public instead of the reverse.

The current era of neoliberal consensus has expanded the scope of the state dramatically, broadening government interference in markets and personal lives, increasing surveillance, incarceration, bureaucracy, and the scope and quantity of legislation. It has shifted who the government serves and funds from a plurality of citizens to a small minority of economic elites – the parasitic classes – but it has certainly not shrunk the state.[135] The ideological specifics evolved over time, adapting to meet political realities. Like the early fascists, this later wave of neoliberal political actors framed their ideology as a 'third way' between extreme libertarian capitalism of earlier ideologues and social democratic policies of the Keynesian period.[136] Neoliberals of all shades have always emphasised privatisation, which involves transforming public commons or public goods into private property for the purpose of extraction. Finally, they resurrected and spread the belief – or at least the stated belief – that self-interested profit-making 'lifts all boats', echoing Enlightenment scholars like Malthus and Turgot who claimed that self-interest is the engine of progress. They would eventually join virtually every other ideology of the time in their faith that infinite compound growth yields continual progress.

Hayek and other leading neoliberals like Milton Friedman and Ludwig von Mises understood the inherent contradiction between economic and political liberalism. Hayek claimed that what he called 'economic freedom' is more important than political freedom. 'Personally,' he told *El Mercurio*, 'I prefer a liberal dictator to democratic government lacking liberalism.'[137] Neoliberals also understood that if the balance of power swung towards economic liberalism and away from political liberalism, the result would be increased economic inequality. The American Enterprise Institute and the Institute of Economic Affairs, two leading neoliberal think tanks, argued, as historian Daniel Stedman Jones renders it, 'that social and economic inequality was necessary as a motor for social and economic progress'.[138] Stedman Jones suggests that the neoliberals' 'fundamental acceptance of substantive inequality' as both inevitable and 'even desirable' is one of the main principles distinguishing it from Keynesianism, which sought to reduce such inequality and understood the destabilising role it played throughout the twentieth century.[139] Hayek justifies the concentration of extreme wealth using a vague invocation of progress, writing, 'in any phase of progress the rich, by experimenting with new styles of living not yet accessible to the poor, perform a necessary service without which the advance of the poor would be very much slower.'[140] This sounds like Herbert Spencer, who wrote, 'the poorest today owe their relative material well-being to the results of past inequality.'[141] One central text puts forward the neoliberal movement's clear ideological position: three years before founding the Mont Pèlerin Society, Hayek published *The Road to Serfdom*. It was popular on publication and is likely the closest thing to a foundational text for neoliberals.

Like Spencer, Hayek starts from a progressive view of history and presumes that the unfolding of time has been a teleological process of ever-increasing freedom:

> It may merely be pointed out that up to the present the growth of civilisation has been accompanied by a steady diminution of the sphere in which individual actions are bound by fixed rules. The rules of which our common moral code consists have progressively become fewer and more general in character. From the primitive man who was bound by an elaborate ritual in almost every one of his daily activities, who was limited by innumerable taboos, and who could scarcely conceive of doing things in a way different from his fellows.[142]

But, like earlier progressive historiography, this example is based on a misunderstanding of history. Countless examples of the 'primitive man' Hayek refers to practised forms of government and social organisation that made them far freer than the states that conquered,

III SYSTEM

displaced or eradicated them, with far fewer laws and 'taboos' to control their behaviour than those the neoliberals themselves imposed.[143]

States trap people. They *depend* for their very existence on trapping people within their boundaries, whether as slaves, workers or conscripts.[144] Recall again all the examples in colonial America of settlers defecting from the colonies to go and live with Native Americans, whose lifestyles promised far more freedom and abundance. The states had to literally kidnap and force colonists to remain. For most of the societies Hayek calls 'primitive', 'freedom' referred to things like being able to decide for yourself how you spent your time, what you would eat, where you would live, with whom you would form relationships, and all the vital decisions one makes in one's day-to-day life. Many such freedoms were associated not only with control of *personal* property – rather than rent-generating *private* property – and liberty to procure sustenance of one's own or to leave an oppressive group, but with civil freedoms like direct democracy and parity across genders and sexualities.

These freedoms stand in stark in contrast with the highly hierarchical and rule-based states that supplanted the societies which guaranteed these freedoms. The freedom to leave a social group and join another was a broadly practised exercise of agency that has been greatly diminished with the advent of bordered states and restriction of movement, and the sorts of structures neoliberals have claimed are engines of greater freedom. Though the greater or lesser, broader or narrower distribution of freedom has never been strictly mediated by the passage of time, it was only with the rise of parasitism in Mesopotamia 5,000 years ago that the systematic curtailment of individual freedoms began, eventually reaching a global scale and zenith. Only with the rise of such parasitic states as Hayek champions did fences and borders begin to enclose land across the world. In fact, many early cities practised egalitarian politics and respected individual freedom to an extent that citizens of many cities today, even ostensibly democratic ones, would consider utopian. This historical reality directly contradicts Hayek's assertion that history has moved teleologically from lesser to greater individual freedom.[145]

As in the ancient empires, the extent to which individuals could claim the autonomy integral to individual freedom mirrored the boom-and-bust nature of the economies themselves: with the dissolution of regimes or the withdrawal of imperial bureaucracies, freedoms would often increase in number and diversity – only to be curtailed again by new regimes or foreign warlords.[146] The conquest of the Americas, to which Hayek – an admirer of Enlightenment liberalism – is certainly alluding in the earlier passage, could not be seen as a historical expansion of individual liberty but rather as its opposite, one of history's greatest curtailments of liberty. The many diverse, often free and democratic nations that were decimated by the arrival of monotheistic European monarchies in the Americas exemplified the literal replacement of freedom-loving democracies with oligarchic autocracies. As this book has demonstrated, as attested by the most up-to-date scholarship, and even by the first-hand accounts of Indigenous peoples captured by colonists – and indeed by the colonists themselves – First Nations in the Americas tended to respect individual autonomy and liberty far more than the colonial governments that arrived to make war on and destroy them.

It is from this fundamental error of history that we can see Hayek's further sense of 'freedom' emerge. Neoliberals following Hayek and the movement directly contributed to the rise of despotic rulers, like Augusto Pinochet in Chile. Pinochet is notorious for his abuse of political rivals. Kidnapping them and dropping them from helicopters to their deaths was one method of execution,[147] while sending dissidents to a death camp run by a former Nazi and cult leader was another.[148] Pinochet's Chile was for Hayek a perfect illustration of the maxim that economic liberalism is more important than political liberalism. On Chile, Hayek had only praise: 'I have seen in some South American countries the most extraordinary progress. . . . In that much condemned country, Chile, the restoration of only economic freedom and not political freedom has led to an economic recovery that is absolutely fantastic.'[149] Milton Friedman, Hayek's most prominent American acolyte, referred to circumstances in Pinochet-run Chile as an economic 'miracle'.[150]

III SYSTEM

The neoliberal movement was neither an immediate nor necessarily an inevitable success.[151] For over twenty years the neoliberals remained marginal in policymaking circles. From the 1940s through the 1960s, Keynesianism enjoyed a political consensus in the US and UK, largely due to the popular redistributive programmes it yielded. The neoliberals waited explicitly for a crisis to undermine this consensus. The oil shock of 1973 and the simultaneous inflation hysteria proved to be the perfect convergence of crises to erode Keynesianism. Stedman Jones observes:

> Neoliberalism was a particularly powerful theory, almost a faith, for true believers in the free market and its possibilities. Friedman's monetarist analysis [that monetary policy that saw central banks controlling money supply was the most important] and market solutions grabbed people's attention as a ready-made and plausible alternative strategy when the economy began to unravel under the weight of stagflation.[152]

He goes on to write,

> The Keynesian economic elite had by then lost credibility with the public and with policymakers. Friedman instead was able to offer a solution with a set of policy prescriptions that could immediately be put into effect. Technical policy ideas burst the dam of Keynesian dominance, and Chicago neoliberal philosophy rushed through.[153]

Keynesian investments in public research institutions yielded technological inventions that became the precursors to much of the high-tech, communications, aeronautics and digital tech of the late twentieth and early twenty-first centuries. But with Keynesian social democracy came limits to what corporations could do with the technologies that universities and research labs had developed: there were environmental limits, cultural limits, and a common good ethic that restrained the invisible – rather, the hidden – hand of corporate commerce.[154] Neoliberalism was a cultural-political revolution whose primary aim was to unleash the extractive potential of the Keynesian technological revolution by building a moral system that could justify removing the limits imposed by public-good government regulations. This is the central reason the neoliberal consensus coincides with increasing rates of deforestation, extinction, greenhouse gas emissions and inequality in the late 1970s; suddenly land in the Global South that had been owned in common was closed to citizens and opened to global markets, suddenly the programmes intended to move wealth from small groups of elites to the public good began an onslaught of defunding and disintegration. The association between growth, progress and neoliberal interventions – global and domestic debt and debt markets, enclosure, marketisation, financialisation, and corporate governance structures – remain hegemonic today.

Environmental campaigner George Monbiot opens his essay on neoliberalism like this:

> Imagine if the people of the Soviet Union had never heard of communism. The ideology that dominates our lives has, for most of us, no name. Mention it in conversation and you'll be rewarded with a shrug. Even if your listeners have heard the term before, they will struggle to define it. Neoliberalism: do you know what it is?[155]

For a long time, neoliberalism has remained cloaked, not taking its rightful place as a distinct ideology, instead disguising itself as the *absence* of ideology. Its proponents claim to be objective, rational observers, devoid of ideological commitments, simply stating the facts, even going as far as to deny that 'neoliberalism' really exists. 'Its anonymity,' Monbiot writes, 'is both a symptom and cause of its power.' But the truth is that neoliberalism is every bit as much an ideology as Nazism or Marxism. Its core tenets arose out of the same complex of ideas as both, and its followers remain as beholden to it as the true believers of any other ideology. It maintains the growth-as-progress narrative that has been so ubiquitous in networked parasitic economies. Despite many declarations of neoliberalism's demise, its basic tenets – an obsession with profit and growth, the marketisation of all realms of life and the devaluing of the non-commercial – remain fully embedded in most of the world's institutions. If anything, this ideology has never been more powerful and all-enthralling. For the half-century following the Second World War, neoliberalism took on mostly grey technocratic forms, donning homogenous corporate aesthetics, all bathed in fluorescent and business casual, smoothly invisible. But as challenges to the ideology have arisen, it has begun experimenting with wilder narratives and tactics for maintaining supremacy.

Heavens of Tomorrow

In 1917, Albert Einstein wrote to his friend Heinrich Zangger, 'All of our exalted technological progress, civilisation for that matter, is comparable to an axe in the hand of a pathological criminal.'[156] It's understandable that, with newly invented aerial bombing, chemical weapons and machine guns rattling in one's ears during the Great War, one might arrive at this conclusion. Einstein was a socialist and a pantheist, holding the belief that the divine can be found in nature and the unity of existence. He considered 'the economic anarchy of capitalist society . . . the real source of evil', and noted wistfully: 'The time – which, looking back, seems so idyllic – is gone forever when individuals or relatively small groups could be completely self-sufficient.'[157] Einstein believed technological and scientific advancements could only be considered achievements insofar as they helped humanity transcend its current 'predatory phase' of development – in which the owning classes preyed upon the working class – and instead produced social goods to fulfil the needs of every member of society. In contrast to Keynes's technological optimism, Einstein was more realistic, lamenting that 'Technological progress frequently results in more unemployment rather than in an easing of the burden of work for all.'[158]

Invention, education, curiosity, the honing of one's unique skills and abilities: these were values Einstein saw as incompatible with capitalist society. As a physicist who understood the laws of nature, he would likely be appalled by the talk today of infinite growth, particularly in the face of ecological emergency. Indeed, contemporary physicists are among the scientists sounding the alarms. It's often *not* the scientists who fetishise the technology that their discoveries frequently enable, but all too often, perhaps unsurprisingly, the businessmen keen to press those technologies into the service of greater extraction.

Today, the progress narrative formula is full of technological jargon spouted by Silicon Valley moguls and myopic visionaries. Confronted with ecological limits and the apocalyptic warnings of climate scientists, those who control the world's institutions and capital have begun to organise a global live-action roleplay game set in a science fiction story of their own invention, full of rockets, robots and infinity. Two billionaires – the world's richest men, at time of writing – illuminate this phenomenon.

In March 2021, SpaceX – the aerospace manufacturing company founded by brand developer Elon Musk – intentionally destroyed a rocket. Re-entering the atmosphere, the debris put on a 'fantastic light show in the Pacific Northwest sky', one headline cooed.[159] A few days later, SpaceX unintentionally destroyed a rocket, this one in Texas.[160] Unlike the military rockets built by, say, Raytheon, SpaceX rockets are typically not meant to explode. They are, allegedly, meant to one day shuttle people to outer space and eventually to colonies on Mars and other planets. Still, their failures and explosions are seen as exciting successes to the devout followers of SpaceX and its CEO, a business prodigy who got his start in business by being born into a family of mine-owning moguls in South Africa and made his fortune buying an early stake in the online payment website PayPal.[161]

Musk is not the only 'richest man in the world' investing in spaceships. Jeff Bezos has also bought shares in the cosmos market with his company Blue Origin. Less flashy and with fewer – but not zero[162] – rockets unintentionally exploded, Blue Origin is two years older than its competitor.[163] While SpaceX is raising furious investment capital and, at time of writing, is valued at around $74 billion,[164] Bezos has shifted his billion-dollar attention from Amazon to Blue

III SYSTEM

Origin, even going so far as blasting himself into space.[165]

Why did the world's two richest men both found spaceship companies within two years of one another at the dawn of the new millennium?

Their – by this point unsurprising – answer: to advance human progress. SpaceX and Musk frequently proclaim that they seek, as the company's shorn-down mission statement puts it, to 'Make Life Multiplanetary'. In response to a CleanTechnica hagiography written in his defence over criticisms of wealth hoarding,[166] Musk blasted off a somewhat messianic tweet: 'I am accumulating resources to help make life multiplanetary & extend the light of consciousness to the stars.'[167] Blue Origin's website copy is less succinct than that of SpaceX, but comes to a similar point: 'Blue's vision is a future where millions of people are living and working in space. In order to preserve Earth, our home, for our grandchildren's grandchildren, we must go to space to tap its unlimited resources and energy.'[168] In some ways, these are reasonable goals: it *is* exciting to imagine the 'light of consciousness' casting its rays upon new worlds and wonders. It *would* be a moral good for humans to spark otherwise dead planets with the miraculous volt of (more-than-human) life. Earth's ecological systems *could* use a break from the pressures of global commerce, industry, and the demands of the multitudes, affluent and otherwise, even if putting those complex processes in space is a quixotic solution.[169]

Musk and Bezos aren't the only parasitic class scions to indulge in this science fiction fantasy, of course. Silicon Valley thinkers, such as they are, have long been dreaming up fantastical futures in which artificial intelligence and bioengineering produce androids and supercomputers who surpass human intelligence and repopulate the world with perfect minds, building permanent and infinite utopias on a plane of human wreckage. Google cofounder Larry Page has, according to an investigation in the *New York Times*, welcomed our new robot overlords. After merging with machines, or being fully replaced by them, Page imagines that, through a social Darwinian process, 'One day there would be many kinds of intelligence competing for resources, and the best would win.'[170] He sounds a lot like the

Italian futurist and proto-fascist Marinetti, with a minor tweak: 'When we are forty, other younger and stronger [robots] will probably throw us in the wastebasket like useless manuscripts – we want it to happen!'

This leaves us with two important questions. First, do these people really believe in the science fiction they apparently live within? This is a corollary question that has run through this whole book: are progress narratives deployed by people who truly believe them, or are they simply tools used cynically to manipulate others into accepting their servitude to emperors, kings, lords, billionaires and millionaires?

The second question this raises is, could their fantasy become a reality and save us? To this point, we have primarily been looking at historical examples of progress myths. Our distance from history gives us a clarity of sight, to see that the promises of progress did not unfold as their prophets foretold. Those promises were revealed to be empty, over and over again. But today's progress narrative makers say this time is different. This empire, unlike all the rest, will *not* collapse and this lineage will continue with limits neither to extent nor duration. Could they be right this time?

Let's start with the first question. Are these scions of today's parasitic classes earnest true believers, zealots of their own version of L. Ron Hubbard's *Battlefield Earth*, cheering on the machines in *The Matrix*?

The *New York Times* article quoted earlier contains a revealing detail about Musk's attitude towards AI. It was in a heated debate with Musk that Page revealed his allegiance to a machine-dominated future, calling Musk a 'specieist' [*sic*] for maintaining a sentimental attachment to human supremacy. Musk voiced a fear of the possibility of a future dominated by AI and out-of-control androids, as well as a fear of climate change and the overpopulation of the planet. While taking the cofounder of AI company DeepMind, Demis Hassabis, on a tour of his SpaceX facility, Musk recounted that his motivation for building rockets was to save humanity by colonising Mars. Hassabis replied that this was a good plan, but only if he could prevent superintelligent robots from following humanity to Mars and destroying them there. 'Mr. Musk was speechless,'

III SYSTEM

according to the *New York Times*. 'He hadn't thought about that particular danger.' He promptly invested in Hassabis's company. This exchange, his heated discussion with Page, and his immediate response of putting his money where his mouth went – hundreds of millions of dollars' worth, which he invested in a different company, OpenAI – suggests that perhaps Musk does believe that he is working for the progress and betterment of humankind. Or at least that he thinks he should be in charge of a potentially catastrophic new technology and choosing who gets on the lifeboat to Mars. One may reply that he invested simply because he saw profit in it, but, to be fair, he pushed for OpenAI to be a non-profit at first – whether in good faith or not, only Musk knows. He eventually changed his mind and attempted, and failed, to seize control of the organisation to make it a commercial entity which it would become anyway after he was ousted.

It should be clear from these exchanges, too, that this belief system is, whether genuine or not, childish.[171] Musk has indulged in other examples of stunted psychological and emotional development. He crowned himself 'Technoking', named his Texas office 'Starbase' and his rockets *Falcon*, *Starship* and *Dragon*, and christened his child 'X Æ A-12 Musk'.[172] He shot a car into space. Combined with cringe-inducing podcast weed-smoking stunts and his compulsive retweeting of dated memes, it seems this isn't all just guerrilla marketing: Musk truly lives in an adolescent sci-fi delusion. His motivations often seem no more complicated than a pedantic teenager's reveries – the protagonist just happens to have tens of billions of dollars and ethically questionable behaviour issues.[173]

Bezos, too, indulges in a degree of teenage narcissism. When he was eighteen years old, he held forth on his fascination with space travel, telling the *Miami Herald*, 'The whole idea is to preserve the earth.'[174] He wished for humanity to leave the planet entirely, keeping Earth a giant wildlife park. More recently Bezos said, 'We need to take all heavy industry, all polluting industry and move it into space, and keep Earth as this beautiful gem of a planet that it is.'[175] His dream of sending '2 or 3 million' people to space (what happened to the rest?) when he was in high school seems to have remained lodged in his mind. Musk claims he was inspired by classic sci-fi author Isaac Asimov's *Foundation* series, while man-of-the-people Bezos anachronistically claims a slightly lower-brow inspiration, the film *October Sky* (1999) starring Jake Gyllenhaal.[176]

But even if Musk and Bezos are sincere in their belief that they can bring consciousness to the stars, there is a dark side of this dream, and there is no doubt that they also deploy it for cynical and self-serving ends. Both Musk and Bezos have mentioned *commoditising* space travel, and not just the small potatoes of selling orbital tickets to rich tourists. Notice again the Blue Origin copy: 'we must go to space to *tap its unlimited resources and energy*.' According to some scientists, celestial bodies like asteroids could contain trillions of dollars' worth of mineral resources (at least, trillions of dollars when they are as scarce as they are on Earth, which presumably would not remain the case once they are all brought to earth from asteroids, flooding supply well beyond demand, a fact which has apparently eluded their keen business minds). Bezos is preparing to become the first cosmic robber baron, monopolising those 'trillions' of dollars sitting out on space rocks.[177] Musk, meanwhile, wants colonies of millions on Mars to 'recreate Earth's entire industrial base', explicitly using the labour of indentured servants to do so (he says workers can pay off the price of their ticket with debt bondage).[178] If the way they run their companies is any indication, this is unlikely to be good either for workers or for newly terraformed environments.[179]

That's because Musk and Bezos use their notion of progress for the same reason it has been used throughout the history analysed in this book: to expand their empires. They are simply bringing parasitism to space. The minds of emperors have always bent towards two aims alone: the expansion of dominions and extraction from land and bodies. Frontier expansionism and extractivism led to the genocide of the Americas that killed 10 per cent of the world's population, it built the Atlantic slave trade, it sent Vikings on quests of slaughter, plunder and collapse (Musk connected his project to this history, tweeting 'Cybervikings of Mars'[180]), and it contributed to the political violence of the twentieth century that killed well over 150 million people.[181] It's no giant leap to imagine some space colonist

III SYSTEM

promising to bring the light of consciousness to the stars and trillions in mineral capital to Earth-based markets in order to justify chattel slavery all over again. Musk is already just one step away with his dreams of Martian serfdom.[182] This view of space as a frontier of extraction is not unique to Musk and Bezos. In President Trump's 2020 State of the Union speech, he justified his desire to settle colonies on the moon, saying, 'We must remember that America has always been a frontier nation [and] embrace the next frontier: America's Manifest Destiny in the stars.'[183] He created a new branch of the military, the US Space Force, to achieve this.

All these men deploy technological narratives of progress for the same reason as all the parasitic classes of the past. We all know the hoarding of wealth is harmful, especially when inequality is so high.[184] Amazon's 2018 carbon footprint was equivalent to Norway's.[185] The company has been credibly found to be forcing employees to urinate and defecate in bags.[186] This is on top of a well-documented history of other abusive practices.[187] Musk, meanwhile, has got into legal trouble for attempting to bust incipient unions at Tesla while workers there suffer overwork and injury.[188] He explicitly justifies employee abuse and overwork at SpaceX with 'greater good' arguments, cajoling workers to work harder and longer for the sake of technological progress.[189] With displays of such behaviour, these men need to prove – to us, to themselves – that they're not villains. Whether they know it or not, they are playing the role of a thespian on the stage of the world's ruling class, putting on an extravagant play, sparing no expense, to create the illusion of progress, for their own moral stardom. Like the Genesis stories of ancient Southwest Asia, the world's richest men offer celestial illustrations of how faith in progress will deliver us to the Promised Land – even if it happens to be on Mars.

One way of doing this is to make us believe that progress is dependent on the ruling class: without them we would never reach the stars. Musk said it himself: he's accumulating capital to 'extend the light of consciousness to the stars'. *You should all be grateful for my wealth hoarding*, he seems to imply. In this light, SpaceX and Blue Origin rockets look less like sincere attempts to push technology forward and more like elaborate public relations campaigns to mitigate the public backlash against their harmful behaviour.[190] Look at the shiny space rocket while we hoard historic magnitudes of wealth. Musk and Bezos are also open about promoting progress narratives. Musk, for instance, tweeted, 'those who attack space maybe don't realize that space represents hope for so many people',[191] explicitly revealing that narratives of space travel are palliatives that make people feel good about problems on Earth. It makes one wonder whether he really believes colonising other planets is possible or desirable for anyone except the parasitic class to which he belongs. If he knows as much as he claims to, and as much as his followers seem to believe he knows, then the answer is he probably doesn't.

There's good reason to believe their public pronouncements are contrary to their private ones. Comedian Jon Stewart discussed on his talk show a private dinner he had at the White House with the Obamas and Jeff Bezos. Bezos described his vision of the future and it did not entail a beautiful pristine earth and happy humanity living in luxury. Instead, he described a vision that essentially entailed most people working to provide services for the wealthy.[192]

Whether Musk and Bezos are true believers living in a sci-fi adolescent fantasy or nihilistic pragmatists giving people false hope to distract from their immense crimes, perhaps does not matter. As hard-won lessons from the twentieth century should have taught us, powerful men promoting niche fantasy worlds can amass huge followings and inflict terrible destruction. But we know from those lessons that men captured by such delusions cannot elevate humanity; they are only capable of debasing it.

Maybe the more important question at this point is whether their technologies could have any positive impact on humanity. Or whether, for that matter, other technologies may be better suited to deliver on such promises. Let's start with rockets and space as a viable frontier.

For all its futuristic glitter, SpaceX's Falcon 9 rockets run on a fuel first trademarked in the nineteenth century: kerosene. And they burn a lot of it. A single launch burns nearly 30,000 gallons of kerosene, emitting more than 336,000 kilograms of carbon dioxide in the process.[193] Space rockets like the Falcon 9 require supply chain complexity, which depends on material inputs like titanium, lithium, copper, glass, rubber and more, on top of labour inputs from countless technicians, miners and transporters in production webs that encircle the globe, and all the resources their bodies need to live and work. Most of the inputs that go into launching just one of these rockets, from kerosene to the petroleum-grown food that fuels the workers who make it all possible, are non-renewable and diminishing.[194] Lithium, copper, rubber, sand, all are running out, along with many other critical minerals and resources.[195] And not only are rockets dependent on dwindling resources, they are *contributing* to that dwindling. Advancing technologies to the complexity necessary to settle other planets requires much more of these inputs and decades of research. But if we are going to sustain complex logistics chains and maintain stable food stocks – let alone ever-improving rockets with a tendency to explode – for decades to come, the intensity of these material inputs must *decrease* rapidly in a variety of sectors.[196] It is fantastical to think we can sustain the complex production processes necessary to build and launch space rockets when the seas are drowning every coastal city and erratic weather disrupting every agricultural zone. What will feed the thousands of workers who build the parts for these rockets? Where will the factories sit? Who will mine the metals? And not only is the increasingly unstable environment bad for launching rockets, launching rockets is bad for the environment. Beyond its resource and carbon footprints, SpaceX negatively affects wildlife. As the Nevada Director of the Center for Biological Diversity noted, SpaceX missions impact 'ocelots, jaguarundis, aplomado falcons, 5 species of sea turtle, the piping plover, and the red knot'.[197] This isn't isolated behaviour. In 2019, the US Environmental Protection Agency penalised Tesla, of which Musk is also CEO, for hazardous waste violations.[198] This paradox appears, for now, unresolvable. Musk and Bezos certainly aren't working to resolve it, given that it entails confronting the ecological and climate crises to which they are disproportionately contributing, and requires massive global cooperation rather than isolated for-profit space businesses.

In addition to the scarcity paradox that is preventing the success of multiplanetary life, the limitations of biological organisms and biospheres present another obstacle. Human bodies cannot live long without biospheres. Our bodies need to respire oxygen, metabolise nutrients, and obtain and synthesise vitamins to live. These processes are complex and depend on inputs that can only be derived ecologically: our bodies evolved in complex ecologies and can only survive in complex ecologies. Sustaining artificial ecologies has never been achieved, especially outside the gravitational and chemical ideals of Earth, and is likely impossible. The 'Biosphere 2' project is the most famous attempt to develop the techniques necessary to build artificial ecologies on other planets.[199] It *did* enjoy the benefits of Earth's gravity, energy, atmosphere and sunlight, and it still failed. It's worth noting that the man in charge of the project at one point was Steve Bannon, who is best known for his role in helping to organise an international fascist movement.[200] Ecologies simply contain too many intricate variables to build them artificially in bubbles, at least with current or anything resembling near-term technology, which again is likely to hit resource walls soon.

The second and more difficult issue is the *macro*-biosphere, i.e., the Earth.

The ecological and climate crises are making infrastructure chains less viable and sea routes more turbulent, threatening food supplies, and jeopardising the stability of both political and economic systems. The increasing number of ecological emergencies is already throwing markets and governments into states of instability.[201] As was demonstrated vividly in 2021 when a single container ship blocked the Suez Canal, disrupting global trade, distribution is more tenuous than we like to imagine: a frail paper rocket sitting atop a pyramid of cards.[202] The coronavirus pandemic, potentially a direct result of either ecological decline or technological hubris,[203] gives a small taste of how the compounding crises will continue to disrupt civilisation into the near future.[204] (Musk and Bezos, incidentally, saw their wealth 'skyrocket' during the pandemic.[205]) Energy scientist Vaclav Smil, in a 2021 interview with *Noema Magazine*, said, 'Managing the biosphere . . . is the main issue, because it's the only biosphere we've got. We are not going to colonize Mars, despite Elon Musk's enthusiasm. . . . It's totally laughable. This is the only biosphere we have, and we have to take care of it here and now.'[206]

The viability of space colonies comes down to one simple question: is the rate of technological progress moving faster than the rate of terminal resource and environmental degradation? The decisive answer is No. Disintegration is moving faster than progress. At the current rate of advancements in space technology, the Earth's capacity to sustain complex civilisations will have collapsed long before building micro- or macro-biospheres on other planets becomes viable (if it ever can be) or before rockets can land reliably without exploding.[207] The fact is, we're not really any closer to a 'multiplanetary' civilisation than Soviet cosmonauts were sixty years ago. Given the degraded condition of the ecological foundation on which all human technology rests, we're probably a lot further away.[208]

All of the technologies dreamed up by Silicon Valley investors and policy think tanks and genuine researchers are dependent on two things: abundant energy and a planet – that is, a climatic, nitrogen, hydrological and biological system – stable enough to allow for all of the activities involved in technological production.[209] All of these necessities are rapidly disappearing.

The Soviet Union's *tsar bomba*, as you may recall, the largest humanity has ever detonated, released 1,500 times more firepower than the combined tonnage of the US's Hiroshima and Nagasaki bombs. The global economy today generates greenhouse gas emissions with the heat energy equivalent of the *tsar bomba every twelve minutes*. This warming not only kills people, by means ranging from disrupted food and water sources to more prevalent infectious diseases, but is also killing the rest of the natural world. The National Oceanic and Atmospheric Administration predict that 'By 2030 . . . the heating imbalance caused by greenhouse gases begins to overcome the oceans' thermal inertia, and projected temperature pathways begin to diverge, with unchecked carbon dioxide emissions likely leading to several additional degrees of warming by the end of the century.' If warming triggers unstoppable feedback loops and yields the so-called 'hothouse earth' scenario, it will almost certainly result in conditions unsuitable for complex economies and, potentially, for human bodies to survive.[210]

We don't know how close we are to tipping into any of these feedback loops – things like ocean anoxia, permafrost melt, forest diebacks and polar ice melt – and some have likely been set in motion already. The most recent Global Tipping Points report, published in 2023, counts *five* thresholds that have already likely been crossed, and three more that are likely to be crossed by the 2030s.[211] The Greenland Ice Sheet is thought to have crossed one such tipping point at the turn of the twenty-first century, which means that even if the atmosphere and oceans were to stop warming immediately, the ice sheet will most likely continue to melt. The Greenland Ice Sheet is the largest contributor to rising sea levels globally; earth scientist Grace Palmer reports that 'Returning to a balanced state would require an extra 60 gigatons per year of snowfall or reduced melting. Yet under essentially all climate change scenarios, the opposite is expected.'[212] The number of species that would go extinct in a hothouse earth scenario would be functionally uncountable. The path from here to there would inevitably be violent as even small disruptions to food systems in recent years have yielded increasing authoritarianism and civil conflict.[213] The Institute for Economics and Peace estimates that by 2050, there will be 1.2 billion climate refugees displaced across thirty-one countries.[214] When many such crises are compounded all over the world, it's hard to imagine anything but some sort of world war resulting. We know from the collapse of past societies that even moderate fluctuations in climatic conditions can lead to violent societal collapse and war. The current nuclear weapons stockpile is about 13,000 known warheads (the real number may be higher).[215] A world war with so many weapons would undoubtedly be catastrophic, likely making large patches of the planet uninhabitable for humans and many other species. Just the fact of a few people controlling whether tens of millions would die painfully from firestorms and radiation poisoning is one of the world's greatest repudiations of progress narratives; what do inflated well-being statistics matter, living ever in the shadow of potential hellfire and oblivion?

III SYSTEM

Yet the broader ecological crisis – of which climate change is one aspect – may present even more pressing challenges. A 2023 study in the journal *Nature* put the urgency of this crisis in terms scientists have traditionally avoided but, out of desperation, are increasingly using:

> A major concern for the world's ecosystems is the possibility of collapse, where landscapes and the societies they support change abruptly. Accelerating stress levels, increasing frequencies of extreme events and strengthening intersystem connections suggest that conventional modelling approaches based on incremental changes in a single stress may provide poor estimates of the impact of climate and human activities on ecosystems. . . . Collapses occur sooner under increasing levels of primary stress but additional stresses and/or the inclusion of noise in all four models bring the collapses substantially closer to today by ~38–81%. We discuss the implications for further research and the need for humanity to be vigilant for signs that ecosystems are degrading even more rapidly than previously thought.[216]

Industrial agriculture is depleting fertile topsoil faster than it can be replenished, portending mass food shortages within fifty or sixty years. Once depleted, it can take between 500 and 1,000 years for healthy topsoil to develop again.[217] Meanwhile, overuse of antibiotics in the meat industry is producing a crisis of 'superbugs', infectious diseases that are evolving defences against antibiotics, which will likely lead to more pandemics and infections that are difficult or impossible to treat. According to one study, these superbugs threaten to take 10 million lives annually by 2050.[218] In 2021, the USDA killed two hundred wild animals *per hour*, ostensibly to 'protect farmers' by annihilating very dangerous animals like 'armadillos, doves, owls, otters, porcupines, snakes and turtles' – a total of 1.75 million creatures (this doesn't include the 70 billion terrestrial animals or the 1 trillion sea creatures killed annually for food). The slaughter also included ecologically vital and critically endangered creatures: hundreds each of wolves, bears, mountain lions and bobcats, and thousands each of coyotes, foxes and beavers.[219] Deforestation is occurring at the virulent pace of about 10 million hectares per year, yielding a pandemic of forest destruction.[220] The species extinction rate is between hundreds and thousands of *times* higher than a natural rate, a breakneck pace not seen since the extinction of the dinosaurs 66 million years ago.[221]

Such statistics can hide some of the more quietly tragic facts of annihilation. As numbers of songbirds dwindle, for instance, even remaining birds begin to lose their songs with fewer elders to teach them, which makes it harder for them to find mates: a small, sad, overlooked feedback loop.[222] On a Discovery Channel programme, you can hear a recording of the last male of a species of songbird singing his side of what was once a duet between pairs that mate for life. He sang his part to silence, for a mate who would never come.[223] 'He is,' the scientist says, 'totally alone.'

Meanwhile, if Musk and Bezos and Page get caught up in their teenage fantasies, others are more clearly deploying faith in technology to cynical ends. Techno-faith and progress propaganda has found allies in the petroleum industry, including proselytes for tech like geoengineering, which might entail, among other ideas, spraying particulates of various composition into the atmosphere to attempt to reflect sunlight (and therefore heat) away from Earth – and its untested, non-viable cousin carbon capture and sequestration. This is no surprise because it would theoretically allow the oil and gas industry to continue profiting from its existing business model for the foreseeable future. BP, meanwhile, has invested heavily in ad campaigns promoting their investment in renewable energy, touting themselves as central actors in making progress on futuristic energy systems technology. But after the ads ran, they removed virtually all their investments in renewables. 'They invested millions in creating a new green sun-based logo, linking it with solar energy. This has remained its logo ever since, despite the promises of "Beyond Petroleum" seemingly having been long abandoned, as BP sold off the token renewable energy projects and refocused on being a fossil-fuel corporation.'[224] Environmental lawyers from ClientEarth filed a complaint against BP's misleading advertising in 2019, reporting,

'While BP's advertising focuses on clean energy, in reality, more than 96% of the company's annual capital expenditure is on oil and gas. According to its own figures, BP is spending less than four pounds in every hundred on low-carbon investments each year. The rest is fuelling the climate crisis.'[225] BP pulled their ad campaigns to stop the complaint proceeding in court.

Shell, meanwhile, puts out press releases with titles like 'Shell accelerates drive for net-zero emissions with customer-first strategy' and copy like: 'Shell is integrating its strategy, portfolio, environmental and social ambitions under the goals of Powering Progress: generating shareholder value, achieving net-zero emissions, powering lives and respecting nature.'[226] Chevron has hopped aboard the corporate progress train, 'pinkwashing' its oil drilling in social media posts: 'Chevron attached a cute little graphic to the tweet, showing cartoon workers above a rainbow band, including two people fixing a pipeline with the caption "acting for allyship."' Gizmodo notes this hypocrisy: 'For years, oil companies' success has depended on having politicians in power who can perpetuate climate denial and fossil fuel apologia in Washington, DC. That has meant donating to the GOP – the party that, incidentally, has also tried to block pretty much every fight for basic rights the LGBTQ community has waged in recent decades.'[227] Major media corporations, including the *New York Times* and *Washington Post*, have abetted oil companies, allowing them to publish paid propaganda: what the industry calls 'native advertising', a common practice that passes off industry press releases as news.[228] These companies have done everything in their power to maintain an oil-based economy. Again, this is just a tip of the iceberg of cynical corporations using progress rhetoric to justify and mask their commitment to a business model that results in nothing less than crimes against humanity and nature.

Maybe the rocket prophets, AI servants and petroleum industry publicists are selling obviously false pledges. What about other technological promises?

A May 2021 interview with John Kerry, President Biden's 'climate czar', revealed another kind of faith in technology. As the *Guardian* reported, Kerry said '50% of the carbon reductions needed to get to net zero will come from technologies that have not yet been invented.'²²⁹ The experimental technology Kerry is talking about includes geoengineering. Such technology is untested and could have catastrophic impacts on agriculture, wildlife and human well-being.²³⁰ We have no reason to believe such technologies will be ever be safe and viable, let alone within the shrinking window of time necessary to avoid catastrophic warming of 1.5–2 degrees Celsius. But such facts don't stop advocates from blithely hand-waving away the risks. John Fialka at *E&E News* wrote that such aerosols are 'relatively harmless', even though there is absolutely *no research whatsoever* from which to make that claim. Meanwhile, when Sámi people in Sweden protested at sun-blocking experiments there and called instead for 'zero-carbon societies in harmony with nature', Fialka dismissed them as 'an Indigenous group of reindeer herders'.²³¹ It's the kind of arrogance common among the techno-utopian crowd that is generally in favour of geoengineering.

During the Great Acceleration of the mid-twentieth century, there was a short-lived acceptance that nuclear energy was part of the technological progress that could deliver free, safe, endlessly available energy. While that dream has been dashed on the hard shore of reality, a group of trans-ideological pundits – their politics lying across a narrow spectrum – have recently resurrected it. Nuclear energy does have some positives as well as many negatives, and these are worth weighing against one another. The positives include (1) the lack of carbon emissions and fine particulates at the point of electricity generation, (2) the reactors' ability to be integrated into existing grid infrastructures, which could theoretically deliver faster decarbonisation of electricity, (3) the stability of power supply, and (4) the scope for improving the technology by, for instance, recycling spent fuel. At its current scale, nuclear energy tends to be a lot less deadly than the fossil fuel-derived electricity that advocates claim it is well suited to replace. Unfortunately, when we start to compile the negatives, they pile up, and when we look more closely at some of the positives, they don't always hold up.

New nuclear power plants are slow and costly to build, taking an average of ten years, with energy costing between $112 and $189 per megawatt hour (MWh), in contrast to $29 to $56 per MWh for wind and $36 to $44 per MWh for solar.[232] Physicist Amory Lovins notes that 'most U.S. nuclear power plants cost more to run than they earn', which means profit-driven markets would not initiate or sustain a nuclear expansion.[233] Scaling nuclear up globally would therefore require mass state investment in a non-profitable industry, which is not likely in a world of near total neoliberal capture. Nuclear power isn't really a solution to climate and ecological problems given that transition must occur quickly; its slowness, costliness and inflexibility is a major hindrance. Meanwhile, technological innovations that could mitigate these issues are still far away, if they're possible at all. The commercial viability and practical scalability of tech solutions loom ever out of reach, with projected timelines stretching years or decades into the future and no guarantees of ever becoming viable. Nuclear power is unpopular in the United States, with just 16 per cent of US adults surveyed believing the country should maintain existing reactors and build new ones. Only 29 per cent of the public view nuclear power favourably, and 49 per cent view it unfavourably.[234] Whether justified or not, its unpopularity presents a major political obstacle to the rapid expansion of nuclear energy.

Continuing scholarly debates about the safety of living near nuclear facilities and its potential for higher cancer risks are unlikely to help nuclear's standing in popularity polls.[235] The nature of nuclear energy means the risk of a catastrophic accident is impossible to bring to zero, human error and natural disaster being ineradicable risks. As weather and political entities become far more erratic, disasters will become far more likely. Furthermore, the fuel on which nuclear energy depends is non-renewable. Engineering professor Derek Abbott has calculated that scaling nuclear production up to meet global demand could leave just five years of uranium supplies.[236] Getting that uranium is dirty and dangerous: environmental journalist David Thorpe calculates that 'To produce the 25 tonnes or so of uranium fuel needed to keep your average reactor going for a year entails the extraction of half a million tonnes of waste rock and over 100,000 tonnes of mill tailings. These are toxic for hundreds of thousands of years.'[237] And while it's true that nuclear reactors do not emit carbon dioxide, all the processes that go into them *do*, from the mining, processing and transportation of uranium to the construction of plants. Nuclear power plants also use emergency (fossil-fueled) diesel generators as back-up sources of energy when reactors need to be shut down for safety, which even conditions like heatwaves can cause.[238] Physics professor Keith Barnham suggests, 'Far from coming in at six grams of CO2 per unit of electricity for Hinkley C,' a new reactor being built in the UK, 'the true figure is probably well above 50 grams – breaching the [Committee on Climate Change's] recommended limit for new sources of power generation beyond 2030.' Barnham argues that 'half of the most rigorous published analyses' find that nuclear power exceeds the limit of carbon dioxide emissions set by the government's climate change advisor.[239] Far from 'carbon-neutral,' it's not even a 'low-carbon' energy source. What's more, there is nothing intrinsic to nuclear power to believe it would *replace* fossil fuel electricity rather than simply supplementing it. For instance, companies like Microsoft are hoping to power their ever-growing electricity demands from AI with expanding nuclear power generation *on top* of their existing fossil-fuelled operations.[240]

Perhaps most relevant to our discussion here, nuclear energy has historically remained an element of the parasitic project of extraction from the marginalised peoples living at the frontiers of expansion. Native American reservations have been central to the entire history of nuclear technology, both power plants and weapons. 'Energy sacrifice zone' was a Nixon-era term coined by the National Academy of Sciences that described the Shoshone and Paiute lands taken for nuclear testing at the Nevada Test Site, a 1,360-square-mile reservation that was bombed 928 times – nearly half of all the world's nuclear explosions.[241] 'The Manhattan Project sought domestic supplies of uranium from the only source of which it was aware, the vanadium mines in and around the Navajo reservation,' writes historian Traci Brynne Voyles. 'With that, uranium went from being a waste by-product of vanadium to the most sought-after ore of the twentieth century.' Today, more than 2,000 abandoned uranium mines sit open on Navajo land, 'leaching radon gas into the air and water and scattering radioactive debris throughout the ecosystem'.[242] Nuclear energy inevitably locks in these extractive relationships and, perhaps more dangerously, creates the need to maintain militarised states simply in order to protect and maintain the highly potent toxicity of waste and operational plants.[243]

Nuclear reactors and all the zones created to sustain them, like mines and waste storage facilities, are extremely dangerous and remain so for tens of thousands of years. No civilisation has ever lasted nearly so long. Few systems of governance remain in place longer than a few centuries before order collapses and wholly new ones replace them. To imagine that *any* civilisation could safely sustain radioactive materials for as long as they remain extremely deadly is a fantasy entirely divorced from reality and historical fact. The world, after all, is not growing more stable, but less so. These sites will continue to become increasingly difficult to maintain in the face of environmental and social instabilities.

Strangely, even as nuclear energy is often touted as the best solution to mitigating climate change, some of its most vocal advocates also write books dismissing the risks of climate change. One such pundit, Bjørn Lomborg, has written unscientific and thoroughly debunked books downplaying climate change while advocating for nuclear energy. Reviewing Lomborg's most recent pro-nuclear climate denial book, the London School of Economics wrote, 'like his previous contributions to this issue, Dr. Lomborg's arguments are based on fantastical numbers that have little or no credibility.'[244] Alex Epstein, meanwhile, has written very similar books; his most recent, *Fossil Future*, argues in favour of increasing use of fossil fuels *and* nuclear energy. To make his argument that fossil fuels are essential to progress, he uses phony poverty statistics and ignores mountains of scientific data.[245] Perhaps unsurprisingly, Epstein is the founder of a think tank called the Center for Industrial Progress.

Another important aspect of this faith in technology to solve all problems is a pattern of political leaders setting dates by which to achieve certain cuts in emissions, justifying them using currently non-existent technologies. World leaders have frequently set deadlines that sound good, like 50 per cent cuts, or the even more ambitious-sounding 'carbon neutral' by 2050, and call it 'climate progress'.

But this is a strategy meant to *delay* progress rather than to achieve it. They do this because it is a way to escape accountability for the crisis that is occurring in the present. The dates they set always lie outside the scope of any participating politician's career timeline, or at least outside the likely scope of their time in a particular current office. As such, there's no way to hold them accountable for the deadlines, and no publicly available mechanism to ensure such policies are maintained between the time they're declared and the deadline that's set. There is, furthermore, no mechanism to prevent those dates from being pushed back as they draw near. At the same time, 'carbon-neutral' only means that carbon emissions are 'offset' rather than eliminated, but the vast majority of offsetting schemes – like tree plantations – are either ineffective in reducing emissions or are eventually uncovered as outright scams.[246] As a result, there has literally been *no* significant 'progress' in mitigating climate change. Total net emissions have risen during the whole span of time that human societies have sought to reduce them.[247] Like the Musk-Bezos rockets, it seems international climate meetings and deadlines are more political theatre meant to obfuscate the issue than genuine aspiration to solve it.

Every day brings record-breaking high temperatures, record-breaking emissions of greenhouse gases, and record numbers of extinctions and habitat destruction. These will continue to accelerate as long as a parasitic mode of extraction, production and excretion continues. There are currently *no technologies* capable of mitigating these problems: no carbon capture and sequestration technology is scalable, no energy production is without unsustainable extractive demands, no form of soil-depleting agriculture at an industrial scale can be sustained into the next century. Vaclav Smil calculates that capturing a quarter of emissions from *only* the world's coal plants would require pipelines capable of transporting a volume of fluid twice the scale of that carried by the entire crude-oil industry.[248] The manufacture, shipping and construction of those pipelines would require immense carbon energy inputs and emissions, and that's just to capture the emissions from coal, a fraction of total emissions, which are primarily coming from oil and gas. These same resource

III SYSTEM

demands plague technologies like nuclear energy and even renewables like solar and wind.[249]

It's worth noting, however, that the latter technologies are far less harmful in their impacts than fossil energy. However dramatic oil spills like Deepwater Horizon and *Exxon-Valdez* may appear, tankers release much more oil-laced ballast water in the normal course of transportation. By the calculations of Michael J. Graetz, 'when all the ships in the tanker fleet are factored in, [it] adds up to five *Exxon-Valdez* spills each and every year.'[250] All that oil contributes to the pollution from fossil fuels that causes nearly 9 million deaths per year, about one-fifth of all deaths.[251] That's more than tobacco and malaria combined. Automobiles, which use much of that oil, kill 1.35 million people per year – not including the deaths that result from pollution-induced climate change, nor the dramatic impact of roads on wildlife.[252] But renewables are only a solution to these problems if they replace fossil energy, rather than supplement it. Currently, the energy they provide is simply being added on top of ever more fossil energy.

Energy transition is much harder than deployment, as Elon Musk discovered. When he was primarily known for building electric cars that people wanted to buy in a time of political hostility towards the idea of electric cars, and when he was talking about 'Gigafactories' making batteries that could solve intermittency problems of distributed renewable electricity generation, he really did look like a 'climate leader'. He contrasted sharply with the dozens of other billionaires who not only don't care about climate and environmental emergencies but are actively working to accelerate those emergencies.[253]

But Musk has moved far away from this original goal, perhaps seeing how entrenched the obstacles to such progress really are. If Musk and Bezos were sincerely interested in making life multiplanetary, or protecting the beautiful gem that our planet is, they would not be investing their billions in private rocket technology; they'd be investing it in the less flashy but more difficult technological challenge of making viable micro-biospheres, and in the much more difficult political challenge of retaining a viable macro-biosphere here on Earth. They would be attempting to address the paradox whereby the rate of progress moves more slowly than the rate of collapse of the resource foundation necessary for technological progress, and society itself, to occur at all. But they're not; they're contributing to the opposite in very real material terms. And for all their entertainment value, Jeff and Elon's Arks will be empty, for the birds and beasts and the rest of us cannot afford the tickets to board.[254] Space billionaires will no more save us from collapse than the myriad emperors of fallen empires saved their own people. Maybe when they founded these companies, Musk and Bezos really did want to save humanity and the Earth. But then they became the world's richest men.

Like the authors of the biblical Genesis and New Testament promising individual ascent to Heaven and collective progress to a New Earth, the latest frontiers of progress today are a secular paradise among the physical heavens. After five thousand years, we are once again admonished to trust our rulers and defer to their judgment based on the ever-elusive promise that we will one day forsake Earth and ascend to an off-world paradise. It's no surprise or accident that the language and aspirations for spacefaring involve frontiers, mining, colonising, or dominion. As in the beginning, the mandate of the parasitic society still dominates. Rather than stewardship, discovery or new possibilities, such endeavours in space are rooted in expanding the metabolism of an insatiable economy.

Progress myths based on a linear momentum of history are impossible to sustain in the face of climate and ecological chaos because they depend on a lineage that can no longer continue. The door to that future is shut no matter what, given the amount of carbon dioxide already in the atmosphere and the near-inevitability of an increase in global temperature of 1.5 or even 2 degrees. If societies, states, economies and communities fail to take climate and ecological collapse seriously, they will fall. And in that fall, they will be so thoroughly changed that the historical precedents on which they're based will cease to have much meaning or impact, any more than other empires' norms continued to affect its colonies after its collapse (which is to say, not none, but not much). The parasitic economy that has been spreading from empire to empire for the last five thousand years is now globally networked in a way it has never been in all of history, meaning it could fully collapse at a global scale and leave a lot of unfamiliar sorts of societies in its wake.

On the other hand, if societies, states, economies and communities succeed in taking climate and ecological collapse seriously, civilisation will *still* be fundamentally changed, because stopping these crises entails dismantling the imperatives and logics of parasitic energy capture, thereby fundamentally reshaping the very idea of civilisation that emerged in Mesopotamia around 3000 BCE. Either way, this progress story will end. This is what technological optimists and other modern progress pundits cannot or will not understand. They think we can keep the story going as long as we make a few technical tweaks here and there, or maintain faith in the miraculous technological breakthrough that shows no evidence that it will ever come. But they fail to grasp the fact that the entire logic on which contemporary economies are based is intrinsically unsustainable. Parasitism is, by nature, a boom-and-bust system. It is encoded in its DNA to collapse. There is no technology that can make this imperative infinite. That doesn't mean technology can't help facilitate a transition to something else, even something better. Maybe it could. But it cannot continue the bloodline of progress that we are currently trapped in; that lineage must end completely. Because techno-utopians cannot imagine the end of that lineage, they are unable to take climate and ecology seriously, they are unable to see these problems with clear eyes. They recede from the world as it is into fantasies of outer space, androids, and carbon capture and sequestration.

But we non-billionaires cannot afford to be teenagers living in fanciful escapist dreams, trailing an eccentric Technoking who names his children after jets. This time is too consequential for those indulgences. The problems of today demand that we be adults and face the facts. Long-term space travel, and enduring technological development itself, will only come to fruition if we learn how to make civilisation both genuinely progressive *and* sustainable, immediately – and if we stop squandering the non-renewable resources being burned for frivolous activities. These goals remain dangerously far away, and maybe they're impossible; we don't know. What we do know is, if we wish to expand the light of consciousness for real, it cannot be the consciousness of emperors. Anyone unwilling to preserve the life that's already here will never invest in promoting life elsewhere.

Yet the future is not destroyed. Until a hothouse earth scenario becomes unstoppable and the planet too hot and volatile for the bodies of *Homo sapiens* – and those of countless others on whom our existence depends – there is yet tomorrow. And there is still an imperative, of the utmost urgency, to turn clear, adult eyes towards that future and imagine now what it ought to look like, and how we can move from here to there. But before we can do so, there is one more obstacle to consider. If we must tear down this lineage of progress and parasitism to build a world worth living in – and capable of being lived in – there is one last force to overcome. A force that has maintained the hegemony of empires for as long as they have existed, and still patrols the frontiers.

Frontier of Ghosts

Toward the end of the twentieth century, the USSR fell and took Russian communism with it. In the chaos of its ruin, corporate fiefdoms and an uneasy neoliberalisation took hold into the twenty-first century. At the time of writing, the country is ruled by an authoritarian executive and corrupt oligarchy. In the 1970s, after Mao's death, the Chinese Communist Party retained control by repositioning its economy to adopt neoliberal tenets, which it has maintained and adapted into the twenty-first century. As a result, China is producing more billionaires than any other country and has pursued with colonial ruthlessness a programme of intensively buying up land, burning fossil fuels and exploiting resources – including those in the Global South. In the West, where neoliberalism arose and dominated first and most fiercely, inequality has reached extremes last seen on the verge of the Great Depression. Consecutive recessions and erosions of human life expectancy and well-being indicators, like increased suicide rates, have accompanied this ideological realignment into the twenty-first century. Most urgently, pollution has warmed the globe to deadly extremes that show no sign of slowing and have eradicated immense quantities of terrestrial and ocean life. The Great Acceleration first sparked with Keynesianism and accelerated again with neoliberalism has opened nearly every imaginable frontier, with declining returns.

The global economy remains in a networked parasitic mode. Growth is the hegemonic imperative, progress the sustained invocation. Even so, the political and ecological crises hastened by neoliberalism have not gone unmarked or unchallenged. But in responding to this ideological consensus, the primary opposition has come from ghosts. The ghost of Marx has risen and drifts where he will. The ghosts of Hitler and Mussolini are haunting the cores of empires and the corners of urban zones. Even the ghost of Keynes has stalked the corridors of the White House and the think tanks in its orbit. None of these ghosts has sought in earnest to challenge the parasitism at the heart of all these insufferable conditions. They grapple for the helm, and the corsair of the past five thousand years sails on towards oblivion.

In the twenty-first century, in addition to the oceans, some of the biological frontiers sustaining this parasitic growth are the last great forests. The Amazon, the Congo and Borneo are among the three greatest.[255] Their transformation from wildlife to agricultural zones are taking biomass and energy from other life and injecting new calories into the human world. These forests are rapidly fragmenting and nearing collapse. Twenty per cent of the Amazon rainforest has been destroyed since the 1970s and is nearing tipping points that could ensure its collapse by the 2060s.[256] The Amazon has become a net emitter of CO_2 rather than a major absorber of it, releasing nearly 20 per cent more CO_2 than it absorbed over the last decade.[257] In the Congo Basin, about 700,000 hectares of rainforest was destroyed *every year* between 2000 and 2010.[258] The forest could be gone by 2100.[259]

In nearby Nigeria, nearly half of the forest cover was destroyed between 2007 and 2017, much of it for charcoal sold in local and international markets.[260] In Borneo – one of the world's only homes to orangutans – an area the size of Belgium was illegally logged between 1985 and 2001.[261] Within a span of twenty-two years, the area given over to oil palm production – a cheap source of vegetable oil present in about half of supermarket products[262] – grew by an order of magnitude, from 600,000 hectares to 6 million hectares. With regard to vegetation loss and carbon sequestration, one study suggests that the forest has reached a point of degradation 'that is beyond a point of no return'.[263] When they have been destroyed, their destruction will bring new challenges for both wildlife and human systems.

New mining and fossil fuel developments continue without an end in sight. Air travel is increasing rapidly, and jet fuel is in great demand. The continued spread of urban zones and the sprawling slums that surround them also provides frontiers of extraction, specifically of abstract energy capture as a source of labour. Some daring pioneers hope to open two more frontiers: deep space, where asteroids tempt the venture capitalists who lust for floating minerals, and the mind, where neurons provide the raw material to be mined or appropriated. Artificial intelligence, too, stimulates the salivary glands of a parasitic class intent on draining organic neuronal activity from any opportunity to extract and consolidate wealth. Space and brains remain frontiers primarily within the imaginations of hollow minds, but how long they will reside mostly skull-bound is impossible to say. Perhaps resource limits will arrive like an impenetrable wall, shutting down the data centres and high-energy manufacturing necessary to maintain machine learning or launch rockets. Perhaps not. Though it would be tempting to place our faith in the folly of emperors to see to their own downfall, there is reason to believe that they will manage to sustain that arrangement well into the future. For there are other ghosts that haunt these frontiers, weaving the first scenes of a nightmare that crisscrosses like a sticky web our pathway into the future.

Many empires of the past – the parasitic economies that dominated vast tracts of land and the majority of the global human population, and perpetually fought enemies on their growing frontiers – used secret police, spies and elite military units to govern their territories and scout out ways of expanding them. Rome, for example, employed the *frumentarii*. Originally engaged in moving grain around the empire – *frumentum* being Latin for grain – the imperial bureaucracy soon put these agents to use in gathering intelligence in faraway colonies. Affiliation with the *frumentarii* appearing on gravestones suggests they may have carried high status in the empire. Rome's *agentes in rebus*, the imperial courier service, served a similar role later on, while the elite Praetorian Guard conducted assassinations on behalf of emperors, and also the assassination of emperors themselves.[264] The *jinyiwei*, or Embroidered Uniform Guard, meanwhile, were a secret police and intelligence force in service to the emperors of China's Ming Dynasty. Feudal Japan's much mythologised *shinobi* ninjas conducted covert functions like espionage, sabotage, psychological warfare and assassination. The ancient Mauryan Empire in India had many secret services. The Arthaśāstra, an ancient Sanskrit text, was a how-to guide for emperors wishing to govern the state and economy while achieving perpetual military victory; one piece of advice admonished leaders to set up clandestine services. Women were frequently used for such purposes. The text suggests that a king should seek to make enemy aristocrats 'infatuated with women possessed of great beauty and youth', who could then manipulate those enemies according to the king's wishes.[265] The Mauryan Empire also used female assassins called *Visha Kanya* who used poison to kill their targets.[266] The Mongols, Hebrews and Egyptians all had clandestine services. The Aztec Empire utilised *pochteca*, a group of travelling commercial professionals similar to Rome's *frumentarii*, who would gather intelligence for rulers and often became very wealthy in the process. The Aztecs' *Quimitchin* were secret agents who could speak local languages and don local dress to infiltrate targets the empire planned to invade.[267]

The United States controls more wealth than any other empire in history; its network of spies is, accordingly, the largest. Today, the clandestine network in the United States sprawls across agencies like the Central Intelligence Agency, the Federal Bureau of Investigation, the National Security Agency (NSA), the Joint Special Operations Command, the intelligence divisions of the five military branches, the Department of Homeland Security, and the Bureau of Intelligence and Research, which number among 1,271 government organisations and nearly 2,000 private companies operating in about 10,000 locations across the country.[268] This network falls under the umbrella of the 'Intelligence Community' (IC), a loose entity formally created in

1981 under Executive Order 12333 by Ronald Reagan.[269] Over the past decade, the Pentagon has recruited some 60,000 people, both soldiers and civilians, to perform domestic and foreign operations under false identities, some embedded in top institutions around the world. This 'secret army' constitutes the largest known undercover force to ever exist.[270] There is no reliable way of tracing what the Pentagon is doing with this secret army or how exactly it is spending money. The Department of Defense is the only major federal agency unable to pass an audit, with billions of dollars unaccounted for, mysteriously disappearing into an ominous 'war budget' (officially the Overseas Contingency Operations account).[271] National security expert William D. Hartung writes, 'Without a reliable paper trail, there is no systematic way to track waste, fraud, and abuse in Pentagon contracting, or even to figure out how many contractors the Pentagon employs.'[272]

The Intelligence Community has ample experience opening or maintaining frontiers of parasitic growth throughout Latin America, with many confirmed cases of direct political involvement on behalf of US corporate assets and interests. Most recently, the US has been involved in attempted coups in Bolivia and Venezuela, in the former case to maintain supplies of lithium. The Intelligence Community has a long confirmed list of places where they've honed their blades helping to maintain the dominance of growth, capital and US economic interests, including Afghanistan, Albania, Angola, Argentina, Bolivia, Brazil, Cambodia, Chad, Chile, Congo, Cuba, Dominican Republic, El Salvador, Ghana, Guatemala, Indonesia, Iran, Iraq, Italy, Laos, Nicaragua, Poland, Syria, Vietnam and elsewhere.[273] High-profile leaks have shown the US Department of Defense, primarily the NSA, surveilling large swathes of the globe, across not only Asia and South America but also Europe, including its allies, and the United Nations, theoretically totalling 193 countries.[274] Their surveillance includes mass spying on US citizens on US soil.[275] It also has extensive experience targeting domestic populations for more than surveillance. Acknowledged past programmes like MK Ultra, the Counter Intelligence Program (COINTELPRO) and Operation Sea-Spray targeted US citizens, while projects like Operation Gladio targeted the civilians of allied nations.

These are not conspiracy theories; they are well-known and well-documented operations conducted by clandestine services against domestic populations, with extensive paper trails.

The executive order by which President Reagan formally created the Intelligence Community offers a clue to the IC's structural significance. The document is an interesting artefact of the neoliberal period, and the global networked parasitism it was meant to sustain. The language of the order places US corporations only after the government – and before US citizens – as subjects to be protected by the Intelligence Community: 'Special emphasis should be given to detecting and countering espionage and other threats and activities directed by foreign intelligence services against the United States Government, or United States corporations, establishments, or persons.'[276] When we look at the countries the Intelligence Community targets and the governments it seeks to overthrow or manipulate, they are invariably states possessing resources that the US government or US corporations seek to access, not ones that constitute any particular threat to US citizens. But capturing resources is not its only role.

The Intelligence Community acts as a psychological deterrent to both domestic and foreign threats to growth-as-progress parasitism. Much of its conduct appears irrational when viewed as a strategy for national defence, or even as a counter to communist regimes or foreign terrorists, which was its original claim to legitimacy. What do testing bioweapons on the population it is meant to protect, conducting expensive interrogation techniques that its own internal research has found to be ineffectual in gathering intelligence,[277] and dumping resources into fantastical 'mind-control' experiments have in common? They present an openly devious face to the world. Recent CIA director Mike Pompeo publicly bragged, 'We lied, we cheated, we stole,' and former director Michael Hayden boasted, 'We kill people based on metadata.'[278] Such behaviour would appear counter-intuitive and counter-productive for an agency formally tasked with quietly carrying out clandestine strategic data collection. One might expect such an agency to be more *covert*.

But when viewed not just as a national security apparatus but instead as a force for maintaining primarily fear-based psychological adherence to parasitic economies both domestically and abroad, its excesses, cruelties and projections of elite competence begin to appear more rational.[279] When viewed as a terroristic force, one whose primary purpose is to instil fear in those who may be tempted to question the imperatives of growth in general, or transgress the norms of capital and neoliberalism in particular – primarily those like radical environmentalists and leftists, who are often the IC's main targets – its actions make more sense. Intimidate, divide, conquer.

They are open about these roles and seem to relish their performance as spooky operators. In 2022, the US Army's 4th Psychological Operations Group published a short film recruitment trailer for their group titled 'Ghosts in the Machine'.[280] Col. Chris Stangle, commander of the 4th, boasted to *Task and Purpose* that 'psychological operations are occurring "literally everywhere, every day, in every component of our lives"'.[281] The trailer reinforces this notion. At one point, text appears over a series of both archival and highly stylised images, asking, 'Have you ever wondered who's pulling the strings?' and accompanied by an archival audio clip of Reagan's 'tear down this wall' speech and the image of a conductor. 'A threat rises in the east' occurs over vaguely communist marching armies and news reports of the Russian invasion of Ukraine. 'Anything we touch is a weapon,' the text notifies viewers ominously. Then: 'We can deceive,' followed by a series of more benign verbs, ending with 'inspire'. 'We are everywhere.' But, more specifically, 'You'll find us in the shadows at the tip of the spear,' mixing clichés with abandon and no sense of irony that a covert force in the shadows is not meant to be found. Though, in the case of *these* forces, it seems they are indeed meant to be found, to be seen, and that we are meant to know vaguely what they are doing. 'Warfare is evolving,' the text informs us, over the conductor again, 'and all the world's a stage.'

The IC's domestic function as a form of social control becomes clearer when we notice how many institutions it has infiltrated, perhaps the most insidious being corporate media. Veteran journalists Matt Taibbi and Glenn Greenwald have both documented Intelligence Community interference with the media. Taibbi reports employees from many major media institutions collaborating with the IC, while

III SYSTEM

Greenwald connects the Trump-era 'Russiagate' conspiracy with the ever-closer relationship between corporate media and the security state.[282] Politico, meanwhile, has reported on the new glut of former spies who are now paid commentators on the biggest television news shows. The aforementioned Michael 'metadata murder' Hayden, for example, is a regular paid contributor on CNN.[283] Neither is world media safe for impartial reporting: recent leaked documents have shown Reuters and the BBC participating in secret UK government programmes to disseminate propaganda targeting Russia.[284]

The US military in general, and the Intelligence Community in particular, have served the same purposes as imperial intelligence forces of the past: to maintain adherence to extractive and expansionist parasitic economies. The post-war period enshrined growth more explicitly and intensively than ever before, and the spear point used to enforce it grew proportionately larger and more intensive than ever before. Militarisation, incarceration and surveillance have increased dramatically during the neoliberal period. The United States has by far the largest prison population and highest imprisonment rate in the world. The number of people under 'correctional supervision' in the United States today – which means either in prison, in jail, on probation or on parole – exceeds the number of people caught in the Gulag system at its height (over 6 million people in the United States, compared to over 5 million in the Soviet Union).[285] The United States' correctional population peaked at over 7 million people in 2007, the equivalent of one in every thirty-two adults in the country.[286] The increase in mass incarceration can be seen as an intensification of tendencies rooted in the nation's history, but imprisonment in the United States has more than tripled between 1980 and 2010, the height of the neoliberal era.[287] Prison populations in the United States began to explode in the early 1970s, around the time neoliberalism became the dominant ideological basis for policy.[288] So much for Hayek's ever-increasing freedom.

Meanwhile, Cold War hysteria and neoliberal governmentality combined with new technological capacities for surveillance – like smartphones and satellites – to produce tightly controlled borders and mass imprisonment. As neoliberalisation made it easier for capital and corporations to move around the globe, it has simultaneously made it harder for people to cross borders.[289] As geographer David Harvey has argued, 'militarisation abroad and at home inevitably go hand in hand, and . . . international adventurism of the neoconservatives, long planned and legitimized after the 9/11 attacks, had as much to do with asserting domestic control over a fractious and much-divided body politic in the US as it did with a geopolitical strategy of maintaining global hegemony through control over oil resources.'[290] Is genuine progress, away from parasitism, possible with such forces on patrol?

Whether anyone – including the ghosts of Keynes, Marx or Hitler – will intercede in the parasitism of the global economy remains to be seen. These frontier guardians like the Intelligence Community have aggressively targeted those who have most explicitly sought to interrupt parasitic economic development: environmentalists and Indigenous land and water protectors. In November 2019, Britain listed the non-violent environmental organisation Extinction Rebellion among 'extremist ideologies' that should be reported to the UK's counter-terrorism programme Prevent in a guide 'aimed at police officers, government organisations and teachers who by law have to report concerns about radicalisation'.[291] Extinction Rebellion appeared alongside neo-Nazi and Islamic terrorist groups. The guide warned recipients to look out for people, including school strikers inspired by Greta Thunberg, who speak in 'strong or emotive terms about environmental issues like climate change, ecology, species extinction, fracking, airport expansion or pollution'.[292] In addition to targeting children and people worried about climate change, the police have also been tasked by the government with targeting those considered to be protesting too loudly: the 2021 Police, Crime, Sentencing and Courts Bill in the United Kingdom introduces ten-year prison sentences for protest activity that constitutes a 'public nuisance', which includes any act from making too much noise, damaging a memorial or marching slowly, to hanging off a bridge. The government policy paper for the Bill specifically cites 2019 Extinction Rebellion and 2020 Black Lives Matter protests as reasons why this new legislation has been created.[293] The first arrests under the Public Order

Act 2023 have recently begun, targeting climate activism group Just Stop Oil. A 57-year-old activist was sentenced to six months in prison for participating in a slow march. As the *Guardian* reported, 'A spokesperson for the campaign said: "Section 7 of the Public Order Act 2023, a law drafted by the fossil fuel lobby, was introduced in April by Priti Patel, and covers 'interference with the use or operation of key national infrastructure'. It seems this government has now made walking down the road, walking on the public highway an illegal act that is worthy of imprisonment".'[294]

And yet, the Intelligence Community assure us they are contributing to progress – and not just growth-as-progress parasitism, but progress in social justice, too. Feminist icon Gloria Steinem used her prominent position to promote the Central Intelligence Agency, where she worked as an agent in the 1950s and 1960s, at the height of some of the Agency's most heinous crimes (that we know of). 'In my experience,' Steinem told the *Chicago Tribune*, 'the Agency was completely different from its image; it was liberal, nonviolent and honorable.'[295] In 2021, the CIA ran a much-pilloried advertisement touting its progressive credentials – minority racial representation, for instance – throwing social justice buzzwords around with little sense of its own humorous excess.[296] In 2019, four of the nation's top five military contractors were run by women, a fact hailed by many as a triumph for gender equality.[297] Raytheon – a company whose weapons have recently been used to kill civilians in Saudi massacres in Yemen[298] – authored a press release titled 'Raytheon named "Best Place to Work" for LGBT equality for tenth straight year'.[299] Meanwhile, the CIA and the US military-industrial complex have unleashed devastation on people of colour, women, LGBT people and children all over the world for decades, in some cases impacting those groups disproportionately. Avril Haines was appointed by the Biden administration as the first woman to serve as Director of National Intelligence, the role to whom CIA and the other Intelligence Community members report. She both broke this glass ceiling and was at the same time centrally involved in covering up past uses of torture by US forces, and integral to making the rule that drones could be used to assassinate US citizens without due process.[300]

Crackdowns on protestors, activists and environmentalists have already accelerated, and there's good reason to believe that as conditions grow worse, and people more desperate, these forces will become even more aggressive. The Intelligence Community has already targeted with impunity the citizens it is ostensibly intended to protect.

When it's not contributing to the suicide of celebrated authors like Ernest Hemingway, the FBI can be found infiltrating and sabotaging activist groups, aiming to smear and discredit their members, and to sow doubt and discord among them.[301] One of the most notorious examples of this was the aforementioned COINTELPRO, a broad and illegal operation lasting from 1956 to 1971 that infiltrated activist organisations, officially in order to disrupt the US Communist Party. The Red Scare, meanwhile, shattered many lives and, combined with cultural neoliberalism, has contributed to the breakdown in working-class solidarity over the last half century. As Astra Taylor notes in *The New Republic*, 'anticommunism is responsible for the deaths of millions of people worldwide.'[302] And as Branko Marcetic writes, 'Just as military equipment and technology is routinely used today against US citizens, the FBI turned the wartime tactics it had used against the Soviet Union on domestic groups.'[303] But the FBI's targets included all manner of 'subversives' including feminist groups, pacifists, civil rights groups, Indigenous independence movements, environmentalists and animal rights activists.[304] Perhaps most infamously, the FBI targeted Martin Luther King, Jr. under COINTELPRO, using blackmail to try to coerce him to commit suicide.[305] Widespread methods used by the FBI included 'burglary, threats, kidnapping, infiltration of groups based on their political beliefs, murder, incitement to criminal activity, illegal wiretapping, intercepting letters and replacing them with changed content, and numerous other tactics (some of the most heinous of which the FBI has not publicly admitted)'.[306] One reason we don't see many of the arguments presented in this book questioning sacred national myths like progress is because those making them have been assassinated. Malcolm X, quoted at the beginning of the book and one of the most eloquent critics of the idea, suffered that fate. He was murdered by a former friend when he was only thirty-nine, but had he survived longer, there's good reason to believe he would have been sidelined by the state. The FBI was surveilling him and recent evidence has emerged suggesting that the NYPD and FBI may have conspired in his murder.³⁰⁷ A deathbed letter by a police officer claimed that he had been indirectly involved in Malcolm X's murder, and was 'told to encourage leaders and members of the civil rights groups to commit felonious acts' by federal handlers. Such practices are still occurring.³⁰⁸

In 2020, the FBI was publicly lauded for thwarting a plot by a right-wing armed militia to kidnap the state of Michigan's Democratic governor, Gretchen Whitmer. A BuzzFeed investigation later revealed that the FBI participated in 'nearly every aspect of the alleged plot, starting with its inception'. Its informants organised and funded the militia's meetings, and gave them 'military-style training', even showing them where to plant explosives. The investigative journalists concluded that the level of FBI involvement suggests that the plot probably would not have happened at all without them.³⁰⁹ This is not the only time the FBI has manufactured terrorist plots. Branko Marcetic connects these tactics to FBI activity under President George W. Bush: 'the FBI's use of informants and undercover agents to effectively manufacture their own terrorist plots – typically by entrapping down-on-their-luck or mentally unwell Muslim men – which they then foiled and publicized, thereby justifying more money, resources, and powers for the "war on terror."'³¹⁰

And still, invocations of progress remain ubiquitous. Progress will follow growth, and growth will follow the development of new technologies and industries. But growth has been yielding declining dividends. No technological change is likely to ever deliver the leap in energy capture that new fossil fuel development did during industrialisation and then the Great Acceleration in the mid-twentieth century. Regardless of technology, thermodynamics will halt growth if ecological fragmentation fails to do so first. Still, parasitism on *a global scale* must end, and may end soon, because all such economies collapse, and this one will collapse. In that collapse, other forms of parasitism may emerge. Old forms like the kind that gripped Nazi Germany and the American West and Rome, or something new, some hybrid, some unimagined form, may rise. Until then, we are trapped by the ghosts on the borders of the frontier, keeping us under surveillance, under thickets of arbitrary bureaucracy, under debt and incarceration. And we seem to have little to accompany us but still other ghosts. Now is the time to imagine and then implement something different.

Coda

The twentieth century and, so far, the twenty-first have maintained a human-ecological relationship of global networked parasitism. In place of gods, and in place of nations, systems contained in ideologies became the vehicles for progress, the ground on which utopian ideals were built. The twentieth century was notable for the rapid expansion of the human world, for the invention of ever more complex technology, but also for famine, genocide and the most destructive wars ever fought. The ideologies born in this era failed to deliver on most of their promises of progress. They dreamt of speed and paradise among the stars, they dreamt of the bloody soil of their fantastical race, they dreamt of industrialised perfection, the perfection of society, of the soul, of the human body. They delivered death and turmoil.

If you have read this book from the beginning and reached this point, I commend you. It has been grim and heavy. We are nearing the final sections. I want to promise you a hopeful end. Doing so, however, would not be in the spirit of the book. Instead, I will deliver a further challenge. For the dark facts in this book are just some of the ones that professional researchers have been paid to investigate. There are far too few researchers and they are paid far too little to understand a world so vast, and phenomena so complex. There are surely challenges currently hiding that will show themselves in the future, or evils that will only become clear in retrospect. The wise, rational and cautious course of action would be to act as if these problems are serious and in need of urgent attention. But even as weather patterns grow increasingly unstable, as scientists voice daily and urgent warnings, and multiple compounding, overlapping and consecutive crises reveal the weaknesses in this mode of living, there are few large-scale, mainstream efforts to address any of the challenges with which we are presented. There are none that are commensurate with the challenges themselves. For those who do not wish to be trapped in an increasingly hot glasshouse, packed with screens and armed gangs, cameras and smoke from wildfires, and always more car exhaust, what would it look like to pursue a non-parasitic mode of human ecology and political economy? What does it take to achieve?

After Progress

For most of my life, I have participated in activism or social reform in some capacity. I've done so because I believed earnestly in the possibility of a future better than the present or the past. I still hold some qualified version of that belief. But the more I have learned about the world, the more that what I imagine as both good and possible has changed. When people are asked to consider the possibility that the future will be worse than the present, and not only worse but potentially horrific, bloodier than anything that has come before, they often evict such a notion from their imaginations. They come up with sci-fi fantasies, or they descend into fearful deliria about imaginary enemies, or they simply avoid any further corroborating information. I can understand and sympathise with those reactions.

We are caught in a difficult trap. If everything that is familiar is torn down and all the structures that govern our day-to-day disintegrated, we risk terrible disorder. We court famines and wars. We invite power vacuums to be filled by even more brutal psychopaths than those who haunt the halls of power now. But if we don't, if we continue on the current path and simply follow inertia, there is a good chance that the outcome will be far worse than the disruption of upending everything today. Maintaining status-quo trajectories in carbon emissions, habitat destruction and pollution, there is a high likelihood of collapse in the existing structure anyway. It will just occur under far worse ecological conditions than if it were to happen sooner, in a more controlled way. At least, that is what all the best science suggests. To believe otherwise requires rejecting science and knowledge itself, which some find to be a worthwhile trade-off. But reality can only be denied for so long. Dream at night we may, the day will ensnare us anyway.

I do not want to suggest that there is nothing to be done. When we do face the facts, there is much we must do. Much to do particularly given that, as I argued at the outset of this book, *never before have so many lives, human and otherwise, depended on the decisions of human beings in this moment of history.*

What is to be done? We must, simply, halt and dismantle the parasitic human ecology in which most of the world lives. In the past, there were myriad forms of human groups. In the present, our task is to lift the weight of parasitism and progress so that in the future, myriad forms may flourish again. Those of us who live in parasitic economies or contribute to their maintenance are directly involved in the problems they cause and so must be involved in their disintegration, even if we have limited tools for achieving that.

If that is what we must do, the harder question is *how?*

As a starting point to answering it, we must explore what a non-parasitic mode of living may look like and has already looked like. This section will examine the vital elements that have been common to non-parasitic human systems for most of human history, and will aim to reconcile those examples with the world as it is in the early twenty-first century and is likely to be in the next. Then we will use two methods for approaching answers to the questions of *what should we do*, and *how should we do it?* One is by examining the big actions that must happen at a collective, systemic level. The other is by confronting how we as individuals engage in those sorts of large-scale actions, which we can only do on an individual and day-to-day basis, since we are simple primates of the great ape variety, living out our precious time hour by hour.

Economies of Reciprocity

The Great Lakes region has been home to many cultures since the last glaciation ended, the ice retreated and formed what are now predominantly called Lakes Michigan, Superior, Huron, Erie and Ontario, and the land masses in which they are nested. The first people to make these regions their home most probably arrived at least 10,000 years ago.[1] These earliest peoples were likely part of the waves of migrants from Asia who crossed the land bridge at Beringia, between Eurasia and North America. They cultivated crops that included corn, beans, peas, squash and pumpkins, among others, and were supplemented with game, fish, and wild rice and fruits. They were nomadic and moved villages periodically, once or twice a generation, as resource limits required. Though they may have been depleting these regions of human-use resources to some extent, they were not causing major extinction events nor intensively, exclusively cultivating human-use crops. Their habitats were able to recover. Scholars Kyle Whyte, Jared L. Talley and Julia D. Gibson have pointed out that 'Indigenous peoples such as the Anishinaabe interpret their history as involving constant migration and transformation, diverging from conceptions of periods characterized by static or stable conditions.'[2] In the same way, these modes of production and human ecology have been fluid through time as well. We can see the commensalism practised by these early migrants as if on a spectrum, located towards the centre, between mutualism and parasitism. This principle is important in guiding how we imagine alternative forms of human organisation, avoiding apparent binaries like sustainable-unsustainable or capitalist-communist or individualist-collectivist. But how do we find the optimal place on this continuum of mutualism-commensalism-parasitism? Where might that place be?

If humans can be said to have a 'natural' state, it is to reside in groups, as much as any school of fish or band of monkeys.[3] We are a society-building species as much as ants and termites are hill-building. Contrary to Hobbes's notion of war of all against all, if you were to toss a hundred individual humans onto an island, all spread out and alone, they would soon find each other and start a society, or, more likely, several. When we speak of human cultures and societies, we are speaking about cohesive assemblages of people of any size, often containing multiple kin groups (so, something bigger than a single immediate family). The number of human cultures and societies that have existed over the past 300,000 years is uncountable, but, extrapolating across many times more millennia from those that have been recorded over documentable history, the number is almost certainly in the tens of thousands. The boundary around what constitutes a distinct culture is not always clear, in the same way that the boundaries between species are sometimes fuzzy and debatable. Cultures, like species, can intermingle and create something new, changing through time in ways that are adaptive and responsive to environmental pressures and are unique to their own internal dynamics, or they can emerge spontaneously and they can disappear forever.

Many of these cultures constitute what I am referring to as '*calculated* commensalist' societies because they engage in concrete energy capture that is non-harmful to the ecologies they inhabit. They do this not in an unthinking, inevitable sort of way, but in a calculated way. That is, they develop strategies for integrating their societies with ecological systems, cultivating those strategies over many generations, honing and refining them over time. This integration entails capturing energy through methods like hunting and catching wild animals, foraging wild plants, nuts and fungi, cultivating domesticated or semi-domesticated organisms in a settled, semi-settled or nomadic way, and semi-wild herding, where people corral or follow migratory ungulates or generational migration patterns.

Although the evidence is slim, given what we know about existing Indigenous peoples and clues from past ones, it wouldn't be unreasonable to assume that for about 99 per cent of our roughly 300,000-year history, the large majority of societies humans have built have been generally commensalistic. Although the societies and cultures that have flourished in that time have been diverse and widely varied, a few patterns emerge from the archaeological and anthropological records such that we can make some assumptions about these sorts of societies.[4]

As we have seen in examples throughout this book, in societies that practise concrete energy capture in a calculated commensalist way, abstract energy capture tends to be similar. That is, the relations between individuals within these societies tend to be either commensalistic or mutualistic, built on reciprocity and non-harmful extraction, and done so in a way that is calculated, honed and refined over generations. The result for life within those societies tends to be a high degree of continuity between nature and the social structures in which life occurred. There are nine primary beneficial outcomes that are worth considering here that have been typically consistent qualities of this sort of society.

1. **less disease:** given lower population densities, no close habitation with livestock, and coevolution in ecological systems, disease burdens tend to be low in such societies, except in places where wild disease vectors have been endemic. They enjoy very low incidences of cancer, heart disease and stroke.[5]
2. **more leisure:** given that such societies need not capture energy as intensely, daily commitments to energy capture tend to be low, as little as an equivalent to 'part-time' work today, or twenty to thirty hours per week, giving people far more time for leisure.[6]
3. **more connection to others:** without the alienating forces of class, unearned hierarchy, warfare and economic competition, it is easier for people to forge strong connections, particularly in smaller and more long-lived groups, with one another.
4. **more connection to nature:** by definition, commensalistic societies must understand the ecological systems they reside in with great sophistication, and integrate their production systems in a way that facilitates close contact with wild plants and animals and maintains biodiversity. The result of this is a deep and sacred relationship with other life.
5. **sometimes more food:** though there can be variability with ecological conditions, typically such societies are able to enjoy natural abundance of food and resources, or are able to move and adapt in times of greater natural scarcity.[7]
6. **less violence:** despite false narratives that have been published on the high levels of violence in Indigenous or prehistoric societies, commensalistic societies in fact tend to be much less violent and do not engage in systematic warfare, which arose with parasitism and large settled states.[8]
7. **more freedom:** as this book and many others have shown,

commensalistic societies have tended to highly value individual personal agency and liberty, and build political systems that enshrine universal representation in decision-making processes.
8. **more equality and fairness:** there tend to be few hierarchical distinctions across variations in human life. That is, freedom and agency are distributed evenly, not withheld from certain groups, and a sense of fairness is embedded. Any natural inequalities are rewarded and mitigated by means of cultural rites and processes.
9. **more cultural diversity:** many more types of cultures, styles of political organisation, languages, arts and histories existed when there was a plurality of commensalistic societies. Diversity in cultures has reduced with the spread of parasitic cultures and empires.

These qualities are not guaranteed in any given society. But they are likelier to be accommodated in societies that practise economies of reciprocity, what I am calling calculated commensalism. The tragic irony is that many of the ecologically sound societies that have been eradicated or greatly curtailed and oppressed by parasitic progressive societies themselves represent a kind of perfect state that so many of the empires that displaced them claim to be pursuing. They achieved, through multigenerational struggle and adaptation, many of the qualities citizens of these states would think of as utopian: low disease rates, low violence rates, high caloric abundance, extensive and equally distributed leisure time, lots of community connection and entertainment, and challenges that bring meaning rather than despair. The *end of history* of prosperity and freedom and the pursuit of happiness was achieved thousands of years ago, over and over again, and over and over it has been destroyed by the very societies that claim to be seeking to achieve it. Of course, those now-marginalised ecological societies may not have seen themselves as 'progressive', or maybe they have at various times. But their pathway to those qualities of freedom, equality, leisure, abundance and peace was not achieved through the conquest of frontiers and the hysterical extraction of life, but through ecological values and knowledge like those detailed at the beginning of this book, through experimentation, through inquisitiveness and courage, and through the willingness to fight for these things rather than give them up to someone who promises them in the future while forever withholding what they had once enjoyed in the present.

Emphasising cultural diversity – that not every alternative iteration of human systems must look the same, that those systems should reflect the biodiversity of the land they inhabit – and forms of energy capture that are commensalist is a first step in imagining large-scale alternatives to the parasitic empires currently ruling the twenty-first century. Progress in this way comes to mean not an intensification and continuation of the parasitic lineage but a fundamental break from it, and the adoption of both very old and new forms of energy capture based on calculated commensalism. We cannot wish ourselves into a world that existed thousands of years ago, however, as it has been destroyed and we live in an entirely different world. Even those still practising commensalist modes have been forced into this very different world. How might we reconcile such principles with the world as it is in the twenty-first century and beyond?

Cities and Farms of Wilderness

The underlying assumption of progress ideologies has been that ever-increasing urbanisation is a human good, particularly the extent to which cities banish other forms of life that have no apparent market utility. Unfortunately, the result of these ideologies has been to fill the world with cities that are ugly and brutal even to their human inhabitants, and much more so to all animals that are not flies and rats and all plants that are not ornamental.

Like the dominant religions, even the most ancient cities are new in the history of our species. The most robust cities have transformed like rivers, shifting over the land they cover, during the last 9,000 years. Cities will undergo massive, rapid changes in the twenty-first century and beyond. Ninety per cent of cities are coastal, and, whether through sea level rise or more intense storms and flooding, every single one of those cities will be impacted.[9] Non-coastal cities are also at risk from climate-related disasters like extreme forest fires, heatwaves unendurable for the human body and infrastructure alike,[10] water shortages, blackouts, and pestilence from increased pest populations and breakdowns in sanitation.[11] In those cities under siege, many will die, quality of life will decline and violence will spread.

Naturally, people will flee – 1.2 billion people by 2050, according to recent estimates.[12] The mass population transfer that will occur in this and subsequent centuries poses an urgent imperative to rethink how we design urban spaces, and to do so on better principles than those that are currently consuming the world to death. This great resettlement need not be slipshod and chaotic; many of the effects of climate breakdown – like sea-level rise and climate zone shifts – occur slowly enough that this movement *can* be more planned and controlled than the migration that occurs as a result of rapid crises like war, invasion or volcanic eruptions (although some climatic impacts are not slow, like forest fires and hurricanes). At the very least, slower-manifesting emergencies may offer openings to shake inertia-bound cities out of their morbid torpor and introduce a new vitality. While it may be too optimistic to presume such a rosy picture as a matter of course, people *are* beginning to think more deliberately about where and how to live in the light of anticipated impacts. This crisis, in some ways, widens the scope of how we envision the city we want now, today. This is important not just because cities are already doomed to the changes, like sea-level rise, that are at the moment likely inevitable, but also because current cities are not working. Most are needlessly inhospitable.

Cities currently account for at least two-thirds of all energy consumption and carbon emissions.[13] Most cities built in the industrial era have been designed to be dependent on fossil fuels, both to construct and sustain them. They have consequently reflected the priorities of the ideologies that built them: primarily, fossil fuel-based parasitism and markets that have no concern for human or animal well-being. The modern city is a fossil city. Fossil cities have produced conditions that are not merely unpleasant to reside in, but are actively harmful to physical health and well-being, deadly for non-human life, and impossible to sustain without dense energy infrastructures, which themselves are impossible to sustain without cataclysmic consequences like ecological disintegration.[14] Cities that are not built on parasitic principles and dependent on fossil fuels, or that are reformed along principles that reject such logics, can offer a much improved way of living in the world, both for humans and other life: an existence not at the edge of bare endurance but thriving.

Cities with more green spaces, like parks and wild zones, green walls and roofs, and street trees, are healthier than cities with less green space.[15] Green spaces can dramatically mitigate heat-island effects in cities and cool them to an extent that would save many lives and make them more pleasant in a hotter world. Furthermore, many city dwellers suffer nature deficit disorder. Creating zones in urban areas that are habitable for wild plants and animals could improve human (and non-human) physical health. Wild urban areas produce healthier microbial life, which positively affects human health in complicated ways, primarily by enhancing immune function.[16] In addition to more wild spaces, parks and green buildings, urban agriculture can play a role in greening cities. Urban agriculture represents a tiny proportion of food produced today, so there is great scope for expansion and the practice can play an important role in reclaiming more space for food production (and less for, say, cars).[17] Neighbourhoods that produce their own food and energy can also benefit from greater self-sufficiency, civic participation, and a certain sense of solidarity that these sorts of collective projects uniquely yield. Plant-rich cities can begin to integrate existing infrastructure with the wild and rural life outside them.[18] This relationship between food and energy production and self-sufficiency shows the interrelated dynamic between material and cultural shifts. Cities are already sites of energy transition, with many focusing on ways of incorporating energy production, like solar-studded roofs.[19] Distributed, community-run energy can be one important way of powering cities (and towns and rural areas) while maintaining local revenue and ownership. Creating a whole civilisational network in which energy is derived this way would make it more likely to be democratically controlled and resilient to climate impacts.[20]

So how do we begin to transform cities in these ways, replacing roadways and cars with parks and trees, swapping ugly towers and boxes with elegant buildings pleasing to the eye and able to produce the food and energy we need where most people need it? Given the extent to which material and cultural change are interwoven in the timeframe we're looking at, it's worth saying a word about how such cities are governed, since their structures of governance are shaped by the physical structures that make them up, and vice versa.

The tale of civilisational progress generally situates cities in a history of hierarchical states. The primordial city is mostly thought of as a gift bequeathed by the glory of kings and conquest, the slave-cut gems set in the crowns at the core of empires; with their bustling agoras, menacing pillars and fetid markets, they can only rise from the spoils of war and subjugation. As mentioned, British archaeologist David Wengrow has convincingly challenged this narrative, suggesting that some of the earliest cities were not the result of empire, but instead were far more egalitarian and free. Such cities emerged outside the frontiers of expansive states and were not necessarily bound by their logic; as such, there is no iron law suggesting that cities must inherently stratify into hierarchical layers for their governance.[21]

But cities erected or governed by fossil empires are neither free nor equal. Of the world's urban population today, nearly a quarter – about 1 billion people – live in slums.[22] Luxury high-rises abut sprawling networks of shacks clad in corrugated aluminium, strewn with mud and trash, where the world's urban poor and workers must battle pollution, heat, disease and destitution. Industrial and transport pollution causes dangerous levels of toxicity in air and water – and thus birth defects, miscarriages, lower life expectancy and early-childhood illnesses – that impact everyone living in urban spaces. If cities continue to follow present trends, the proportion of people living in slums will increase and cities will become epicentres of armed conflict, impacting how new cities are built and existing ones reformed. The US military is anticipating this: the Pentagon developed a leaked internal video that presents a chilling, dystopian vision of the future of urban life.[23] In it, a dramatic voiceover declares the need to militarise cities and deploy elite units to crush networks of gangs, terrorists and the increasingly angry working poor: 'this is the world of our future . . . and it is unavoidable.'

Geographer Mike Davis writes:

Thus the cities of the future, rather than being made out of glass and steel as envisioned by earlier generations of urbanists, are instead largely constructed out of crude brick, straw, recycled plastic, cement blocks, and scrap wood. Instead of cities of light soaring toward heaven, much of the twenty-first-century urban world squats in squalor, surrounded by pollution, excrement, and decay.

He muses further that 'the one billion city-dwellers who inhabit postmodern slums might well look back with envy at the ruins of the sturdy mud homes of Çatal Hüyük in Anatolia, erected at the very dawn of city life nine thousand years ago.'[24]

Only cities built and run on principles that are democratic and participatory, focused on liberty, egalitarianism and mutual integration with ecologies – that is, that are commensalistic – will produce an antidote to the spreading scourge of slums and skyscrapers, and halt the grim future the Pentagon deems unavoidable. Challenging planning decisions and dismantling existing infrastructures that cement economic castes into the fabric of the city is necessary, from highways carving up neighbourhoods, to toxic landfill sites, to extraction processes adjacent to the homes of many. The causation in this intervention is non-linear: people inside and outside those marginal zones are shaped by city governments, their rules, norms and laws, which in turn are shaped by those residents. The infrastructure of the city impacts their participation, and is itself shaped by their participation. The design process requires interventions both cultural and material. At the cultural level, there is simply no alternative to individuals and groups intervening in the political process of governing cities and making planning decisions. Citizens must disrupt and interject themselves into that process, and overpower opposition. Precisely how will depend on the details of each locality.

On the more material side, one significant and relatively easy, low-cost intervention is to set aside space for wildlife and then introduce that wildlife to urban, suburban and rural areas. 'Rewilding', or the process of reclaiming land for wildlife, is broadly popular. Some of the most important animals to reintroduce in the rebuilding of healthy wildlife are predators, and, unfortunately, many people retain an irrational fear of predators. For example, as of 2020, only 36 per cent of the British public would like to see wolves and lynx reintroduced, while just under a quarter would welcome bears back to the wild.[25] The reintroduction of such predators would have positive ecological and humanitarian effects: reintroducing over a hundred grey wolves in the United States has had tremendous benefits in Yellowstone National Park, where they have played an extensive, vital and well-documented role in restabilising ecosystems and bringing biodiversity.[26] Reintroducing predators like eagles, hawks, wolves and lynx would almost certainly *save* human lives, as they would prey on many of the deer and mice that carry dangerous disease vectors like ticks, which are increasing in number and range across Europe, in addition to the broader health benefits of retaining more intact ecological systems. While the public is skittish, it is primarily those with economic interests in controlling land and wildlife, like avaricious livestock businesses and generational landowners, who actively move the levers of power to prevent predators from living here. The moral problem with preventing these species' reintroduction is the simple fact that they have as much right to inhabit this island as humans, or perhaps even more right given that they were here first: bears, lynx, wolves and countless other species, like pine martens, wildcats, wild pigs, and dozens of species of birds.[27]

Fear of predators, moreover, is deeply irrational. There are no documented cases of lynx killing people, nor even any anecdotes or rumours of lynx killing people. The threat they present to livestock, meanwhile, is low.[28] In the United States, where there are many more wolves, there are also no documented accounts of humans being killed by wild wolves during the entire twentieth century.[29] In that time, there have been only sixteen cases of wolves biting humans and none of those bites were life-threatening. By contrast, domesticated dogs kill around thirty Americans every year (not counting rabid dogs, where the number is closer to 25,000 deaths annually around the world).[30] Domesticated dogs may offer some benefit to human beings, but none to ecological systems, and are instead taking biomass from wild dogs who do have great value to ecological systems – and, by extension, to far more human beings than individual house dogs.

Bears, the most feared animal of the group, also rarely harm humans. For those 76 per cent of British holdouts who are afraid of bears, I will say that when I lived in New York State, I was taking a walk in the woods, minding my own business, when a black bear charged me, exploding from a nearby meadow, thick black legs devouring the sixty or so metres between himself and me in a few seconds. He (or she) stopped within about three metres of me, stood on his hind legs, sniffed the air and then bounded off into the forest. That moment was among the most intense I've experienced, standing dead still hoping he did not wish me harm, really fearing I could be killed, living at the mercy of a forest god. He scampered off, and soon after, he came close again further down the path; this time he seemed only to be playing. Even having met face-to-face with what at first had seemed to be an aggressive bear (and maybe was, I don't know what he was thinking) – or *because of* that experience – I would still very much welcome wild bears back to the forests of Britain. Little did I know then that I had almost nothing to fear. Black bears are rarely dangerous to humans; only sixty-three people have been killed by black bears in over one hundred years in North America.[31] The fact is, we shouldn't just pick and choose the species we feel comfortable with. That was the whole problem in the first place. Even large predators are at least as afraid of humans as we are of them. A sad study found that playing human voices in the wilderness caused animals to flee in fear; mountain lions and bobcats showed signs of terror at the sound of a nearby human being.[32] In reality, these species are much, much less dangerous, and far more beneficial, than the average car, dog or person, or the rapidly expanding populations of ticks and mosquitoes that carry debilitating diseases.[33] The progress narrative formula, as noted at the beginning, has sought to convince people that banishing wilderness leads to happiness and health. As we have found, the opposite is true, and we can restore some health and happiness by bringing these creatures back.

* * *

Like cities, progress ideologies have cast agriculture as a force for improvement to the extent that it strips life out of land and replaces it with market-valuable commodities. The very simplistic progress narrative has claimed that more and bigger farms, growing more – but not necessarily better – products equals progress. The results of this attitude have been catastrophic. Today, agriculture is a massive source of greenhouse gas emissions, globally accounting for nearly 20 per cent.[34] The methane emissions from synthesis of ammonia fertilisers are a hundred *times* higher than reported by the industry.[35] Methane, meanwhile, is dozens of times more potent in capturing heat than carbon dioxide. Industrial agriculture is the greatest cause of the deforestation and habitat destruction that's responsible for the greatest extinction since the dinosaurs.[36] Beyond the environmental harms, which are already catastrophic, industrial food processes have a lot of other harmful effects. The food system in the United States sickens around 48 million Americans per year.[37] Chronic illnesses, like diabetes, heart disease and cancer, often caused by unhealthy foods, are among the country's greatest killers.[38] The air pollution caused by growing corn alone kills about 4,300 Americans per year.[39] The consolidation of agriculture into a few megacorporations is leading to an epidemic of suicide among smaller farmers. Farmers kill themselves at a rate double that of veterans, a rate higher than any other occupation in the United States.[40] India has often been touted by technological fetishists as a great success story in the industrialisation of agriculture. One of the results is that an average of twenty-eight Indian farmers take their own lives per day.[41] Industrial agriculture destroys soil fertility far faster than it can be replenished, even with synthetic means. At current rates, this will almost inevitably lead to collapses in food supply this century.[42] As the best research shows, industrial agriculture is not only *not* a solution to the myriad environmental and humanitarian problems bearing down on us, it is in fact one of the major causes of those problems.[43]

Yet determining reasonable solutions to the problem remains a major challenge. The first and most fundamental way human-ecological relationships form is through the technologies humans develop to feed themselves. Ingestion is the most direct form of concrete energy capture, with one organism building its body out of the body of another. Perhaps the most vital front line in transitioning from parasitism to commensalism is how human systems get food. Aside from a few examples of urban agriculture, this process happens almost entirely in rural areas.

As geographer Mike Davis notes, urbanisation is a process that interacts with the rural world. People in the countryside are forced by economic circumstances to migrate to cities and, 'in many cases,' he writes, 'rural people no longer have to migrate to the city: it migrates to them.'[44] This process cuts people off from the ecologies they've inhabited, in some cases for many generations, even millennia, even as it leeches resources from those same people and ecologies. How polities and individuals design rural landscapes is inextricable from the challenge of urban design. Given that so much of the human world is now a network of interconnected metropolitan zones, knitted together by road systems and waterways, and woven into the rural areas from which they extract, it is reasonable to aspire to future settlement patterns in part as a matter of intentional design, even if that process generally occurs sporadically and organically. As with urban design, there are a few goals and principles that should guide which land policies are implemented.[45]

Human population density today is dependent on equivalently dense agriculture. Human beings are a large mammal. No other large mammal, or at least no omnivorous predator, has achieved such population volume and density, and it would have been impossible to do so without organised agriculture. While there are 8 billion human beings, our next closest evolutionary kin, the genus *Pan* (chimpanzees and bonobos), number only 350,000 maximum.[46] As much as we might find joy and liberty from some sort of return to mass foraging, wild ecologies simply cannot yield caloric density sufficient to feed the numbers of humans now on the planet. Some form of agriculture is necessary while there remains anything close to this population size. But economic and state incentives have driven a process of consolidation in agribusiness. These incentives come from IMF loans and multinational corporations and from state subsidies like those that emphasise commodity crops and large consolidated industrial farms. Rural areas, both food systems and settlements, are heavily dependent on fossil fuels for transportation and production. Virtually every edible calorie consumed by someone in the industrialised world depends on many calories derived from fossilised carbon.

In the United States, packaged lettuce is grown almost entirely in two states – California and Arizona – and the process of getting a leaf from the farm to the rubbish bin can entail as many as thirty inputs of fossil fuels, from precision seeders, petroleum fertilisers and pesticides, to multiple trucks, trains and refrigerators.[47] And these are just simple leaves. More intensive processed foods require many other inputs, and meat production is heavily land, water and energy intensive as well. The clearest solution to this problem is agroecology, an agricultural technique that integrates food production into wild ecologies.[48] Agroecology forgoes the intensive fossil fuel inputs of synthetic pesticides, fertilisers and long-distance transport for growing locally, using organic fertilisers like compost and other waste, and employing more targeted, sophisticated and organic pest-killers like predators instead of spraying everything with poison. Generally speaking, agroecological farming makes food production more resilient to climate disruption – to impacts like less predictable weather patterns and invasive pests – while maintaining vital habitats for wild creatures. A paper in *Agronomy and Sustainable Development* states in terms quite unequivocal (particularly for scientists):

> The biggest and most durable benefits [for climate adaptation] will likely result from more radical agroecological measures that will strengthen the resilience of farmers and rural communities, such as diversification of agroecosytems in the form of polycultures, agroforestry systems, and crop-livestock mixed systems accompanied by organic soil management, water conservation and harvesting, and general enhancement of agrobiodiversity . . .

Observations of agricultural performance after extreme climatic events (hurricanes and droughts) in the last two decades have revealed that resiliency to climate disasters is closely linked to farms with increased levels of biodiversity.[49]

In an agro-market dominated by major corporations that are subsidised by wealthy governments, agroecological farms will be at a disadvantage until large farms are broken up and government incentives are shifted to support small and medium-sized farms that implement agroecological principles. While it's desirable for individuals and neighbourhoods to start up such projects in both rural and urban areas, concerted political campaigns to shift incentives and structural advantages are necessary. With agroecological farms serving local areas, food would not have to be transported halfway around the world; our reliance on the trucks and shipping vessels that make up precarious global supply chains would be reduced, further reducing carbon emissions. While some may lament the loss of far-flung foods available year round, eating seasonally and cultivating a greater diversity of local foods could offer delicious alternatives under an agroecological model. For example, there are over 2,500 species of apples grown today in the United Kingdom, but the two most popular varieties sold in supermarkets, Gala and Braeburn, are imported from New Zealand. An emphasis on local food production not only means saving on energy and carbon emissions, but exploring overlooked flavours.[50]

Such a shift would likely entail a greater proportion of the labour force working on farms. This isn't necessarily negative: a recent Gallup poll found that, when asked where they want to live, the largest percentage of Americans (27 per cent) chose rural areas.[51] Indeed, common career aspirations are to be one's own boss, to work outside or to work with animals, desires that converge in an economy of agroecologists. One way such a transfer of labour could be safely and equitably achieved, without the forced relocations that have proved so disastrous in the past, is through the implementation of a subsistence-level universal basic income (an income that, importantly, is not used as a weapon to dismantle other social safety nets).[52] Short of a society in which all basic necessities are simply supplied, a guaranteed income is a necessary precondition for abolishing forced labour. A subsistence-level version paid for by a wealth tax would be an easy way to transfer wealth from a few consolidated bank accounts and investment portfolios to whole populations – mitigating the urban violence the Pentagon is salivating over and freeing people from the tyranny of management teams who often hold the power of life and death over employees dependent on wages in order to afford basic necessities.[53] Combined with other incentives and subsidized agroecological education, this would be a way of allowing people the freedom to cultivate small farms and the means to move to or between rural areas to do so. Agroecology is sophisticated and requires lots of training, time and experimentation to master. Something like a guaranteed income could help to fund such endeavours at a mass scale.

In many places, rewilding can be implemented in concert with another vital change in rural design, one that is potentially just as powerful and replicable: giving land stolen from Indigenous peoples back to their descendants. Such a policy can be implemented in every single country across the Americas, to the Sámi in Scandinavia, to Siberian peoples in Russia, to peoples across the Global South. There is precedent for this; in 2021, the US Department of the Interior transferred 18,000 acres of the National Bison Range to the Confederated Salish and Kootenai Tribes (the land, it should be noted, was already within the boundary of their reservations).[54] Returning land that was stolen clearly contains a forceful moral component, every bit as urgent as righting the wrong of personal property stolen from someone who demands the return of that property. And, as mentioned in previous chapters, the crimes committed against Indigenous peoples around the world extend far beyond theft. This should be self-evident to anyone with basic moral intuition.

But there is also a practical benefit to handing the management of land to people who are, generally, better equipped to steward that land through what is likely to be an unstable Anthropocene. Most Indigenous groups are, or are most recently, descended from societies that practised commensalistic modes of production and human ecology – though, of course, they are not universally better equipped because Indigenous groups are never monolithic, containing within them disagreements and disparities in capacity. In the Lower Murrumbidgee Valley in New South Wales, Australia, over 200,000 acres of land has been transferred to the tribal council of the Nari Nari, the land's traditional inhabitants for 50,000 years. Under its new management, the old settler irrigation system was torn out and replaced with a more traditional water regime; when water began to flow again in 2018, 'species such as golden perch and southern bell frogs, along with spoonbills, egrets, black swans and other birds, grew more abundant'.[55] The Nari Nari have protected the land and its native creatures – hunting out invasive species like deer and foxes – and have also been able to defend their ancestral burial grounds and other significant cultural sites.

These examples are inspirational but are still a pittance: there are millions of hectares of land that should be formally returned to the nations and peoples from whom they were taken. The Red Deal is a new manifesto drafted by Indigenous and non-Indigenous activists laying out various demands, including the return of land to Indigenous nations, and is one place to start.[56] Until this combination of urban and rural designs, or something like them, are integrated it is unlikely that we will begin to build a non-parasitic mode of living in the world. But even if the rudiments of these changes are implemented, there is one great boulder still to be moved.

Equilibria of Energy

Parasitic societies strip down the genetic diversity of an area until only, or primarily, the species that remain are those useful to human reproduction and production. When the human group reproduces up to the limits of available energy in that realm, or an environmental cataclysm reduces the availability of energy to – or below – the reproduction rate, they must expand to new territories and again strip down the genetic diversity of that region. If someone is already occupying the newly conquered land, they also strip down its cultural diversity by eliminating or enslaving and assimilating the conquered people. The invading parasitic society repeats the process once they have again reproduced to the limit of that region.

The greatest problems with parasitism are the result of an imbalance in energy capture. When it comes to concrete energy capture, there is too much human biomass and human-use biomass and energy for the rest of the biome to be able to live and inorganic systems to function stably. One species – *Homo sapiens* – and its organised life, the anthroposphere, has captured too much relative energy to maintain a stable biosphere.[57] This is bad for both intrinsic and instrumental reasons. The intrinsic reasons are that diversity in species creates beauty in an otherwise dead solar system, and other species have a right to life on this planet equal to *Homo sapiens*. The instrumental reasons are that all life, including human beings, depends on diversity, and no human society can exist without a healthy biosphere. A world consisting only of humans, livestock and commodity crops is a world without pollinators to maintain soil health, without forests and living oceans to absorb carbon dioxide, without predators to maintain healthy ecological systems or keep disease-carrying pests in check; it is a dead world in which human minds cannot long remain sane, and potentially, one day, a world without clouds, without breathable air, without drinkable water, without life. Human systems have monopolised too great a proportion of the planet's bioenergy and have accumulated too much heat energy in the atmosphere.

The first imbalance can be addressed through the interventions such as those described in the previous chapter, in urban and rural design, by reassigning land to non-human species. One way of doing this is through policies like those advocated by the Land Back movement, in which land is placed under Indigenous control. Another is through reform of large-scale urban and rural policy and human settlement patterns. Many locally developed tactics and strategies may be successfully deployed to those ends. But there's only one way to fix the second problem.

Fossil fuels are often seen as passive tools that human systems, like institutions and political structures, utilise for their own ends. But the material realities of fossil fuels actively *shape* those institutions and political structures. The 'oil curse' is perhaps the clearest example of this, in which states that gain a disproportionately large share of their economy from petroleum tend to be more corrupt, unequal, authoritarian, belligerent and unstable. The nature of petroleum itself – its susceptibility to being concentrated and monopolised, the ease with which it can be converted into liquid capital, its ability to power weapons of war – all contribute to its impact on politics. In the same way, the transition to decentralised energy production can have similar cascading benefits that solve a lot of problems, from reducing air pollution to granting more civic autonomy to disparate communities, which can lead to greater democracy and equality, to the reshaping cities of and rural areas, to – like agroecology – improving resilience in the face of climate emergencies.

The first and most important step in such an energy recalibrating is a keep-it-in-the-ground style policy that prohibits the exploitation of fossil fuels. While fossil fuel divestment campaigns have seen some successes, reliance on the fossil fuel industry continues to expand.[58] As long as fossil fuels have market utility, financial institutions and corporations cannot be in charge of mediating whether they get used or not. States, intergovernmental institutions and non-state actors, including activists, must instead use force to ensure that remaining fossil fuel resources remain unexploited – a course of action which would pressure institutions to increase their funding for the invention of alternatives. Of course, at the moment this seems unlikely, to say the least, particularly given that state-owned oil companies control three-quarters of the world's oil reserves.[59] Even if there's full divestment of private oil companies, and BP and ExxonMobil have to shutter, there's no reason to believe the United States, the United Kingdom, Canada, Norway and other countries with access to oil wouldn't simply nationalise the industry in order to continue the flow. Their current governments are certainly more likely to maintain an oil economy through nationalisation than to dismantle those companies. There are few examples in the world of governments and companies wilfully forgoing readily accessible mineral resources, given their massive market and strategic advantages. Militaries represent the largest consumers of fossil fuels in the world, with the U.S. military by the far the largest. It is hard to a imagine a scenario in which all the world's militaries agree to forego using this strategically advantageous resource. But some governments have done surprising things.

One positive case is New York State's permanent moratorium on the use of hydraulic fracturing (fracking) to access gas shales.[60] Scotland, too, has blocked the use of fracking indefinitely.[61] Fracking causes cancer, respiratory problems, birth defects and many other public health burdens; it causes animals to die off, and contributes dramatically to global warming, through both combustion of carbon and the abundant methane leaks that occur throughout gas pipeline infrastructure, the latter of which make it about as bad as coal from a global warming standpoint.[62] Even so, it took a massive public campaign to compel New York's governor to sign an initial executive action to ban the practice, which was recently codified more permanently in a budget law. Scotland and New York's political environments are conducive to such a move; many other states and countries will not be so lucky. But such policies are necessary to maintain a liveable world and they must start now.

At some point, energy expenditure will have to decrease. This is simply due to the laws of thermodynamics. Even if we abandon fossil fuels and carbon-emitting energy sources entirely, and even if somehow we could continue generating energy in great densities without the demise of life-sustaining ecological well-being, there is still a hard limit on how much energy can be produced and used by the human economy. Physicist Thomas Murphy illustrates this well in the journal *Nature*:

> Selecting a mathematically convenient growth rate of a factor of ten each century (corresponding to 2.3% per year; roughly commensurate with the human enterprise in recent times . . .), our present-day expenditure at the level of 18 TW (18 × 10^{12} W) extrapolates to about 100 TW in 2100, 1,000 TW in 2200, and so on. In a continued progression, we would exceed the total solar power incident on Earth in just over 400 years, the entire output of the Sun in all directions 1,300 years from now, and that of all 100 billion stars in the Milky Way galaxy 1,100 years after that. This last jump is made impossible by the fact that even light cannot cross the galaxy in fewer than 100,000 years. Thus, physics puts a hard limit on how long our energy growth enterprise could possibly continue.[63]

Learning to live within the boundaries of available energy set by both the ecological realms of the planet and the physical laws of the universe, and in a way that is not catastrophically violent or unfair, will be the great task of the twenty-first century. Doing so, and doing all the other things that must be done, will not be easy and they will not be peaceful, but they will be worth the struggle.

War of Life

By now many will have read these proposed interventions and thought, *Sure, but how?*[64] These proposals in urban and rural design are good not just for their own sake and for the immediate health and wellbeing benefits they deliver, but also because they offer local means for individuals to intervene in ways whose wider effects defy the logics of parasitism. But, while there are plenty of solutions that sound good, the more difficult question is, how do these interventions accumulate in a coordinated and meaningful way? How do we come to understand progress not as the ever-increasing domination of the globe by human systems, but as the opposite, and then implement that?

All effective strategy and all long-term solutions must be hyperlocal. They must be decided and imposed at the most temporally and geographically granular level possible. As such, I cannot tell you one model, one ideology or one solution that can be effectively imposed and implemented globally. No one can, and no one should.

Many of those in leadership positions have tried to convince my generation that change only happens gradually. Reform, they say, is a slow incrementalism that happens over the course of generations, administrations and decades. 'You just have to have faith and, sure, the arc of the moral universe, it'll bend toward, you know, justice.' But this is not factually true: virtually all positive change has occurred rapidly, in short spurts. For example, in the United States, slavery went from being a legitimate sector of the economy to illegal in the span of one presidential administration. It did not end through piecemeal policy shifts, it took a war. During the Progressive Era, although some reform had begun under Presidents McKinley and Cleveland, President Theodore Roosevelt signed a wave of reform legislation, despite a Congress hostile to reform and a dominant conservative wing. In the span of two terms, bolstered by a labour movement, muckrakers and an activist constituency, Roosevelt signed into law trust-busting Elkins and Hepburn Acts that broke up monopolies, established the Department of Labor and Commerce, passed the Pure Food and Drug Act and set aside 230 million acres of National Park land. Meanwhile, the sweeping New Deal, packages of legislation and executive orders passed over the course of five years and one president, Theodore's cousin Franklin, included the creation of the Wagner Act, which gave employees the legal right to organise into unions and engage in collective bargaining; it introduced Social Security and inspired a major, enduring expansion in social welfare programmes. This was one of the most consequential policy shifts in the modern era and was neither piecemeal nor gradual. Sweeping environmental legislation like the Clean Water Act, the Endangered Species Act, and the creation of the Environmental Protection Agency all passed in the course of a few years and within President Nixon's administration. (These are American examples because this is where parasitism is currently most intensively practised; the US is still the technological and political leader of the world, and it may be there where pioneering possibilities remain. Other examples throughout the world bear this out, from independence movements in the South, to European revolutions, to the collapse of regimes.)

But even if it were historically accurate to say major policy shifts occur incrementally over time, there is little time for incrementalism to address the ecological crisis. The simple maths of the problem demands that only massive, rapid changes will deliver anything close to a worthwhile outcome for a liveable world.[65] If tipping points of climate change and ecological destruction have not already been crossed, they are likely to be crossed within a few years, certainly within the twenty-first century. But even as there have been large, sweeping policy changes for the better, they have failed to dismantle parasitism and in some cases, like the New Deal, have done more to exacerbate it. We must not cling to the myth that we are nestled in a warm, comfortable heritage of ever-increasing justice, that we ride some historical crest towards a tranquil, sunny beach. We are trapped in a vast desert of imperial graves and our only hope for the future is to get out and spread seeds for forests as we go.

So perhaps reformism and incrementalism are insufficient. What about the more radical end of historical struggles?

Unfortunately, even revolutionaries too often base their point of reference on a false narrative of linear progress. Many today look for lessons from a progressive past, reading the tea leaves of revolutions in Russia and China, Cuba and Haiti, divining lessons from the French and American Revolutions, or the decolonising independence movements of the mid- to late twentieth century. But like most of the rebellions, revolts and civil wars of the medieval and ancient worlds, these revolutions ultimately succumbed to the swapping of one oligarchy for another, exchanging one parasitic class with another and continuing the endless march of expansionism, war and collapse. Russia is now just another market kleptocracy; China is state capitalism with shallow pretences to socialism and no pretences to democracy; Cuba has recently welcomed economic liberalisation without any of the apparent benefits of political liberalisation;[66] soon after its revolution, Haiti descended into dictatorship, feudalism and extractivism.[67] The American Revolution ejected the British aristocracy and monarchy and entrenched its own largely hereditary power system in the void; the French went through a bellicose monarchic empire before finally settling on a technocratic liberal empire. Work fields, mines and armaments changed hands. Revolutionary progress has never yet overthrown the *logic* of parasitism itself; it has only reiterated parasitism in other forms. Even if all the revolutionary dreams were to come true – the billionaires toppled, labour unions ascendant, governments accountable, education and healthcare guaranteed to all – if those dreams remain funded by the spoils of industrial production, expansion and extraction, we will *still* be doomed to a lifeless world.

What about technological breakthroughs coming as a *deus ex machina* at the last moment? Can societies that produce compounding advancements in technological sophistication do so without exceeding ecological limits? There is no solid base of evidence to feel confident in this. So far, resource limits have always been placed on technological advances. Foragers cannot build rocket ships because the wild resources they depend on cannot yield the raw materials and labour time necessary for building them. We required the accumulated biomass of our deep past in the form of fossilised carbon – millions of years of energy trapped in rock – to access the raw energy, materials and labour time necessary to build rockets that have, compared with the vastness of the universe, provided us with barely a grasshopper's leap away from Earth. Most past societies that have advanced their technology in successive leaps have also been ecologically extractive and expansive. This is why most compounding technological advancements are transboundary: centres of invention and innovation hop from city to city, from empire to empire, following concentrations of resources as past centres of research and invention decline with the falls in ecological output that arise from hoarding, overproduction and overconsumption (or conquest by foreign invaders).

Historically, technological innovation has mostly depended on increasing, compounding growth in energy expenditure, which means an increase in capture from ecological sources. To illustrate the principle concretely: developing a bow and arrow requires experimenting with lots of wood and flint, trying new techniques and designs over time, even generations. What seems a simple technology today has nevertheless required countless hours of invention and mastery, including surpluses in food, shelter and entertainment that sustained the minds and bodies that did the inventing. Moving from bows and arrows to gunpowder cannons required a leap in productive power, from relying on wood and knapped stone to iron smelting and the ability to find, mine and synthesise combustive chemicals. Now, leaping from a technology based on burning carbon in combustion engines to one that isn't so based – what people call, perhaps misleadingly, 'renewable' technology – still requires burning a lot of carbon in combustion engines to invent, produce, build, ship and maintain that technology.

Is there some other way we could overcome this dynamic between the organic limits to technological progress and the desire for compounding discoveries and inventions? No one knows. To believe so is purely an act of faith. Given how unpredictable human behaviour and societies can be, and given that history is not always the best source for predicting where technology can go, it is possible – to imagine at least – that steady-state economies may rise in future which invest heavily in research and maintain a careful balance between ecological energy expenditure within their sociotechnical metabolism and the careful compounding of technological advances. The problem is that we need to both reduce energy expenditure *and* produce revolutionary leaps in technology – like that from carbon to entirely non-carbon electricity production, for instance – in the near term if mitigating climate and ecological crises is to be possible without total collapse. Technological progress cannot mean, therefore, the simplistic and almost childish logic of calling greater sophistication or greater quantities of tech 'progress'. Instead, working towards the resolution of the paradox of sustainability can be the only measure for any true sort of technological progress.

Resolving this paradox is not easy and there is no obvious answer to it. The only way may be through a guided decline in some forms of production – wasteful and superfluous kinds like massive personal vehicles and extreme individual wealth – in the near term while we figure out how to mitigate these crises and design an ecologically viable civilisation; then, once such an ecological economy has been stabilised, we might begin to increase technological intensity while decreasing intensity in some other area.[68] But even that scenario, quite prudent and conservative, would require a degree of coordinated action that has never been done on this scale, anywhere. When have such large groups of people ever agreed on what to do and how to do it – how many readers just thought 'no one's taking *my* car!'? And currently, for better or worse, there is simply no institution capable of coordinating and enforcing such a scenario.

If we cannot have faith in revolution and technology, and the urgency of parasitism simply does not allow for long-term trial and error incrementalist policy, where does that leave us?

The tired truth is that all of this change requires mass

participation in effective movements: activists, organisers, theorists, political commentators and leaders have been saying as much for many years. There are movements working on such issues: there are technocratic liberal organisations, leftist and worker organisations, climate organisations, environmental groups, and all of these overlap in various ways.[69] Even so, such movements rarely seem capable of achieving even the slightest victories necessary to avert disaster. There are few examples from the recent past we can look to since most tend to suffer rollbacks or are co-opted by the instruments of parasitic logics – capitalism, neoliberalism, corporations, states. Or, as mentioned earlier, revolutionary changes devolve into something worse than what revolutionaries fought to change. Pointing to historical working-class militancy, peasant revolts, or social movements is understandable. Taking pride in some of those successes, and feeling kinship with those who failed, is reasonable. But we must not see the political change we fight for today as yet another step – whether in victory or defeat – along the lineage of progressive history within which we so often situate our struggle. We must fight for a total break with that lineage, because it is false and only meant to ensnare us.

Indigenous resistance stands outside this lineage of parasitism. Indigenous warriors inhabit their own histories, cosmic timelines separate from Whig or Spencerian or otherwise imperial historiographies of progress that have sought to entrap them, and this present struggle is one among many that have been narrated throughout their many interconnected generations. In every period and every iteration of progress, Indigenous and commensalistic groups have fought against expansive empires either from within them or from the outside. Few have ever gained the upper hand in the last 5,000 years, and they have failed to stem the tide. But they are among the only modern examples fighting the literal engines of empires, and they should be at the forefront of any strategy for global change. A recent study found as many as 649 land-based resistance movements focused on thwarting ecologically disruptive energy infrastructure, and indicated many to be effective.[70] That is where we can find the beginning of the end of parasitism.

But environmental defenders will no more succeed in peacefully protesting against the destruction of wilderness than European Jews could have peacefully protested against the destruction of their people by the Nazi war machine. It is an eliminationist evil that they face. The work remains extremely dangerous; states, corporations and their informal mercenaries are happy to mete out violence against any who stand in their way. Even with its 'progressive' reputation, Canada implements ruthless, violent policies against Indigenous resistance. The *Guardian* reported that Canadian police were 'prepared to shoot Indigenous land defenders blockading construction of a natural gas pipeline in northern British Columbia'.[71] Meanwhile, the Canadian military used a 'counter-intelligence unit' to monitor an anti-fracking camp led by the Mi'kmaq Warrior Society.[72] From the Wet'suwet'en in Canada to the Standing Rock Sioux in the United States to the Mapuche in Chile, Indigenous groups are on the front lines of the battle against parasitism, as they have been for millennia. Defenders of land and forests around the world are being murdered at record rates, records that are broken every year; four were killed per week in 2019, and that number has grown dramatically since.[73] In addition to murder, prison and harassment, Indigenous people globally are faced with the extinction of their ways of life – the only ways of life human societies have made capable of thriving sustainably for tens of millennia.[74] 'Where indigenous groups have control of the land, forests and biodiversity flourishes,' according to the United Nations Development Program.[75] They are giving their lives to protect that land, which benefits us all. And indeed, not only is it the bravest, Indigenous resistance is the most effective means of preventing harmful pollution and environmental destruction. As an OilChange International study found:

> Victories in infrastructure fights alone represent the carbon equivalent of 12 percent of annual U.S. and Canadian pollution, or 779 million metric tons CO_2e. Ongoing struggles equal 12 percent of these nations' annual pollution, or 808 million metric tons CO_2e. If these struggles prove successful, this would mean

Indigenous resistance will have stopped greenhouse gas pollution equivalent to nearly one-quarter (24 percent) of annual total U.S. and Canadian emissions.[76]

And while they represent such a small portion of humanity, they also represent more than a third of those directly impacted by extractive industries. The destruction of Indigenous ways of life is still taking place today.[77] Even if such groups lead the fight, they must not be left on the front to die alone, or die first, or in the greatest numbers, or die at all; the opposite must be the case, particularly for those who have gained material wealth from the lineage of accumulation wrought by parasitic societies. Support and solidarity from all other groups claiming to be waging the same battle is absolutely necessary, at every level. Technocratic liberals inhabiting powerful institutions; socialists organising unions; environmental radicals sabotaging infrastructure: it is essential for all these to work closely with Indigenous resistance groups if the 'how' of the policies and goals mentioned previously is to be attained, and the many other necessary interventions are to take place.

The extractive industries destroying the living world, including that of humans, know they are waging a war. They are armed and dangerous. Those who wish to stop this war on life must also realise as much, and fast. When someone wages war against you, you either arm yourself and fight back, or you die. When a parasite invades your body, you don't negotiate with it, you kill it. When illegal loggers, miners and builders come to steal homes and lives, a doctrine of self-defence gives those in their path the right to use whatever means necessary to prevent their own demise. Those of us behind the frontiers, being exploited in the cores of parasitic societies, would do well to remember this doctrine.

There have probably been many examples, throughout the hundreds of millennia in which humanity has existed, of groups that did not learn or choose to restrain their expansion, and groups that did not place these biocentric and egalitarian ethics at their core. Many of these are unknown to us because they left no oral histories to span so many thousands of years; they would have inevitably collapsed or splintered.

There is also evidence from contemporary proxies of such societies that those individuals who attempt to abandon commensalistic or mutualistic ethics for parasitic ones are dealt with harshly. Writer and botanist Robin Wall Kimmerer of the Citizen Potawatomi Nation describes how, in a commons-based society, 'sharing was essential to survival and greed made any individual a danger to the whole. In the old times, individuals who endangered the community by taking too much for themselves were first counseled, then ostracized, and if the greed continued, they were eventually banished.'[78] Archaeologist Ian Morris also speaks of the tendency of communities to punish overconsumption and stifle 'upstarts' – individuals attempting to hoard too much wealth, power or status for themselves. After a series of interventions, like the ostracism, criticism and abandonment mentioned by Wall Kimmerer, some communities would go even further: 'While [they] rarely explicitly condone violence,' Morris writes, 'they do typically recognize multiple situations in which men are expected to use force to solve problems,' sometimes taking the form of 'an entire community agree[ing] that the only way to put down an upstart is to work together to kill him'.[79]

Societies of Tomorrow

What specific shape future societies may take does not concern me as such. People have chosen to build their societies in myriad ways, with thousands of permutations in governing processes, gender relations, resource distributions and cultural mythoi, each working through ingenious solutions to balance the complex mix of personalities and motivations that make up any human group. These ways have often conformed to the ecological realities from which they sprang and have taken on the hard policy of balancing collective good and individual freedom. The future offers a broad canvas for developing creative means of building alternative sorts of social groups. The lesson these many examples impart is that, as we think about how to reshape society into something that can survive morally in the coming years, we should take heart from our diverse and creative past. We must neither let our imaginations be confined by a narrow sense of what is possible, nor let them ignore the realities of nature that demand we do what is necessary.

What is clear is that we need the preponderance of societies to practise commensalism or mutualism, and allow only a small minority, if any, to practise parasitism. This shift does not come with a single policy mechanism or a model. It requires those patterns in policy I have outlined and many more. It also requires that we, as citizens or prisoners of parasitic human ecologies and their harsh political economies, begin to refashion our ways of living by whatever agency we have. There are principles and practices that can help us begin to realign our individual selves in such a way that we may realign the political realities of the societies we inhabit. They are prerequisites for becoming the kinds of people we must be to dethrone the administrators who are keeping us on course to oblivion. Instead of clinging to faith in some false doctrine of progress handed down from towers in capital cities, we have to take our destiny into our own hands.

Progress narrative formulae have been used for millennia to promote parasitism across the world. They have created certain kinds of individuals, subjects who follow fixed and narrow patterns in living out their lives. Following such ruts has allowed some to extract disproportionate amounts of energy from our lives, and from the environmental context of life. One vital step towards interrupting that process is by stepping outside those narrow frames that confine the contours of life.

The individual and the society in which one lives are mutually creating. The society shapes the individual and the individual shapes the society. As obvious as that may sound, it has less obvious implications for the question of how we can change society. How, for instance, do we reshape our individual selves in such a way that we improve beyond the limits of the society in which we were nurtured and honed? How do we, as individuals bounded within the confines of the societies we inhabit – the cognitive biases, the ideologies, the false histories and narratives – exercise the will necessary to refashion society in our image, or our image at our best? These are lifelong tasks, not feats that we achieve one day and sit back and enjoy for the remainder of our lives. They are daily practices. Even a few such practices can take us closer, as individuals and societies, to outcomes that forsake parasitism for superior relations, to each other and to our ecological homes. The following general principles and practices should be adapted to specific conditions, but they are a good starting point for guiding that process of mutual co-creation of the kinds of societies that must emerge if we are to avoid catastrophic shocks:

1. **Principle:** *universal personhood.* **Practice:** *recognise kin, human and beyond.*

Like commensalistic and mutualistic societies past and present, the rest of us need to accept and cultivate an expansive understanding of personhood, of who deserves the dignity and protection that we primarily extend only to human beings, or certain privileged groups of humans. Not only must this be expansive to more human beings, but to animals with which we share biomass on this planet, and even beyond them: the rivers, the trees, the very air that gives life. This respect must be enshrined in law and must be enacted on a daily basis.

2. **Principle:** *mutual extraction.* **Practice:** *balance your extraction and creation.*

It's natural to put limits on ourselves as individuals. This is the definition of maturity. Being able to forgo some base impulse for the sake of others or for the future is a mark of being worthy of respect. It's no different at the level of a society: most societies for most of human history have imposed limits on themselves against over-exploiting the ecologies in which they are embedded. Commensalistic societies across the world have used force to stop members of their communities from taking too much land or power for themselves. We must do so again. Extraction must become far more mutualistic than it is today, and get closer to what it has been for most of human history. The fetish of profit is untenable and must be crushed. No one should take significantly more or less from the labour of others than they give back, just as no society should take significantly more from the ecology it inhabits than what it gives back. Give out as much as you take, or more and demand the same of others.

3. Principle: *there is no future.* **Practice:** *fight for today.*
It may sound like I'm contradicting the rest of the book to say that we ought not to fight for the future. After all, is it not the future that is so imperilled by the ecological emergency? First, no: vast death is already enveloping the world like a fetid fog. The seduction of progress comes from the desire to dream of the future at the expense of fighting for life today. The future is the great allure and danger of the progress myth. If we ease the sting of defeats today with thoughts that our work will carry into some future victory, then no victory will ever come, because the future does not exist. Instead, we must fight for today and not stop fighting until the present is as we want it. Today, after all, was yesterday's future, and if we accept that victory can only reside in the future, then we accept defeat.

4. Principle: *generosity of spirit.* **Practice:** *trust your peers.*
Building solidarity movements, new communities and cities, and rebuilding civilisation in the image of better values will require practising a generosity of spirit that the psychopathic competition of market economies and parasitic states do not incentivise. As much as possible we must remind ourselves to practise patience and conscientiousness, and trust in our peers and compatriots. This does not mean that we should be overly ready to forgive or forget the harms done by the pathological and careless, who extract from us and from the world more than they have earned. Nor does it mean we should go out of our way to see the humanity of those who only wish to destroy other life for the sake of their own or their families. They deserve the harshest treatment. But this parcelling out is hard work that must be done.

5. Principle: *irreplaceable earth*. Practice: *protect your ecology*.
No matter how far our fingers extend beyond the Kármán line separating Earth from beyond, we must never forget that there is unlikely to be any other place in the universe as suitable to our minds and bodies as this planet and these ecological systems. We can expect any harm we do to the environments on this planet to return to ourselves and our futures, perhaps with even greater violence. Technology should be deployed first and foremost with the goal of protecting ecologies and the planet at large. Protecting your land and forests, the wildlife that resides there, and the communities that you belong to that are dependent on them, sometimes means being willing to use force and resilience to fight human enemies. This is something we need the courage and strength to do.

In the transition from parasitic economies to commensalistic ones, there is much we can learn from cultures that have practised commensalism or mutualism in the past and continue to do so today. But many of the conditions in which historically they practised such a mode of human ecology are gone for eternity. Pandora's box is open. The Apple is eaten. The Flame stolen and given. Comparable conditions may arise again far in the future, but we cannot have faith in that now. Instead, we have to reconcile the necessities of the moment with those of the past and the conditions of the future.

* * *

There is no one single model for how best to organise a society, since what's 'best' will depend on local ecologies and circumstances. If we avoid a hothouse Earth, the Anthropocene is likely to be an epoch of rediversifying human cultures and societies, even as coastlines shift, deserts expand and climate zones climb latitudes. What I am most concerned with is that the faulty relationship between human systems and ecological systems at the heart of the last 5,000 years – a relationship built on expansion and exploitation – should be forever destroyed. Freedom from parasitism will ensure long-term survival of some sort – not the dreadful survival of eking out an existence like a tormented mouse scurrying back and forth along a floorboard, but long-term prosperity, abundance and joy. This we can be confident in: we know commensalistic human ecologies can deliver health, happiness and wealth because they have done before. Linear progress is a plunge into the unknown. This can be exciting and meaningful, but it can also be perilous. In a time when peril can mean the permanent extinction of all that we are and know, that drive must be checked. Commensalism is known, it is tried and true. But it is not dull or dreary. The reverse is true: the progress of parasitism is dull and dreary. The diversity that emerges from commensalist or mutualist human ecologies leaves open the space for incorporating existing and evolving technology and knowledge to create something new and superior. This future is theoretically attainable based on the tools currently at our disposal and the knowledge we have about how humans have lived for a vast majority of our history – around 99 per cent of it in fact. But we have to have the courage to take this step.

Hope is a treacherous emotion. It quiets rage and it settles the mind into comfortable stasis. Recent studies have suggested that a feeling of hope is contrary to taking action to fight for your life and dignity and that of others.[80] Righteous anger is a far more motivating emotion. This has been the central promise and peril of progress. It will smother rational collective rage, act as a blanket on a fire, by granting us the image of a beautiful future at the cost of a perpetually grim present. To grow and become the people we will have to be to thrive in what will likely be a hard future, we must set aside that hope, and look elsewhere to find peace or salvation or contentment. Maybe it's in something old, far older than any ideologies or narrative lineages dominant today. Or maybe it's new, something that must be collectively patched together in the midst of the catastrophes of this century. Maybe it is something that will only thrive in song or whisper, coaxed into a blaze in a close huddle, and even there will remain as ineffable as a shared dream.

Meaning Beyond Progress

As we began, let us end with a word from Thomas Jefferson. In his most famous composition, his 1776 Declaration of Independence, Jefferson proclaimed, 'That whenever any Form of Government becomes destructive of these ends' – those of Life, Liberty, and the Pursuit of Happiness – 'it is the Right of the People to alter or to abolish it, and to institute new Government, laying its foundation on such principles and organizing its powers in such form, as to them shall seem most likely to effect their Safety and Happiness.'[1]

This is not an invocation of hope for a better future. It is a call to action in the present, to remove those forces that obstruct our ability to achieve freedom, fairness and joy, now. For all the crimes of Jefferson and the other Founders, and for all the hypocrisy present in the United States' founding, even in this document itself, that message remains a vitally important principle. 'Happiness' comes not from placing all of our faith in a future that may never come, nor in cultivating a numbing exercise – of prayer or meditation or collective delusion – meant to reconcile us to insufferable conditions. Happiness, not fleeting pleasure or bewildered contentment, but some deeper sort of balance, comes from elsewhere, from a different practice, a different sort of action. If our *right* to claim it is uncomplicated, what it means in day-to-day experience is not so straightforward.

The basic content of a life – creation, obtaining food and water, exploration, recreation, enduring and enjoying the vicissitudes of weather and season, maintaining a dwelling, procreation – is being fundamentally disrupted by current conditions in a way that no other force, from war to recession to everyday trials, can disrupt. As the stability and predictability of the ecological context in which all life occurs diminishes, so does the meaning, stability and security of every single life, and so, therefore, does any individual's ability to find significance in the quotidian pleasures and pains of that life. And while wars may last years, recessions may skip from decade to decade, this cataclysm threatens to linger for centuries, millennia, or even, as far as humans and many other species are concerned, for eternity. How do we contend with this fact in daily life? How do we find meaning and purpose in a world so brutal and so deeply compromised, and a future likely to be even less stable than the present?

In his autofictional hexalogy *My Struggle*, novelist Karl Ove Knausgård pursues an analysis of the self, finding meaning through daily challenges, and an ethnography of his time and place through this self, that penetrates in great depth and detail the question of how an individual may pursue a meaningful life. Philosopher Martin Hägglund interprets and summarises Knausgård's 'struggle' as the imperative to own one's life, to avoid disowning or detaching from the sorrows and joys of daily existence, to, in Knausgård's words (or rather, those of a priest he records), 'focus your gaze by attaching yourself to what you see'.[2] Hägglund connects this ethic to a mindfulness practice, but a secular rather than a Buddhist one because the latter seeks the opposite, to ultimately detach oneself from the world. He argues that meaning comes not from seeking an off-world afterlife or living for a promised eternity, but from immersing oneself in the impermanence inherent to mortal life, appreciating things for their finitude, not striving for an impossible and meaningless infinitude. That is, we must find meaning in the present rather than deferring it to an imagined future of infinite expansion – though, I would add, there is meaning to be found in deferring the fulfilment of certain present impulses in the interest of future balance, and reconciling these is not always easy. Ultimately what fixing one's gaze leads to, for Hägglund, is a struggle less individual or familial than Knausgård's, less rooted in one's individual experience, and more in a larger mutual struggle. This struggle focuses its gaze by attaching itself to a particular programme. For Hägglund, that programme is democratic socialism:

the key to democratic socialism is to have institutions (including educational institutions and forms of political deliberation) that enable individuals to lead their lives in light of recognizing their dependence on others and on collective projects [and] to have institutions in which we participate because we *recognize ourselves and our freedom* in their form.[3]

It is not sufficient to immerse oneself in the experience of daily life: we may be drowned in the habits of our days, in the chores of raising children, the rhythms of employment, the escape of entertainment, the deep lows – illnesses, accidents, deaths – and soaring highs – births, ceremonies, romances, extreme experiences – that punctuate the day-to-day. But unless we are exercising our agency within the world in some way to improve it, crafting institutions that govern the world in our image at our best, in order to prevent or predict or survive the cataclysms that are accelerating into the future, then a certain meaninglessness that sits as a black emptiness the size of an acorn in our core will take root and its branches will slowly spread, petrifying our limbs and stupefying our minds. Participation in that higher good is a necessary remedy to such an ailment. There are social ways of participating: joining organisations, building unions, starting community projects, making media and performance art; and there are solitary ways of participating: scholarship, writing, creating art and other cultural artefacts. The vital thing is to engage with power as it is expressed, in whatever way possible, in the same fixed and attached way as we engage with our life.

People love to sneer at the word *politics*. The most commonly accepted thing to complain about, if not the weather or the traffic, is that big mausoleum we call simply 'politics'. Even so, the inescapable fact remains: politics is the process by which power is exercised and the parameters of life get decided. By 'politics', I don't mean the horse-trading football-hooligan practice of politics that many commentators, politicians and journalists attempt to infuse in electoral and bureaucratic procedure – that is indeed worthy of contempt. By *politics* I mean the exercise of hard and soft power within society and the exchange and distribution of the resources – concrete and abstract – that make a society and an economy. Politics is the means by which the contents of your life – your time, your food, your clothes, your house, your travels, your relationships, your hobbies, your death, and everything else – are parcelled out and accessed. If our earthly time is all we have, then politics is the primary determinant of how much of that time we get to spend and how we get to spend it. Power grants or rescinds life. It is the mediator of what our jobs are, and whether there are jobs, where our houses sit, who gets to live in them, or whether they remain standing at all. Politics is not an alien sphere outside of the domain of life; it is, or it has become, the foundation of the domain of life, the greatest decider of what the domain becomes. For better or worse, participating in politics *is* participation in this life and the precious time we have on earth.

But we must go further. It is true that immersion in our own individual struggle to reconcile ourselves with being is important to living with meaning; meanwhile, immersion in some grander mutual struggle to reshape the world in the image of our highest nature is necessary to live with meaning. But just as vital as both of these is immersion in the non-human world. A purely human world is insufficient to deliver calm, contentment, meaning, or even physical health. What's missing from much political advocacy and popular philosophy today is acknowledgement of our deep need for the beyond-human. Humanity alone is nothing. Rather, it is worse than nothing; alone in a once-diverse world, it is a grotesque, bloated ape seated on a throne of corpses. The parasitic delusion that only the human world matters has separated us from other life and sought to make us forget that it is everything that we depend on for our own lives; your life is inseparable from the lives of the plants and animals that share the world with us, even if you've been tricked into forgetting it. It is only by cutting down the forest that your love for it could be stolen. But it's not just a matter of practical dependence on other life, not just about where you get your food, clothes and fresh air. It is an ethical, even spiritual dependence: a life devoid of the more-than-human is not a life worth living. And a life that actively and knowingly aids in the diminishment of the more-than-human is a life not worthy of being.

Our struggle must be one of appreciating life in its precious shortness, dipping in and out of living pleasure, not striving for some future eternity of benign comfort that progress promises. But, maybe as a paradox, the struggle must be one of working together to build more democratic structures of life, to fight alongside our compatriots for the same goal of a society in which we are free to spend our own life time, whose fruition may only come in a non-existent future. And the struggle must also be one of working to expand the time for living of those others who live outside the human world – i.e., not your dog or cat – whose own worlds are on the precipice of non-existence and who lack the means to fight for their own futures. There is intrinsic value to having diversity in the kinds of minds that hold the world in the spheres of their comprehension. The world is made richer by its being perceived in diverse multitudes: only by cohabiting a world populated by other minds, ones that grasp the world both deeply unlike, but also recognisably like our own, can we claim any value or meaning in the world. The highest purpose of our species is that we can protect and nurture the beings of other species. And so diminishing this diversity of minds is an act of the darkest possible evil.

Karl Marx described his critique of religion as 'pluck[ing] the imaginary flowers from the chain, not in order that human beings shall bear the chain without caprice or consolation but so that they shall cast off the chain and pluck the living flower'.[4] Hägglund uses this language to explain his own call for democratic socialism: 'Such democratic emancipation requires that we remove "the imaginary flowers" which make our current chains seem bearable. Whether the imaginary flowers are the promises of the free market or the religious promises of eternal life, they serve to make us accept or forget the injustices of *this life*.'[5] The narrative formula of progress is the water that all such flowers on the chain have drunk, bequeathed through the lineage of desolation within which we are enmeshed. And for the same reason we must resist such false promises, so that we may recall not merely the perversions of justice we should hate, but the ever-accelerating total destruction of this life and all others to come. Faith in progress prevents salvation today by promising salvation in some far-off tomorrow. Just as we must find meaning in the present rather than deferring it to the future, we must also demand political victory – for freedom, fairness, wilderness – in the present rather than deferring it to the future.

We may all be building a future together *metaphorically*, but in *practical reality* there's only now, and all we have is this present time we currently inhabit. If all we have is this time, then any society ordered in such a way that we must hand most of it over to someone else is not a society worth having, nor is one that reduces the agency over time and the amount of time we have. The abstract energy capture of parasitism is another way of measuring the theft of living time. The parasitic classes can only commit this act by promising you time in the future, through retirement or through an improved situation for your children or promotions or the promise that this toil is working towards some glorious golden age, while hoping you don't live long enough to collect on their debt to you. If we wish to escape that trap, we need to seize the present. 'Seize the day' has come to mean something like 'live every day as if it is your last': a better meaning is 'seize your present moment back from those forces who want to steal it from you in exchange for nothing'. It may be cliché to say it, but it bears repeating in days of imperial decline and paradoxical future fetish: the future never arrives. We are imprisoned or liberated in a perpetual present.

Casting aside faith in progress is not easy. I have suffered personal crises from doing so, and I have no doubt others have and will as well. It is a faith that offers comfort in the face of fear of nonexistence. In Book Five of *My Struggle*, Knausgård depicts his grandfather in hospital, nearing death in his eighties. His grandfather muses that many of the ailments that afflicted people in his generation are now being cured regularly. He considers that perhaps one day soon, people will live for a thousand years, extrapolating from all the medical progress in his lifetime. Knausgård wonders if perhaps this musing on the abolition of universal death is his grandfather's way of ameliorating his fear of his own personal death.

But there is comfort to be found in accepting the reality of the great repopulation of more-than-human life and the spiritual contentment that can emerge from an animistic worldview, comfort even to face the great horror or succour of nothingness in death. Knausgård concludes the first book in his series with a depiction of his emerging reconciliation with grief. He reminds us that 'humans are merely one form among many, which the world produces over and over again, not only in everything that lives but also in everything that does not live, drawn in sand, stone, and water.'[6] The marriage between an empirical understanding of the world as it is and a love of the divinity within the world has been a vital condition of long-term societies. If we are to endure the mercurial world of coming centuries, we will likely need to cultivate some version of this science and divinity of natural existence. What this will look like in practice today or tomorrow may be far different from what has been practised for the past tens of millennia, but the essential orientation of biophilia and biocentricity, the love of living things and their diversity, must be there, and they must infuse how we design the material civilisation that we now call home. Getting there is what progress must mean now.

The ecological crisis, which encapsulates the climate emergency, threatens to make the possibility of pursuing both the individual struggle to immerse oneself in daily life and collective struggle in politics increasingly difficult, nearing impossible.[7] In a sense, we have no choice but to engage in something like the collective, egalitarian struggle that Hägglund champions in order to maintain any semblance of the individual struggle that Knausgård profoundly documents; at no other time in history has such an ecological cataclysm struck human life in such a way, with such a threat of permanence. The way we begin this process is by questioning our own deeply held assumptions about morality, history and the future: as this book has attempted to inspire, to rethink our place in a larger timeline, to recast our current trajectory not as some zenith of human potential, but as the nadir in a dark age that started five thousand years ago and, for better or worse, in one form or another, may be coming to some sort of end that we must manage so that a better world can be built.

Climate change and ecological collapse are the product of a dysfunctional relationship between human life and more-than-human life. The question that we each must answer, every day, as individuals and as communities, is whether we are automata, living only to gratify immediate impulses or social imperatives imposed from above, or whether we have the will, the self-mastery, the dignity to rise above that and exercise our natural agency. If humanity is but a pestilence that consumes like a vast, unthinking swarm, then what makes us exceptional? What makes us any better than bacteria? What makes us particularly deserving of life? It is our agency, and only our agency, that marks us out as a sometimes exceptional species, for better or worse. Unless we use that agency, individual and, to whatever extent is possible, collective, to master our own destructive impulses, then we will do what many other parasites have done: destroy our host to our own demise.

I can't give you hope, wouldn't delude you with it if I could, but I can just about guarantee that there will come a time when there is too little energy, both living carbon and fossilised carbon, to run the engine of this vast parasitic global network. There will be no more space to expand into and no more resources to extract. When that day comes, this network will fracture. And when it does, there is no reason to believe it won't resemble, in some ways, the collapses of regional empires of the past. Though chaos will claim many lives, so too vistas of possibility will open: possibilities as to how societies are ordered and governed, how people relate to one another, how they fill their time, and how they relate to the rest of life on earth, diminished as it may be. But this will only be true if it begins soon: it won't be true in a world too hot to inhabit. Nothing will reside there.

Humans were once as lions and wolves and orcas: social hunters moving with confidence through the ecological systems in which they were in joy. Few remain so. For so many have become as invasive ants and termites and locusts, building hills and towers and farming other life in a dense collectivity, swarming over land in a hysteria of feeding, roving in gangs driven by vengeful hunger. In his most extreme moments, in this extreme time, a man may become as a plague. What will we become next? We will not return to the life of wolves, not for a long time anyway, for we have destroyed the density and abundance of ecological diversity necessary for that. But nor will we remain as pests and tormentors. Will we retain our remarkable capacity to create great things while shedding our hideous capacity to destroy on a grand scale? Will we show our immense capacity for love and wonder in greater measure than doling out our unparalleled capacity for capricious cruelty or mindless indifference?

For five millennia, progress has offered a paradise of tomorrow, a new frontier that would finally bring everlasting joy, peace and contentment, not revealing that someone else's paradise had to be destroyed to open that frontier. In order to have any future, or any worth enduring, we need a new conception of our place in our ongoing history. We might not build paradise, but at least we may craft branching slivers of peace and contentment, arteries coursing through the world along which joy, life and beauty may still pass, even if they are bounded on all sides by chaos and violence. Many of those reading this book now may see some new world that emerges with their own eyes, build it with their own hands. As builders of this new civilisation, we, you, have to do the intellectual work, the social work, the spiritual work and the physical work of making new cultures, value systems and infrastructures that are capable of existing without parasitism and progress, without the need to expand and extract. This is a more difficult challenge than it sounds. It means reconciling our thirst for frontiers and excess and novelty, for exploration and reproduction, with the need to temper our impulses and allow ourselves, and each other, to be content and satiated and rediscover the joys of abundance already there in the wild and in the creative act. If we achieve such a thing, we may find on the other side of it a form that can mingle with the heavens while nurturing life in the wilderness of Earth.

Acknowledgements

For their incisive interventions, my infinite thanks go to my editors, Grace Pengelly, George Witte, Martin Toseland, Shoaib Rokadiya, Steve Gove, Eva Hodgkin, Katy Archer and Freya Alsop. For the integrity to publish books that push against conventional notions, my enduring appreciation goes to the teams at William Collins and St. Martin's Press. And my heartfelt thanks go to my agent, Julia Eagleton, and the team at Janklow & Nesbit UK.

Some of this work was done while pursuing my doctoral research at the University of Oxford and while writing analyses for various publications. I give warm gratitude to my academic colleagues, particularly Aoife Brophy, Tim Schwanen, Nick Eyre, and Giorgos Kallis, who have improved my research, and to the many editors who have improved my writing over the years.

Friends generously gave their time to read early chapters and offer feedback, which undoubtedly improved the book and bolstered my confidence. Thank you, Shane, Beth, Dan, Matt, and Meg – I owe you. Special thanks to Nathan J. Robinson for his in-depth edit on a portion of the book and for publishing and editing so many of my essays over the years. Many ideas in these pages were first nurtured in issues of *Current Affairs*.

My parents, John and Donna, have always given their unwavering support to me and to all those around them. Their sincere commitment to education, integrity, and personal and societal improvement – *progress!* – has laid a strong foundation for this work. Finally, my love, Heather, has been involved in this project since the beginning, commenting on drafts, helping keep my words tidy, and collaborating to make a sane refuge in an increasingly mad world. Thank you. Many stars in the sky have gone dark, but yours shines out the brighter.

Endnotes

WHAT IS PROGRESS?

1 Thomas Jefferson, letter to William Ludlow, 6 Sep 1824, Founders Online, National Archives, founders.archives.gov/documents/Jefferson/98-01-02-4523
2 Robert A. Nisbet, *History of the Idea of Progress* (London: Routledge, 1980; 1994), 4. There have been few recent books on the idea of progress. Some of those are Matthew W. Slaboch, *A Road to Nowhere: The Idea of Progress and Its Critics* (Philadelphia: University of Pennsylvania Press, 2018); Peter Wagner, *Progress: A Reconstruction* (Cambridge: Polity Press, 2016); Ronald Wright, *A Short History of Progress* (Edinburgh: Canongate, 2004); and Christopher Lasch, *The True and Only Heaven: Progress and Its Critics* (New York: W. W. Norton, 1991).
3 There are frequent claims to the contrary of this fact, but it is indeed true. See Tom Crouch, 'Yes, the Wright Brothers Really Were the First to Fly', *AIR & SPACE Magazine*, Sep 2013, www.airspacemag.com/history-of-flight/who-flew-first-290750/
4 Here's one: Ana Swanson, 'Charted: The tallest buildings in the world for any year in history', *Washington Post*, 11 Mar 2015, www.washingtonpost.com/news/wonk/wp/2015/03/11/charted-the-tallest-buildings-in-the-world-for-any-year-in-history/. And another: Iman Ghosh, 'These are the world's tallest structures throughout history', World Economic Forum, 5 Sep 2019, www.weforum.org/agenda/2019/09/tallest-historical-structures/
5 Figures from Vaclav Smil, *Energy and Civilisation: A History* (London: MIT Press, 2017), 374.
6 Stephen Dowling, 'The monster atomic bomb that was too big to use', BBC Future, 16 Aug 2017, www.bbc.com/future/article/20170816-the-monster-atomic-bomb-that-was-too-big-to-use

7 Here are some bomb charts: Casey Chan, 'The True Scale of Nuclear Bombs Is Totally Frightening', Gizmodo, 7 Oct 2016, gizmodo.com/the-true-scale-of-nuclear-bombs-is-totally-frightening-1787538060
8 'Economic Growth', Sustainable Development Goals, United Nations, www.un.org/sustainabledevelopment/economic-growth/
9 For some of the most prominent examples, see Gapminder, Our World in Data, and the works of Hans Rosling and Steven Pinker.
10 Jason Hickel, 'Progress and its discontents', *New Internationalist*, 7 Aug 2019, newint.org/features/2019/07/01/long-read-progress-and-its-discontents
11 Jason Hickel [@jasonhickel], 'Those who measure "progress" against global poverty by using the $1.90 per day (PPP) line should keep in mind that this is lower than the consumption level of enslaved people in the United States in the 19th century', Twitter, 27 Mar 2021, twitter.com/jasonhickel/status/1375765315397763076. Based on data in Robert C. Allen, 'Poverty and the Labor Market: Today and Yesterday', *Annual Review of Economics*, 12 (2020), 107–34, doi.org/10.1146/annurev-economics-091819-014652
12 Jason Hickel, 'Bill Gates says poverty is decreasing. He couldn't be more wrong', *Guardian*, 29 Jan 2019, www.theguardian.com/commentisfree/2019/jan/29/bill-gates-davos-global-poverty-infographic-neoliberal. Another analysis can be found by Roge Karma, '5 Myths About Global Poverty', *Current Affairs*, 26 Jul 2019, www.currentaffairs.org/2019/07/5-myths-about-global-poverty
13 There are a few recent books that offer a very in-depth debunking of these sorts of data visualisation websites and progress arguments. Those I would recommend include Rodrigo Aguilera, *The Glass Half-Empty: Debunking the Myth of Progress in the Twenty-First Century* (London: Repeater Books, 2020); Danny Dorling, *Slowdown: The End of the Great Acceleration – and Why It's a Good Thing* (London: Yale University Press, 2021); Stephen J. Macekura, *The Mismeasure of Progress: Economic Growth and Its Critics* (Chicago: University of Chicago Press, 2020); and Christopher Ryan, *Civilized to Death: The Price of Progress* (New York: Avid Reader Press, 2019).
14 'Life Expectancy by Country 2024', World Population Review, 2024, worldpopulationreview.com/country-rankings/life-expectancy-by-country
15 J. D. Montagu, 'Length of Life in the Ancient World: A Controlled Study', *Journal of the Royal Society of Medicine*, 87 (1994), 25–6.
16 Cited in Amanda Ruggeri, 'Do we really live longer than our ancestors?' BBC: 100 Year in Life, 3 Oct 2018, www.bbc.com/future/article/2018 1002-how-long-did-ancient-people-live-life-span-versus-longevity

ENDNOTES

17 Ian Morris, *Foragers, Farmers, and Fossil Fuels: How Human Values Evolve* (Princeton: Princeton University Press, 2015), 57.
18 Walter Scheidel, 'The Roman Slave Supply', in Keith Bradley and Paul Cartledge (eds), *The Cambridge World History of Slavery: Volume 1: The Ancient Mediterranean World* (Cambridge: Cambridge University Press, 2011), 287–310.
19 Juliana M. Horowitz, Ruth Igielnik and Rakesh Kochhar, 'Trends in U.S. income and wealth inequality', Pew Research Center, 9 Jan 2020, www.pewsocialtrends.org/2020/01/09/trends-in-income-and-wealth-inequality/#income-inequality-in-the-u-s-has-increased-since-1980-and-is-greater-than-in-peer-countries
20 Michael Gurven and Hillard Kaplan, 'Longevity Among Hunter-Gatherers: A Cross-Cultural Examination', *Population and Development Review*, 33:2 (June 2007), 321–65.
21 Zachary Gidwitz, Martin Philipp Heger, José Pineda and Francisco Rodriguez, 'Understanding Performance in Human Development: A Cross-National Study', United Nations Development Programme, Human Development Research Paper 42 (November 2010), hdr.undp.org/content/understanding-performance-human-development
22 Interview clip of Malcolm X in Mar 1964, 'MALCOLM X - If You Stick A Knife In My Back', uploaded by finifinito, YouTube, 5 Nov 2011, www.youtube.com/watch?v=XiSiHRNQlQo
23 This is just accounting for those in the United States, with the first slaves landing in Virginia in 1619 and emancipation occurring in 1865. Slaves were first taken in the Atlantic slave trade by the Portuguese to Brazil in 1526. Ten times more slaves were brought from Africa to Brazil than to the United States. See Nicolas Bourcier, 'Brazil comes to terms with its slave trading past', *Guardian*, 23 Oct 2012, www.theguardian.com/world/2012/oct/23/brazil-struggle-ethnic-racial-identity
24 Cited in George Breitman (ed.), *Malcolm X Speaks: Selected Speeches and Statements* (New York: Merit Publishers, 1965), 36.
25 Kate Hodal, 'One in 200 people is a slave. Why?' *Guardian*, 25 Feb 2019, www.theguardian.com/news/2019/feb/25/modern-slavery-trafficking-persons-one-in-200
26 Rekha Rani Sharma, 'Slavery in the Mauryan Period (C. 300 B.C.–C. 200 B.C.)', *Journal of the Economic and Social History of the Orient*, 21:2 (1978), 185–94, doi.org/10.2307/3632125
27 Stephanie Nebehay, 'Executions, torture and slave markets persist in Libya: U.N.', Reuters, 21 Mar 2018, www.reuters.com/article/idUSKBN1GX1L5/
28 'Highest to Lowest – Prison Population Total', World Prison Brief, Institute for Crime & Justice Policy Research, www.prisonstudies.org/highest-to-lowest/prison-population-total?field_region_taxonomy_tid=All

29 Leah Sakala, 'Breaking Down Mass Incarceration in the 2010 Census: State-by-State Incarceration Rates by Race/Ethnicity', Prison Policy Initiative, 28 May 2014, www.prisonpolicy.org/reports/rates.html
30 Adam Gopnik, 'The Caging of America', *New Yorker*, 22 Jan 2012, www.newyorker.com/magazine/2012/01/30/the-caging-of-america
31 D. L. Davis, 'American Family CEO makes a point about black male enslavement and today's incarceration', PolitiFact, 1 Nov 2019, www.politifact.com/factchecks/2019/nov/01/jack-salzwedel/american-family-ceo-makes-point-about-19th-century/
32 Jared A. Brock, 'As California wildfires raged, incarcerated exploited for labor', *USA Today Opinion*, 11 Nov 2020, eu.usatoday.com/story/opinion/policing/2020/11/11/california-wildfires-raged-incarcerated-exploited-labor-column/6249201002/
33 Stian Rice, 'Convicts Are Returning to Farming – Anti-Immigrant Policies Are the Reason', *UMBC Magazine*, 7 Jun 2019, magazine.umbc.edu/convicts-are-returning-to-farming-anti-immigrant-policies-are-the-reason
34 Wendy Sawyer and Peter Wagner, 'Mass Incarceration: The Whole Pie 2020', Prison Policy Initiative, 24 Mar 2020, www.prisonpolicy.org/reports/pie2020.html
35 M. Marit Rehavi and Sonja B. Starr, 'Racial Disparity in Federal Criminal Sentences', *Journal of Political Economy*, 122:6 (2014), 1320–54, repository.law.umich.edu/cgi/viewcontent.cgi?article=2413&context=articles; Christopher Ingraham, 'Black men sentenced to more time for committing the exact same crime as a white person, study finds', *Washington Post*, 16 Nov 2017, www.washingtonpost.com/news/wonk/wp/2017/11/16/black-men-sentenced-to-more-time-for-committing-the-exact-same-crime-as-a-white-person-study-finds/
36 'Report to the United Nations on Racial Disparities in the U.S. Criminal Justice System', The Sentencing Project, 19 Apr 2018, www.sentencingproject.org/publications/un-report-on-racial-disparities/
37 Mohamed Abdelfattah, 'Study: White Americans still vastly overestimate racial progress', *Northwestern Now*, 21 Sep 2020, news.northwestern.edu/stories/2020/09/study-white-americans-still-vastly-overestimate-racial-progress/
38 Heather Long and Andrew Van Dam, 'The black-white economic divide is as wide as it was in 1968', *Washington Post*, 4 Jun 2020, www.washingtonpost.com/business/2020/06/04/economic-divide-black-households/
39 Grace Panetta, 'How Black Americans still face disproportionate barriers to the ballot box in 2020', *Business Insider*, 18 Sep 2020, www.businessinsider.com/why-black-americans-still-face-obstacles-to-voting-at-every-step-2020-6?r=US&IR=T

40 Jennifer A. Richeson, 'Americans Are Determined to Believe in Black Progress', *The Atlantic*, Sep 2020, www.theatlantic.com/magazine/archive/2020/09/the-mythology-of-racial-progress/614173/
41 See a strong argument against relying on elite representation in Sudip Bhattacharya, 'The Limits of Descriptive Representation', *Current Affairs*, 11 Feb 2021, www.currentaffairs.org/2021/02/the-limits-of-descriptive-representation
42 Luke Savage and Nathan J. Robinson, 'The Fraudulent Universalism of Barack Obama', *Current Affairs*, 23 Dec 2020, www.currentaffairs.org/2020/12/the-fraudulent-universalism-of-barack-obama, emphasis in original.
43 See Michael Gray, 'Obama orchestrated a massive transfer of wealth to the 1 percent', *New York Post*, 17 Jan 2016, nypost.com/2016/01/17/occupy-obama-he-orchestrated-a-massive-transfer-of-wealth-to-the-1-percent/; Hillary Hoffower, 'The Obamas are worth 30 times more than when they entered the White House', *Financial Review*, 12 Sep 2018, www.afr.com/work-and-careers/management/the-obamas-are-worth-30-times-more-than-when-they-entered-the-white-house-20180912-h159bg
44 Jon Swaine, Paul Lewis and Oliver Laughland, 'Troops roll in to Baltimore as Obama urges US to start "soul-searching"', *Guardian*, 28 Apr 2015, www.theguardian.com/us-news/2015/apr/28/baltimore-obama-troops-riots-police-protesters; Sam Levin, 'Obama's Dakota pipeline response "puts lives in danger", government official says', *Guardian*, 2 Nov 2016, www.theguardian.com/us-news/2016/nov/02/dakota-access-pipeline-protests-barack-obama-response
45 Matt Bruenig and Ryan Cooper, 'How Obama Destroyed Black Wealth', *Jacobin*, 12 Jul 2017, jacobinmag.com/2017/12/obama-foreclosure-crisis-wealth-inequality
46 Margaret Kimberley, 'The Obama Presidential Center Will Displace Black People', *Black Agenda Report*, 13 Oct 2021, www.blackagendareport.com/obama-presidential-center-will-displace-black-people
47 Jessica Purkiss and Jack Serle, 'Obama's covert drone war in numbers: ten times more strikes than Bush', Bureau of Investigative Journalism, 17 Jan 2017, www.thebureauinvestigates.com/stories/2017-01-17/obamas-covert-drone-war-in-numbers-ten-times-more-strikes-than-bush; Serena Marshall, 'Obama Has Deported More People Than Any Other President', ABC News, 29 Aug 2016, abcnews.go.com/Politics/obamas-deportation-policy-numbers/story?id=41715661; Chelsey Cox, 'Fact check: Obama administration approved, built temporary holding enclosures at southern border', *USA Today News*, 26 Aug 2020, eu.usatoday.com/story/news/factcheck/2020/08/26/fact-check-obama-administration-built-migrant-cages-meme-true/3413683001/

48 Jason L. Riley, 'Why Obama's Presidency Didn't Lead to Black Progress', Manhattan Institute, 17 June 2017, www.manhattan-institute.org/html/why-obamas-presidency-didnt-lead-black-progress-10380.html
49 Glenn Harlan Reynolds, 'Africans are being sold at Libyan slave markets. Thanks, Hillary Clinton', USA Today, 27 Nov 2017, eu.usatoday.com/story/opinion/2017/11/27/clinton-ponders-2020-run-lets-not-forget-her-real-libya-scandal-glenn-reynolds-column/895853001/
50 I develop this argument further here: Samuel Miller McDonald, 'Democrats Believe in "Freedom Gas" Too', *Current Affairs*, 7 June 2019, www.currentaffairs.org/2019/06/democrats-believe-in-freedom-gas-too.
51 Jeff Gammage, 'Kamala Harris follows Kaw Nation's Charles Curtis as the second person of color to become vice president', *Philadelphia Inquirer*, 17 Nov 2020, www.inquirer.com/news/kamala-harris-vice-president-charles-curtis-kaw-native-american-indigenous-indian-20201117.html
52 M. Kaye Tatro, 'Curtis Act (1898)', *Encyclopedia of Oklahoma History and Culture*, 20 July 2010, web.archive.org/web/20100720014537/http://digital.library.okstate.edu/encyclopedia/entries/C/CU006.html
53 Encyclopedia of the Great Plains, Benjamin Y. Dixon, University of Oklahoma, https://plainshumanities.unl.edu/encyclopedia/doc/egp.na.054
54 'American Indian Biography: Vice-President Charles Curtis', Native American Roots, 19 Aug 2010, nativeamericanroots.net/diary/642
55 Berlin B. Chapman, 'Charles Curtis and the Kaw Reservation', *Kansas Historical Quarterly*, 14:4 (1947), 337–51, www.kshs.org/p/charles-curtis-and-the-kaw-reservation/13069
56 Or poor men: throughout history, queens have more often waged war than kings. See Gwynn Guilford, 'Throughout history, queens were more likely to wage war than kings', Quartz, 26 Apr 2017, qz.com/967895/throughout-history-women-rulers-were-more-likely-to-wage-war-than-men/
57 Alan Travis, 'Thatcher pushed for breakup of welfare state despite NHS pledge', *Guardian*, 25 Nov 2016, www.theguardian.com/politics/2016/nov/25/margaret-thatcher-pushed-for-breakup-of-welfare-state-despite-nhs-pledge
58 Damien Gayle, 'Women disproportionately affected by austerity, charities warn', *Guardian*, 28 May 2015, www.theguardian.com/society/2015/may/28/women-austerity-charities-cuts-gender-inequality; Stewart Lansley, 'Women lose out from government cuts', PSE:UK, 2010, www.poverty.ac.uk/articles-government-cuts-gender-welfare-system-tax-government-policy-uk/women-lose-out-government

59 Simon Rogers, 'How Britain changed under Margaret Thatcher. In 15 charts', *Guardian*, 8 Apr 2013, www.theguardian.com/politics/datablog/2013/apr/08/britain-changed-margaret-thatcher-charts
60 'Increasing Environmental Pollution (GMT 10)', European Environment Agency, 18 Feb 2015, www.eea.europa.eu/soer/2015/global/pollution; G. Shaddick, M. L. Thomas, P. Mudu, G. Ruggeri and S. Gumy, 'Half the world's population are exposed to increasing air pollution', *npj Climate and Atmospheric Science*, 3:23, (2020), www.nature.com/articles/s41612-020-0124-2
61 Ed Yong, 'Wait, Have We Really Wiped Out 60 Percent of Animals?', *The Atlantic*, 31 Oct 2018, www.theatlantic.com/science/archive/2018/10/have-we-really-killed-60-percent-animals-1970/574549/
62 Louise Boyle, 'Two-thirds of world's original tropical rainforest cover degraded or destroyed by people, report finds,' *Independent*, 8 Mar 2021, www.independent.co.uk/climate-change/tropical-forest-destroyed-b1814083.html
63 Alister Doyle, 'Ocean Fish Numbers Cut in Half Since 1970', *Scientific American*, 16 Sep 2015, www.scientificamerican.com/article/ocean-fish-numbers-cut-in-half-since-1970/
64 Ann Gibbons, 'Are We in the Middle of a Sixth Mass Extinction?', *Science*, 2 Mar 2011, www.science.org/content/article/are-we-middle-sixth-mass-extinction
65 Peter S. Ross *et al*., 'Pervasive distribution of polyester fibres in the Arctic Ocean is driven by Atlantic inputs', *Nature Communications*, 12:106 (2021), www.nature.com/articles/s41467-020-20347-1
66 'New Report: The global decline in democracy has accelerated', Freedom House, 3 Mar 2021, freedomhouse.org/article/new-report-global-decline-democracy-has-accelerated; Zia Qureshi, 'Tackling the inequality pandemic: Is there a cure?', Brookings, 17 Nov 2020, www.brookings.edu/research/tackling-the-inequality-pandemic-is-there-a-cure; 'Goal 1: End poverty in all its forms everywhere', United Nations, 2021, unstats.un.org/sdgs/report/2021/goal-01/ ; Patrick Heuveline, 'Global and National Declines in Life Expectancy: An End-of-2021 Assessment', *Population and Development Review*, 48:1 (2022), 31–50, doi.org/10.1111/padr.12477; Sara Srygley *et al*., 'Losing More Ground: Revisiting Young Women's Well-being Across Generations', *Population Bulletin*, 77:1 (2023), www.prb.org/wp-content/uploads/2023/11/Losing-More-Ground_Population-Bulletin-Vol-77-No-1.pdf; Institute of Medicine (US) Forum on Microbial Threats, 'Infectious Disease Emergence: Past, Present, and Future', *Microbial Evolution and Co-Adaptation: A Tribute to the Life and Scientific Legacies of Joshua Lederberg: Workshop Summary* (Washington: National Academies Press, 2009), www.ncbi.nlm.nih.gov/books/NBK45714/ ; 'Trust in Public

Institutions: Trends and Implications for Economic Security', United Nations Policy Brief 108 (June 2021), 1–4, www.un.org/development/desa/dpad/wp-content/uploads/sites/45/publication/PB_108.pdf; 'Stagflation Risk Rises Amid Sharp Slowdown in Growth', World Bank, 7 June 2022, www.worldbank.org/en/news/press-release/2022/06/07/stagflation-risk-rises-amid-sharp-slowdown-in-growth-energy-markets; Xi Song et al., 'Long-term Decline in Intergenerational Mobility in the United States since the 1850s', PNAS, 117:1 (2019), 251–8, doi.org/10.1073/pnas.1905094116; Some species are thriving in this necrotic age, like mosquitos, ticks, viruses and other parasites.

67 For a contemplative look at this particular possibility, see Samuel Miller-McDonald, 'Bifröst to Nowhere', The Trouble, 31 July 2018, www.the-trouble.com/content/2018/7/31/bifrst-to-nowhere

68 Craig Welch, 'Earth now has 8 billion people—and counting. Where do we go from here?', National Geographic, 14 Nov 2022, www.nationalgeographic.com/environment/article/the-world-now-has-8-billion-people. To situate the writing of this book in time, I only noticed the coincidence when I went to double-check the current human population numbers for this introduction. This is the new introduction of the third draft of this manuscript; I started the first draft of the book in early 2020 during the COVID-19 pandemic, and it will be published in 2025, being a product of the first half of the 2020s. The issues it deals with will certainly be with us through the twenty-first century and probably beyond.

69 'Historical Estimates of World Population', United States Census Bureau, 5 Dec 2022, www.census.gov/data/tables/time-series/demo/international-programs/historical-est-worldpop.html

70 'World Cities Database', Simple Maps, 19 Mar 2024, simplemaps.com/data/world-cities#:~:text=Up%2Dto%2Ddate%3A%20It,every%20country%20in%20the%20world

71 David Farrier, *Footprints: In Search of Future Fossils* (London: 4th Estate, 2020).

72 'Agricultural land (% of land area)', The World Bank, 2021, data.worldbank.org/indicator/AG.LND.AGRI.ZS; A. Mood and P. Brooke, 'Estimating Global Numbers of Fishes Caught from the Wild Annually from 2000 to 2019', *Animal Welfare*, 33 (2024), doi.org/10.1017/awf.2024.7

73 Agence France-Presse, 'More than a third of world's population have never used internet, says UN', *Guardian*, 30 Nov 2021, www.theguardian.com/technology/2021/nov/30/more-than-a-third-of-worlds-population-has-never-used-the-internet-says-un?CMP=share_btn_url

74 Max Roser and Esteban Ortiz-Ospina, 'Literacy', Our World in Data, 2013, ourworldindata.org/literacy

75 See Samuel Miller McDonald, 'Empire of Same', *Current Affairs*, May/June 2020, www.currentaffairs.org/2020/08/empire-of-same
76 Bill McKibben, 'Our Stuff Weighs More than All Living Things on the Planet', *New Yorker*, 16 Dec 2020, www.newyorker.com/news/annals-of-a-warming-planet/our-stuff-weighs-more-than-all-living-things-on-the-planet; Ellie Elhacham *et al.*, 'Global Human-made Mass Exceeds All Living Biomass', *Nature*, 588 (2020), 442–4, doi.org/10.1038/s41586-020-3010-5
77 See Elizabeth Kolbert, *The Sixth Extinction: An Unnatural History* (London: Bloomsbury, 2014).
78 David Wallace-Wells, 'The Uninhabitable Earth', New York Magazine, July 2017, nymag.com/intelligencer/2017/07/climate-change-earth-too-hot-for-humans.html
79 Diagram made by the author with data from Hannah Ritchie, 'Wild mammals make up only a few percent of the world's mammals', Our World in Data, 15 Dec 2022, ourworldindata.org/wild-mammals-birds-biomass. Note that it does not include biomass of insects, plants, reptiles, birds and other wild species. Nevertheless, the starkness of this biodiversity destruction cannot be denied.
80 For an in-depth treatment of this, see Rutger Bregman, *Humankind: A Hopeful History* (London: Bloomsbury, 2020).

BEFORE PROGRESS

1 Thomas Hobbes, *Leviathan* (Andrew Crooke, 1651).
2 This narrative is captured most seriously in Steven Pinker's book, *The Better Angels of Our Nature: A History of Violence and Humanity* (London: Allen Lane, 2011), which has been systematically debunked by multiple scholars. See Douglas P. Fry, 'Peace in Our Time', *Book Forum*, Dec/Jan 2012, www.bookforum.com/print/1804/steven-pinker-offers-a-curiously-foreshortened-account-of-humanity-s-irenic-urges-8575; Pasquale Cirillo and Nassim Nicholas Taleb, 'The Decline of Violent Conflicts: What Do The Data Really Say?' Nobel Foundation Symposium 161: The Causes of Peace, 27 Nov 2016, www.fooledbyrandomness.com/violencenobelsymposium.pdf; R. Brian Ferguson, 'Pinker's List: Exaggerating Prehistoric War Mortality', in Douglas P. Fry (ed.), *War, Peace, and Human Nature: The Convergence of Evolutionary and Cultural Views* (Oxford: Oxford University Press, 2013), oxford.universitypressscholarship.com/view/10.1093/acprof:oso/9780199858996.001.0001/acprof-9780199858996-chapter-7. But versions of the narrative pervade political discourse and many people's basic assumptions. For a productive discussion of the impact of neo-Hobbesian thinking, see Bregman, *Humankind*.

3 Helen C. Rountree, 'Powhatan Indian Women: The People Captain John Smith Barely Saw', *Ethnohistory*, 45:1 (1998), 3, doi.org/10.2307/483170
4 Rountree, 'Powhatan Indian Women', 8.
5 James I, legal charter of New England, avalon.law.yale.edu/17th_century/mass01.asp. I have left most of the original spelling. I nod to Naomi Klein, *Doppelganger: A Trip Into the Mirror World* (London: Allen Lane, 2023), where a modified portion appears.
6 The region housed many different cultures, the most important in this history being the Miami, Illinois and Wendat.
7 Robert Ridgeway quoted in Susan Sleeper-Smith, *Indigenous Prosperity and American Conquest: Indian Women of the Ohio River Valley, 1690–1792* (Williamsburg, VA: The Omohundro Institute of Early America History and Culture and University of North Carolina Press, 2018), 17.
8 Sleeper-Smith, *Indigenous Prosperity and American Conquest*, 146.
9 Henry Knox quoted in Sleeper-Smith, *Indigenous Prosperity and American Conquest*, 1.
10 Sleeper-Smith, *Indigenous Prosperity and American Conquest*, 5.
11 Ibid.
12 Sleeper-Smith, *Indigenous Prosperity and American Conquest*, 19.
13 Ibid.
14 Ibid. See also Samuel Miller McDonald, 'The Last Great Forests', *Current Affairs*, 10 Nov 2021, www.currentaffairs.org/2021/11/the-last-great-forests
15 Colleen Connolly, 'The True Native New Yorkers Can Never Truly Reclaim Their Homeland', *Smithsonian Magazine*, 5 Oct 2018, www.smithsonianmag.com/history/true-native-new-yorkers-can-never-truly-reclaim-their-homeland-180970472/
16 Benjamin Franklin, letter to Peter Collinson, 9 May 1753, Founders Online, National Archives, founders.archives.gov/documents/Franklin/01-04-02-0173
17 J. Hector St. John de Crèvecoeur, *Letters from an American Farmer: Describing Certain Provincial Situations, Manners, and Customs, Not Generally Known; and Conveying Some Idea of the Late and Present Interior Circumstances of the British Colonies of North America.* (London: Thomas Davies and Lockyer Davis, 1782), 294–6.
18 Franklin, letter to Peter Collinson, 9 May 1753.
19 Traci Brynne Voyles, *Wastelanding: Legacies of Uranium Mining in Navajo Country* (Minneapolis: University of Minnesota Press, 2015), 7.
20 Alex Ross, 'Scientific breakthrough confirms how early man left Africa 84,000 years ago', *Independent*, 3 Oct 2023, www.independent.co.uk/news/science/early-man-left-africa-jordan-b2423115.html
21 Jean-Jacques Hublin *et al.*, 'New Fossils from Jebel Irhoud, Morocco

and the Pan-African Origin of *Homo sapiens*', *Nature*, 546 (2017), 289–92; Israel Hershkovitz *et al.*, 'The Earliest Modern Humans Outside Africa', *Science*, 359:6374 (2018), 456–9, doi.org/10.1126/science.aap8369; Scott Hucknall, 'Humans did not drive Australia's megafauna to extinction – climate change did', *Guardian*, 19 May 2020, www.theguardian.com/science/2020/may/19/humans-australia-megafauna-to-extinction-climate-queensland. See also Mathew Stewart, W. Christopher Carleton and Huw S. Groucutt, 'Climate Change, Not Human Population Growth, Correlates with Late Quaternary Megafauna Declines in North America', *Nature Communications*, 12:965 (2021), www.nature.com/articles/s41467-021-21201-8

22 Building upon the work of British archaeologist Ian Morris.
23 See David Wengrow, 'Cities before the State in Early Eurasia', Goody Lecture 2015, Max Planck Institute for Social Anthropology, www.eth.mpg.de/4091237/Goody_Lecture_2015.pdf; David Graeber and David Wengrow, *The Dawn of Everything: A New History of Humanity* (London: Penguin, 2021).
24 See Karl August Wittfogel, *Oriental Despotism: A Comparative Study of Total Power* (Forge Village: Yale University Press, 1957); Marvin Harris, *Cannibals and Kings* (London: Vintage, 1991); Ian Morris, *Foragers, Farmers, and Fossil Fuels: How Human Values Evolve* (Princeton: Princeton University Press, 2015); Karl Marx and Friedrich Engels, *The Communist Manifesto*, translated by Samuel Moore (London: Penguin Classics, 2015).
25 Wengrow, 'Cities before the State in Early Eurasia'.
26 Mimi Kirk, 'The Case for a Nature-Rich Urban Future', Bloomberg, 13 June 2017, www.bloomberg.com/news/articles/2017-06-13/the-case-for-a-nature-rich-urban-future; Meg St-Esprit McKivigan, '"Nature Deficit Disorder" Is Really a Thing', *New York Times*, 23 June 2020, www.nytimes.com/2020/06/23/parenting/nature-health-benefits-coronavirus-outdoors.html
27 Lorena Becerra-Valdivia and Thomas Higham, 'The Timing and Effect of the Earliest Human Arrivals in North America', *Nature*, 584 (2020), 93–7, www.nature.com/articles/s41586-020-2491-6; Nick Howe and Shamini Bundell, 'When did people arrive in the Americas? New evidence stokes debate', *Nature*, 22 July 2020, podcast, www.nature.com/articles/d41586-020-02200-z
28 Jason Hickel, *Less is More: How Degrowth Will Save the World* (London: Penguin Random House, 2020), 257.
29 Hickel, *Less is More*, 255, emphasis in original.
30 Bruce Pascoe, *Dark Emu* (London: Scribe Publications, 2014), 2.
31 Lisa Brooks, 'Indigenous Oral Traditions of North America, Then and Now', in Paul Lauter, ed., *A Companion to American Literature and Culture* (Chichester: Wiley-Blackwell, 2010), 126.

32 Ibid.
33 Robin Wall Kimmerer, *Braiding Sweetgrass: Indigenous Wisdom, Scientific Knowledge and the Teachings of Plants* (London: Penguin Books, 2020), 307.
34 David Bowman, Jason Gibson and Toshiaki Kondo, 'Aboriginal Myth Meets DNA Analysis', *Nature*, 520:33 (2015), doi.org/10.1038/520033a
35 Susie Cagle, '"Fire is medicine": the tribes burning California forests to save them', *Guardian*, 21 Nov 2019, www.theguardian.com/us-news/2019/nov/21/wildfire-prescribed-burns-california-native-americans
36 William Ramsay Smith, *Myths and Legends of the Australian Aborigines* (New York: Dover Publications, 2003), 228.
37 Hobbes, *Leviathan*.
38 Eze Paez, 'Wild Animal Ethics: A Freedom-Based Approach', *Ethics, Policy & Environment*, 26:2 (2023), 159–78, doi.org/10.1080/21550085.2023.2200728
39 Philip Hoare, 'Sperm whales in 19th century shared ship attack information', *Guardian*, 17 Mar 2021, www.theguardian.com/environment/2021/mar/17/sperm-whales-in-19th-century-shared-ship-attack-information?CMP=share_btn_tw
40 Nicholas J. Mulcahy, 'An Orangutan Hangs Up a Tool for Future Use', *Scientific Reports*, 8 (2018), www.nature.com/articles/s41598-018-31331-7
41 Thibaud Gruber, Ian Singleton and Carel van Schaik, 'Sumatran Orangutans Differ in Their Cultural Knowledge but Not in Their Cognitive Abilities', *Current Biology*, 22:23 (2012), 2231–5.
42 Désirée Brucks and Auguste M. P. von Bayern, 'Parrots Voluntarily Help Each Other to Obtain Food Rewards', *Current Biology*, 30:2 (2020), 292–7; Schuyler Velasco, 'Parrots learn to make video calls to chat with other parrots, then develop friendships, Northeastern University researchers say', Northeastern Global News, 21 Apr 2023, news.northeastern.edu/2023/04/21/parrots-talking-video-calls/
43 Dominique Potvin, 'Altruism in birds? Magpies have outwitted scientists by helping each other remove tracking devices', The Conversation, 21 Feb 2022, theconversation.com/altruism-in-birds-magpies-have-outwitted-scientists-by-helping-each-other-remove-tracking-devices-175246
44 Peter Godfrey-Smith, 'The Mind of an Octopus', *Scientific American*, 1 Jan 2017, www.scientificamerican.com/article/the-mind-of-an-octopus/
45 Kumagusu Minakata, 'An Intelligence of the Frog', *Nature* 50:79 (1894), www.nature.com/articles/050079d0; David Graeber, 'What's the Point If We Can't Have Fun?' *The Baffler*, Jan 2014, thebaffler.com/salvos/whats-the-point-if-we-cant-have-fun; Scarlett R. Howard et al., 'Numerical Cognition in Honeybees Enables Addition and Subtraction', *Science Advances*, 5:2 (2019), doi.org/10.1126/sciadv.aav0961

46 The treatment of animals in industrial-scale feedlots around the world are enough to make one sick. Some examples can be found in the following sources: Yuval Noah Harari, 'Industrial farming is one of the worst crimes in history', *Guardian*, 25 Sep 2015, www.theguardian.com/books/2015/sep/25/industrial-farming-one-worst-crimes-history-ethical-question; Matthew C. Halteman, 'Varieties of Harm to Animals in Industrial Farming', *Journal of Animal Ethics*, 1:2 (2011), 122–31, www.jstor.org/stable/10.5406/janimalethics.1.2.0122; Jordan O. Hampton *et al.*, 'Animal Harms and Food Production: Informing Ethical Choices', *Animals (Basel)*, 11:5 (2011), doi.org/10.3390/ani11051225; Nancy Perry and Peter Brandt, 'A Case Study on Cruelty to Farm Animals: Lessons Learned From the Hallmark Meat Packing Case', *Michigan Law Review*, 106, (2008), 117–22, repository.law.umich.edu/mlr_fi/vol106/iss1/7

47 Hobbes's methodology for arriving at his claims about how society and politics work was to use his *reason* – that is to say, to imagine himself in a 'state-of-nature' and consider how he might act in this 'condition of war of all against all', as he famously put it. No wonder when more empirical methodologies are applied to the question, the answer is quite different from the imagination of a man who spent much of his life in the intimate employ of an aristocratic family. To be fair, Hobbes also probably drew conclusions from the conflicts that were raging between European settlers and Native North Americans during his lifetime, and made the leviathanic presumption that the behaviour of First Nations under military siege by an aggressive invader was reflective of all societies without European states and monarchies to rule them.

48 Eric D. Galbraith *et al.*, 'High life satisfaction reported among small-scale societies with low incomes', *PNAS*, 121:7 (2024), doi.org/10.1073/pnas.2311703121

49 Anthropologist David Graeber and archaeologist David Wengrow, among other archaeologists, have noted the curious burials found dotted around Eurasia in which apparently deformed bodies are buried with great riches, perhaps suggesting that not only were they well looked after, despite their lack of physical fitness, but even worshipped.

50 David Graeber, *Debt: The First 5,000 Years* (New York: Melville House, 2014), 29.

51 Paula Gunn Allen, *The Sacred Hoop: Recovering the Feminine in American Indian Traditions* (Boston, MA: Beacon Press, 1992).

52 Maya Wei-Haas, 'Prehistoric female hunter discovery upends gender role assumptions', *National Geographic*, 4 Nov 2020, www.nationalgeographic.com/science/2020/11/prehistoric-female-hunter-discovery-upends-gender-role-assumptions/

53 Hannah Devlin, 'Early men and women were equal, say scientists', *Guardian*, 14 May 2015, www.theguardian.com/science/2015/may/14/early-men-women-equal-scientists

I HEAVEN

1 Though it may be tempting to accuse me of accidentally vindicating Jefferson for comparing historical early agrarianism to that which could be found in middle America during his time, the important point is that this mode of production is not a teleological feature, not an indication of humanity moving upward through time or, given the prosperity of those Ohio River Valley tribes, of their being at a more primitive stage of development. Nor is the narrative indicating that times before this were more barbaric or violent or impoverished.
2 See, for instance, Yuval Noah Harari, *Sapiens: A Brief History of Humankind* (London: Vintage, 2015), James C. Scott, *Against the Grain: A Deep History of the Earliest States* (New Haven: Yale University Press, 2017) and work by Jared Diamond, and many of the classics in both trade and scholarly archaeology and anthropology.
3 David Wengrow, *What Makes Civilisation? The Ancient Near East and the Future of the West* (Oxford: Oxford University Press, 2018), 72–3.
4 Wengrow, *What Makes Civilisation?*, 80.
5 Ian Johnstone, 'Dogs' innate sense of fairness being eroded by humans, study suggests', *Independent*, 8 Jun 2017, www.independent.co.uk/news/science/dogs-wolves-unfairness-acceptance-human-owners-pets-behaviour-research-vienna-a7779076.html; Jennifer L. Essler, Sarah Marshall-Pescini and Friederike Range, 'Domestication Does Not Explain the Presence of Inequity Aversion in Dogs', *Current Biology*, 27:12 (2017), 1861–5, doi.org/10.1016/j.cub.2017.05.061
6 Some examples of inequity aversion and other sorts of fairness and altruism have already appeared in the text, but others exist, among canines, for instance, and other non-human primates. See Sarah F. Brosnan, 'Justice- and Fairness-Related Behaviors in Nonhuman Primates', *Proc Natl Acad Sci USA*, 110:2 (2013), 10416–23, doi.org/10.1073/pnas.1301194110; Deutsches Primatenzentrum (DPZ)/German Primate Center, 'Insights into the evolution of the sense of fairness', ScienceDaily, 2 Mar 2023, www.sciencedaily.com/releases/2023/03/230302114205.htm; and Claudia Wascher, 'Animals know when they are being treated unfairly (and they don't like it)', The Conversation, 21 Feb 2017, theconversation.com/animals-know-when-they-are-being-treated-unfairly-and-they-dont-like-it-73152
7 *The Epic of Gilgamesh*, trans. Maureen Gallery Kovacs (Wolf

Carnahan, 1998), uruk-warka.dk/Gilgamish/The%20Epic%20of%20Gilgamesh.pdf
8 Yoshinori Yasuda, Hiroyuki Kitagawa and Takeshi Nakagawa, 'The Earliest Record of Major Anthropogenic Deforestation in the Ghab Valley, Northwest Syria: A Palynological Study', *Quaternary International*, 73–74 (2000), 127–36, doi.org/10.1016/S1040-6182(00)00069-0
9 'Enuma Elish: The Epic of Creation', trans. L. W. King, *The Seven Tablets of Creation* (London: Evinity Publishing, 1902), www.sacred-texts.com/ane/enuma.htm
10 'Enuma Elish', Tablet Seven.
11 'Enuma Elish', Tablet Four.
12 'Enuma Elish', Epilogue.
13 Richard G. A. Buxton, 'Functions of myth and mythology', Encyclopedia Britannica, 3 Nov 2020, www.britannica.com/topic/myth/Functions-of-myth-and-mythology#ref386978
14 H. D. Baker, 'Degrees of Freedom: Slavery in Mid-First Millennium BC Babylonia', *World Archaeology*, 33:1 (2001), 18–26, www.jstor.org/stable/827886
15 Morris, *Foragers, Farmers, and Fossil Fuels*, 72.
16 Though these connections are difficult to prove. Some assert the possibility that there was no direct influence between them, or even that Zoroastrian eschatology was inspired by these other faiths: 'Eschatology i. In Zoroastrianism and Zoroastrian Influence', Encyclopædia Iranica, 19 Jan 2012, www.iranicaonline.org/articles/eschatology-i
17 Revelation 21:1–4, www.biblica.com/bible/niv/revelation/21/
18 Daan Nijssen, 'Cyrus the Great', World History Encyclopedia, 21 Feb 2018, www.worldhistory.org/Cyrus_the_Great/
19 Genesis 1:26–28, King James Version, Bible Gateway, www.biblegateway.com/passage/?search=Genesis%201:26-28&version=KJV
20 Genesis 12, King James Version.
21 Mordechai Cogan, 'Israel's Incomplete Conquest of Canaan', The Torah, 25 Jan 2022, www.thetorah.com/article/israels-incomplete-conquest-of-canaan
22 Genesis 46:2-4, King James Bible (KJV)
23 Clifford R. Backman, *Worlds of Medieval Europe* (Oxford: Oxford University Press, 2002), 25.
24 See e.g., Matthew W. Slaboch, *A Road to Nowhere: The Idea of Progress and Its Critics* (Philadelphia: University of Pennsylvania Press, 2018), 4; Peter Wagner, *Progress: A Reconstruction* (Cambridge: Polity Press, 2016); and Christopher Lasch, *The True and Only Heaven: Progress and Its Critics* (New York: W. W. Norton, 1991), 54.

25 Robert A. Nisbet, *History of the Idea of Progress* (London: Routledge, 1994). It's important to note that Nisbet is unabashedly and self-avowedly an advocate for the idea of progress as a positive force in societies, so we should read some of his work with scepticism. I have mostly used it here as a guide for tracing the lineage of the idea through other thinkers in the Western canon, which is not substantially impacted by a positive or negative bias.
26 Nisbet, *Progress*, 13.
27 Nisbet, *Progress*, 26.
28 Benjamin Isaac, 'Proto-Racism in Graeco-Roman Antiquity', *World Archaeology*, 38:1 (2006), 32–47, www.jstor.org/stable/40023593
29 Isaac, 'Proto-Racism in Graeco-Roman Antiquity'.
30 Nisbet, *Progress*, 41.
31 Nisbet, *Progress*, 38.
32 Ashley Dawson, *Extinction: A Radical History* (London: OR Books, 2016), 34.
33 Dawson, *Extinction*, 32–3.
34 Ali Parchami, *Hegemonic Peace and Empire: The Pax Romana, Britannica and Americana* (London: Routledge, 2009), 26.
35 'Desert' has sometimes been translated as 'solitude', but the former is both more literal and more appropriate. Cornelius Tacitus, *The Complete Works of Tacitus*, translated by William J. Brodribb and Alfred John Church (New York: Random House, 1942).
36 Richard Nicolaus Coudenhove-Kalergi, *Crusade for Pan-Europe: Autobiography of a Man and a Movement* (New York: G. P. Putnam, 1943), 299–304.
37 Parchami, *Hegemonic Peace and Empire*, 26.
38 Theodor E. Mommsen, 'St. Augustine and the Christian Idea of Progress: The Background of the City of God', *Journal of the History of Ideas*, 12:3 (1951), 346–74, www.jstor.org/stable/2707751
39 Ian Morris, *Geography is Destiny* (London: Profile Books, 2022), 116.
40 Tacitus, *Agricola*, chapter 21. Quoted in Morris, *Geography is Destiny*, 109.
41 See Reza Aslan, *Zealot: The Life and Times of Jesus of Nazareth* (London: Penguin Random House, 2014).
42 Theodor E. Mommsen, 'St. Augustine and the Christian Idea of Progress: The Background of the City of God', *Journal of the History of Ideas*, 12:3 (1951), 346–74, www.jstor.org/stable/2707751
43 Garth Fowden, *Empire to Commonwealth: Consequences of Monotheism in Late Antiquity* (Princeton: Princeton University Press, 1993), 88.
44 Matthew 28:18–20, New International Version.
45 Fowden, *Empire to Commonwealth*, 91.
46 Fowden, *Empire to Commonwealth*, 139.
47 Nisbet, *Progress*, 53.

48 For details of Constantine's 'Pro-Christian Legislation', see Chapter 6, no. 5 in Timothy D. Barnes, *Constantine: Dynasty, Religion and Power in the Later Roman Empire* (Chichester: Wiley-Blackwell, 2011).
49 Libanius, Oration 30, *Pro Templis*, 'To the Emperor Theodosius, for the temples', 8–10, in *Libanius: Selected Orations*, ed. and trans. A. F. Norman, Loeb Classical Library (London: Heinemann, 1977).
50 Libanius, Oration 30. For more on the destruction of pagan temples, see Edward Watts, 'Fiddling while Rome converts', *Aeon*, 27 Oct 2020, aeon.co/essays/pagan-complacency-and-the-birth-of-the-christian-roman-empire
51 Nisbet, *Progress*, 76.
52 Quoted in Nisbet, *Progress*, 54.
53 Hickel, *Less is More*, 65.
54 Timothy Bolton, *Cnut the Great* (New Haven: Yale University Press, 2017).
55 James H. Barrett et al., 'Ecological Globalisation, Serial Depletion and the Medieval Trade of Walrus Rostra', *Quaternary Science Reviews* 229 (2020), 1–15, doi.org/10.1016/j.quascirev.2019.106122
56 Andrewd Lawler, 'Kinder, Gentler Vikings? Not According to Their Slaves', *National Geographic*, 28 Dec 2015, www.nationalgeographic.com/news/2015/12/151228-vikings-slaves-thralls-norse-scandinavia-archaeology/
57 Jared Diamond, *Collapse* (London: Penguin, 2005), 276.
58 Kevin Crossley-Holland, *The Norse Myths* (London: The Folio Society, 1989), xvii.
59 Crossley-Holland, *The Norse Myths*, 200.
60 Tharik Hussain, 'Why did Vikings have "Allah" embroidered into funeral clothes?' BBC News, 12 Oct 2017, www.bbc.co.uk/news/world-europe-41567391
61 Caitlin Ellis, 'Vikings didn't just murder monks and pillage monasteries – they helped spread Christianity too', The Conversation, 23 Dec 2019, theconversation.com/vikings-didnt-just-murder-monks-and-pillage-monasteries-they-helped-spread-christianity-too-128910
62 Ellis, 'Vikings didn't just murder monks'.
63 James Gormon, 'The Vikings Were More Complicated Than You Think', *New York Times*, 16 Sep 2020, www.nytimes.com/2020/09/16/science/vikings-DNA.html
64 Francis Joy, 'The Disappearance of the Sacred Swedish Sámi Drum and the Protection of Sámi Cultural Heritage', *Polar Record*, 54:4 (2018), 256, doi.org/10.1017/S0032247418000438
65 Ibid.
66 Ibid.
67 Morris, *Geography is Destiny*, 164.
68 Fowden, *Empire to Commonwealth*, 139.

69 Some may balk at a 'golden age' framing, especially when contrasted with a subsequent 'dark age', but there can be no doubt that this period produced an abundance of scholarship and learning – and also gold. For a contrary opinion, see Asad Q. Ahmed, 'Islam's invented Golden Age', openDemocracy, 28 Oct 2013, www.opendemocracy.net/en/openindia/islams-invented-golden-age/
70 Matthew E. Falagas, Effie A. Zarkadoulia and George Samonis, 'Arab Science in the Golden Age (750–1258 C.E.) and Today', *The FASEB Journal* 20:10 (2006), 1581–6, doi.org/10.1096/fj.06-0803ufm
71 Falagas *et al.*, 'Arab science'.
72 Dennis Overbye, 'How Islam Won, and Lost, the Lead in Science', *New York Times*, 30 Oct 2001, www.nytimes.com/2001/10/30/science/how-islam-won-and-lost-the-lead-in-science.html
73 Jim Al-Khalili, 'It's time to herald the Arabic science that prefigured Darwin and Newton', *Guardian*, 30 Jan 2008, www.theguardian.com/commentisfree/2008/jan/30/religion.world
74 Mihai Andrei, 'Meet Islam's Da Vinci: Al-Biruni, father of geodesy, anthropology, and master of pharmacy', ZME Science, 27 Nov 2020, www.zmescience.com/other/feature-post/meet-islams-da-vinci-al-biruni-father-of-geodesy-anthropology-and-master-of-pharmacy/
75 Bernard K. Freamon, *Possessed by the Right Hand: The Problem of Slavery in Islamic Law and Muslim Cultures* (Leiden: Brill, 2019), 197–8.
76 Freamon, *Possessed by the Right Hand*, 199.
77 See Ibn Khaldun, *The Muqaddimah: An Introduction to History*, trans. Franz Rosenthal, ed. N. J. Dawood (Princeton: Princeton University Press, 2015). Or an online version: delong.typepad.com/files/muquaddimah.pdf
78 Khaldun, *The Muqaddimah*.
79 Khaldun, *The Muqaddimah*.
80 Khaldun, *The Muqaddimah*.
81 Dieter Weiss, 'Ibn Khaldun on Economic Transformation', *International Journal of Middle East Studies*, 27:1 (1995), 29–37., doi.org/10.1017/S0020743800061560
82 'Ibn Khaldun and Thomas Hobbes both present the pre-political state as characterized by intolerable violence, and view the establishment of a strong political authority as the best hope of stability.' From Alexander Orwin, 'Dawla and Leviathan: Ibn Khaldun and Hobbes in Defense of State Authority', *İbn Haldun Çalışmaları Dergisi*, 3:1 (2018), 47–64., doi.org/10.36657/ihcd.2018.35
83 Khaldun, *The Muqaddimah*, 123.
84 Khaldun, *The Muqaddimah*.
85 Cited in Freamon, *Possessed by the Right Hand*, 220.
86 *The Qur'an*, 35:39.

87 *The Qur'an*, 40:79–80.
88 See Morris, *Geography is Destiny*, 118, and Robin Fleming, *Britain after Rome: The Fall and Rise, 400–1070* (London: Penguin, 2011), 357. Fleming writes that life expectancy likely increased in Britain with the collapse of Rome and that 'many people had healthier childhoods once urbanisation, taxation, and galloping social inequalities had disappeared.'
89 Marvin Harris, *Cannibals and Kings* (London: Vintage, 1991), 325.
90 Harris, *Cannibals and Kings*, 328.
91 Karl Marx, 'A Contribution to the Critique of Hegel's Philosophy of Right', *Marx: Early Political Writings*, ed. Joseph J. O'Malley (Cambridge: Cambridge University Press, 1994), 57–70.
92 Harari, *Sapiens*, 234.
93 Peter Turchin *et al.*, 'War, Space, and the Evolution of Old Word Complex Societies', *Proceedings of the National Academy of Sciences*, 110:41 (2013), 16384–9, doi.org/10.1073/pnas.1308825110
94 See Dean Hamer, *The God Gene: How Faith is Hardwired Into Our Genes* (New York: Doubleday, 2004), largely dismissed in M. Goldman, 'The God Gene: How Faith is Hardwired Into Our Genes', *Nature Genetics*, 36:12 (2004), 1241, doi.org/10.1038/ng1204-1241

II NATION

1 Alan Riach, 'Dúthchas: The word that describes understanding of land, people and culture', The National, 16 Mar 2020, www.thenational.scot/news/18306403.duthchas-word-describes-understanding-land-people-culture
2 Hickel, *Less is More*, 44.
3 Kate Ravilious, 'Europe's Chill Linked to Disease', BBC News, 27 Feb 2006, http://news.bbc.co.uk/1/hi/sci/tech/4755328.stm
4 Hickel, *Less is More*, 42.
5 Spencer Dimmock, 'Expropriation and the Political Origins of Agrarian Capitalism in England', in Xavier Lafrance and Charles Post (eds), *Case Studies in the Origins of Capitalism* (London: Palgrave Macmillan, 2019), 39–62.
6 Roxanne Dunbar-Ortiz, *An Indigenous Peoples' History of the United States* (Boston, MA: Beacon Press, 2014), 33.
7 Dimmock, 'Expropriation and the Political Origins of Agrarian Capitalism'.
8 Alexandra M. de Pleijt and Jan Luiten van Zanden, 'Two Worlds of Female Labour: Gender Wage Inequality in Western Europe, 1300-1800', EHES Working Papers in Economic History no. 138 (2018), 1–31, http://www.ehes.org/EHES_138.pdf

9 *The Journal of Christopher Columbus (During His First Voyage), and Documents Relating to the Voyages of John Cabot and Gaspar Corte Real*, ed. and trans. Clements R. Markham (London: Hakluyt Society, 1893), 41.
10 Columbus wrote, 'As soon as I arrived in the Indies, on the first Island which I found, I took some of the natives by force in order that they might learn and might give me information of whatever there is in these parts.' Cited in Howard Zinn, *A People's History of the United States: 1492 to Present* (London: HarperCollins, 2005), 2.
11 Zinn, *A People's History of the United States*, 3.
12 Jason Hickel, *The Divide: A Brief Guide to Global Inequality and its Solutions* (London: Heinemann, 2017), 71.
13 Juan López de Palacios Rubios, 'El Requerimiento' (1513), English translation, kdhist.sitehost.iu.edu/H105-documents-web/week02/Requerimiento1513.html
14 Bartolomé de las Casas, *A Brief Account of the Destruction of the Indies* (London: R. Hewson, 1689), www.gutenberg.org/cache/epub/20321/pg20321.html
15 Aaron O'Neill, 'Annual share of slaves who died during the Middle Passage 1501–1866', Statista, 2 Feb 2024, www.statista.com/statistics/1143458/annual-share-slaves-deaths-during-middle-passage/
16 Alexander Koch et al., 'Earth system impacts of the European arrival and Great Dying in the Americas after 1492', *Quaternary Science Reviews* 207 (2019), 13–36, www.sciencedirect.com/science/article/pii/S0277379118307261
17 Kyle Powys Whyte, 'Our Ancestors' Dystopia Now: Indigenous Conservation and the Anthropocene', in Ursula Heise, Jon Christensen and Michelle Niemann (eds), *The Routledge Companion to the Environmental Humanities* (London: Routledge, 2017), 206–15.
18 Dunbar-Ortiz, *An Indigenous Peoples' History*, 43.
19 John Dorney, 'The Great Irish Famine 1845–1851 – A Brief Overview', *The Irish Story*, 18 Oct 2016, www.theirishstory.com/2016/10/18/the-great-irish-famine-1845-1851-a-brief-overview/
20 Ocean Malandra, 'EarthRx: The Irish Potato Famine Was Caused by Capitalism, Not a Fungus', *Paste*, 13 Mar 2017, www.pastemagazine.com/science/irish-potato-famine/earthrx-the-irish-potato-famine-was-caused-by-capi/
21 Chris Bambery, *A People's History of Scotland* (London: Verso, 2014), 73.
22 Graeber and Wengrow, *The Dawn of Everything*, 31
23 Graeber and Wengrow, *The Dawn of Everything*, 44–59.
24 Graeber and Wengrow, *The Dawn of Everything*, 38.
25 Graeber and Wengrow, *The Dawn of Everything*, 39.
26 Graeber and Wengrow, *The Dawn of Everything*, 40.
27 Graeber and Wengrow, *The Dawn of Everything*, 41.

28 Nisbet, *Progress*, 171.
29 Karl Jaspers, who coined the term 'Axial Age', puts the date of the era from around 800 BCE to 200 BCE, a period in which many of the world's major philosophical debates arise and the world's major religions have their roots. The anthropologist David Graeber extends the era from 800 BCE to 600 CE in his book *Debt*, to include the founding of Christianity and Islam. Ian Morris discusses the absorption of Axial Age ideals into dominant regimes in *Foragers, Farmers, and Fossil Fuels*.
30 And where, incidentally, I wrote this book.
31 Charlie McKinnon, 'The radical Robert Burns', *International Socialism*, 9 Jan 2018, http://isj.org.uk/the-radical-robert-burns/
32 From Moncure Daniel Conway (ed.), *The Writings of Thomas Paine, Vol. 1 (1774–1779)* (New York: Knickerbocker Press, 1894).
33 Ronald L. Meek (ed.), *Turgot on Progress, Sociology and Economics* (Cambridge: Cambridge University Press, 1973), 42.
34 Ibid.
35 As Nisbet puts it, 'Turgot's experience within a single year can be seen as the very epitome of processes of intellectual development . . . which take us . . . from Providence-as-progress to progress-as-Providence.' Nisbet, *Progress*, 182.
36 Meek, *Turgot on Progress*, Introduction, 29.
37 J. Baillie, *The Belief in Progress* (Oxford: Oxford University Press, 1950), 113, quoted in Meek, *Turgot on Progress*.
38 Golo Mann, 'The Belief in Progress by John Baillie', *Commentary*, Dec 1951, www.commentary.org/articles/golo-mann/the-belief-in-progress-by-john-baillie/
39 Meek, *Turgot on Progress*, 27.
40 Nisbet, *Progress*, 180.
41 Meek, *Turgot on Progress*, 43.
42 The Mises Institute, 'Online Library of Liberty', and Libertarianism.org all revere Turgot. For instance: mises.org/profile/arj-turgot; oll.libertyfund.org/page/turgot-and-enlightened-progress; www.libertarianism.org/articles/turgot-first-theory-progress
43 Nisbet, *Progress*, 117.
44 Edward Gibbon, *The History of the Decline and Fall of the Roman Empire*, ed. J. B. Bury (New York: Fred de Fau and Co., 1906), volume 1, chapter 9.
45 Gibbon, *History of the Decline and Fall*.
46 Ibid.
47 Marie Jean Antoine Nicolas de Caritat, *Outlines of an Historical View of the Progress of the Human Mind*, translated from the French (Philadelphia: M. Carey, 1795).
48 Quoted by Bregman in *Humankind*, 248.

49 Dunbar-Ortiz, *An Indigenous Peoples' History*, 38–9.
50 Sikata Banerjee, *Muscular Nationalism: Gender, Violence, and Empire in India and Ireland, 1914–2004* (New York University Press, 2012), 30.
51 Cited in Banerjee, *Muscular Nationalism*, 31.
52 Full disclosure, there's a good chance that relatives on my father's side were victims of both the Highland Clearances and the Irish famine.
53 Kevin Keane, 'Study into impact of land ownership in Scotland', BBC News, 28 Sep 2017, www.bbc.co.uk/news/uk-scotland-highlands-islands-41414706
54 Bambery, *People's History of Scotland*, 111.
55 Dunbar-Ortiz, *An Indigenous Peoples' History*, 36.
56 It is important to note that many Irish and Scottish people, while themselves from a colonised or at least regularly brutalised population, went on to commit atrocities in service of parasitic economies and imperial ambitions. As Sikata Banerjee notes, Irish people contributed to colonisation in India as missionaries, soldiers and bureaucrats: 'As both colonized and colonizer, the Irish, although deemed barbaric and ape-like at certain times, at other times were seen as worthy allies in the British imperial project.' Racial hierarchies are almost infinitely flexible and easily repurposed to serve the imperatives of changing regimes. Banerjee, *Muscular Nationalism*, 44.
57 John Cottingham, '"A Brute to the Brutes?": Descartes' Treatment of Animals', *Philosophy*, 53:206 (1978), 551–9, www.jstor.org/stable/3749880
58 Stanley E. Hedeen, 'From Billions to None: Destruction of the Passenger Pigeon in the Ohio Valley', *Ohio Valley History*, 10:3 (2010), 28.
59 Hobbes, *Leviathan*.
60 Frans de Waal, *Our Inner Ape: The Best And Worst Of Human Nature* (Granta Books, 2014); see also 'veneer theory', Emily Rios, 'De Waal Sides with Darwin: Morality is Instinctual, Evolved', *Emory Report*, 59, 27 (2007), www.emory.edu/EMORY_REPORT/erarchive/2007/April/April%2016/DeWaal.htm
61 Christopher Bertram, 'Jean Jacques Rousseau', in Edward N. Zalta (ed.), *The Stanford Encyclopedia of Philosophy* (Winter 2020), plato.stanford.edu/archives/win2020/entries/rousseau/.
62 Paul Oslington, 'God and the Market: Adam Smith's Invisible Hand', *Journal of Business Ethics*, 108 (2012), 429–38.
63 Paul Sagar, 'The real Adam Smith', *Aeon*, 16 Jan 2018, aeon.co/essays/we-should-look-closely-at-what-adam-smith-actually-believed
64 William Easterly, 'Progress by Consent: Adam Smith as Development Economist', *The Review of Austrian Economics*, 34 (2019), 179–201, doi.org/10.1007/s11138-019-00478-5

65 For a recent treatment of Malthus, see Giorgos Kallis, *Limits: Why Malthus Was Wrong and Why Environmentalists Should Care* (Stanford: Stanford University Press, 2019).
66 Thomas Robert Malthus, *An Essay on the Principle of Population* (London: John Murray, 1826).
67 Samuel M. Levin, 'Malthus and the Idea of Progress', *Journal of the History of Ideas*, 27:1 (1966), 92–108, www.jstor.org/stable/2708310
68 Malthus, *Essay on the Principle of Population*.
69 For a good look at the difference between 'liberals' and 'leftists', see Nathan J. Robinson, 'The Difference Between Liberalism and Leftism', *Current Affairs*, 7 June 2017, www.currentaffairs.org/2017/06/the-difference-between-liberalism-and-leftism
70 The company captured nearly a quarter of a million people, more than 40,000 of whom died before reaching markets. Locke's involvement in the Royal Africa Company was at least in part politically motivated and he sold the stocks within a couple of years. There is reason to believe his views evolved over time, as Holly Brewer writes in 'Slavery-entangled philosophy', *Aeon*, 12 Sep 2018, aeon.co/essays/does-lockes-entanglement-with-slavery-undermine-his-philosophy
71 John James Audubon, cited in Stanley E. Hedeen, 'From Billions to None: Destruction of the Passenger Pigeon in the Ohio Valley', *Ohio Valley History*, 10:3 (2010), 28.
72 Mark Kurlansky, *Cod: A Biography of the Fish that Changed the World* (London: Vintage Books, 1999), 49.
73 Kurlansky, *Cod*, 70.
74 Kurlansky, *Cod*, 69.
75 Danielle Taschereau Mamers, 'Historical photo of mountain of bison skulls documents animals on the brink of extinction', The Conversation, 2 Dec 2020, theconversation.com/historical-photo-of-mountain-of-bison-skulls-documents-animals-on-the-brink-of-extinction-148780
76 J. Weston Phippen, '"Kill Every Buffalo You Can! Every Buffalo Dead Is an Indian Gone"', *The Atlantic*, 13 May 2016, www.theatlantic.com/national/archive/2016/05/the-buffalo-killers/482349/
77 Colin Woodard, 'Determining America's National Myth Will Determine the Country's Fate', Zocalo Public Square, 22 Feb 2021, www.zocalopublicsquare.org/2021/02/22/america-nationhood-george-bancroft-frederick-douglass/ideas/essay/.
78 Woodard, 'America's National Myth'.
79 John O'Sullivan, 'Annexation', *United States Magazine and Democratic Review*, 17:1 (1845), 5–10.
80 While Jackson and Lincoln are not exactly 'Founding Fathers', they are early presidents who administered, in the former case, major growth and conquest of Indigenous peoples' land, and in the latter,

both conquest and a sort of 'second founding' with the culmination of the Civil War and Reconstruction.
81 See Graeber and Wengrow, *The Dawn of Everything*.
82 Zinn, *A People's History of the United States*, 118.
83 Ibid.
84 Thomas Jefferson, letter to Henry Dearborn, 28 Aug 1807, Founders Online, National Archives, founders.archives.gov/documents/Jefferson/99-01-02-6267
85 Thomas Jefferson, letter to David Bailie Warden, 29 Dec 1813, Founders Online, National Archives, founders.archives.gov/documents/Jefferson/03-07-02-0046
86 Zinn, *A People's History of the United States*, 126.
87 Patrick Wolfe, 'Settler Colonialism and the Elimination of the Native', *Journal of Genocide Research*, 8:4 (2006), 401.
88 Nisbet, *Progress*, 193.
89 The Thomas Jefferson Foundation, an institution aimed at maintaining a generally positive legacy for the former president, has written a rather chilling account of his actions: 'Thomas Jefferson's Enlightenment and American Indians', *Thomas Jefferson's Monticello*, www.monticello.org/thomas-jefferson/louisiana-lewis-clark/origins-of-the-expedition/jefferson-and-american-indians/jefferson-s-enlightenment-and-american-indians/. Some more accounts of this history can be found in Donald L. Fixico, 'When Native Americans Were Slaughtered in the Name of "Civilisation"', History, 2 Mar 2018, www.history.com/news/native-americans-genocide-united-states
90 Zinn, *A People's History of the United States*, 126.
91 Jeffrey Ostler, *Surviving Genocide* (New Haven: Yale University Press, 2020).
92 For raw numbers, which are staggeringly high, see André B. Rosay, 'Violence Against American Indian and Alaska Native Women and Men', National Institute of Justice, 1 June 2016, nij.ojp.gov/topics/articles/violence-against-american-indian-and-alaska-native-women-and-men. Rosay notes that 'in Oliphant v. Suquamish Indian Tribe (1978), the U.S. Supreme Court ruled that tribes did not have criminal jurisdiction over non-Indians', which has led to many non-Natives acting with impunity against Natives. There is an 'epidemic' of Native women suffering violent and sexual attacks: Giulia Marchiò, 'The Deafening Silence on The Massacre Of Native Women in The United States Of America', Human Rights Pulse, 12 Aug 2021, www.humanrightspulse.com/mastercontentblog/the-deafening-silence-on-the-massacre-of-native-women-in-the-united-states-of-america
93 Thomas Jefferson, letter to Edward Carrington, 16 Jan 1787, Founders Online, National Archives, founders.archives.gov/documents/Jefferson/01-11-02-0047. Parenthesis in the original letter.

94 Quoted in Levi Rickert, 'U.S. Presidents in Their Own Words Concerning American Indians', Native News Online, 16 Feb 2020, nativenewsonline.net/currents/us-presidents-words-concerning-american-indians/
95 George Washington, letter to Major General John Sullivan, 31 May 1779, Founders Online, National Archives, founders.archives.gov/documents/Washington/03-20-02-0661. See also Barbara Alice Man, *George Washington's War on Native America* (Lincoln, NE: University of Nebraska Press, 2009).
96 Stuart Leibiger, '*George Washington's War on Native America*, and *The Political Philosophy of George Washington* (review)', *Journal of the Early Republic*, 31:3 (2011), 529–33, muse.jhu.edu/article/448135/summary
97 Gillian Brockell, 'George Washington owned slaves and ordered Indians killed. Will a mural of that history be hidden?' *Washington Post*, 25 Aug 2019, www.washingtonpost.com/history/2019/08/25/george-washington-owned-slaves-ordered-indians-killed-will-mural-that-history-be-hidden/
98 George Washington, letter to Lafayette, 15 Aug 1786, Founders Online, National Archives, founders.archives.gov/documents/Washington/04-04-02-0200
99 George Washington, letter to Lafayette, 29 Jan 1789, Founders Online, National Archives, founders.archives.gov/documents/Washington/05-01-02-0198
100 It is possible that the Mesoamerican empires deployed their own versions of the progress narrative formula. The scope of this book is limited to covering the cultural artefacts of history's expansionist states, but it would be an interesting study to determine whether non-Western cultural lineages independently deployed similar political rhetorics. 'Capturing society' was coined by anthropologist Fernando Santos-Granero; see 'Amerindian Torture Revisited: Rituals of Enslavement and Markers of Servitude in Tropical America', *Tipití*, 3:2 (2005), 147–74.
101 Graeber and Wengrow, *The Dawn of Everything*, 187–90.
102 Graeber and Wengrow, *The Dawn of Everything*, 187.
103 Zinn, *A People's History of the United States*, 127.
104 Gale Courey Toensing, 'Indian-Killer Andrew Jackson Deserves Top Spot on List of Worst US Presidents', *Indian Country Today*, 10 Sep 2017, indiancountrytoday.com/archive/indian-killer-andrew-jackson-deserves-top-spot-on-list-of-worst-us-presidents-q-Qg-O3lJUCE1bdhzyeS-A
105 Dunbar-Ortiz, *An Indigenous Peoples' History*, 97.
106 Andrew Jackson, 'On Indian Removal', speech to Congress, 6 Dec 1830, Records of the United States Senate, 1789–1990, Record Group 46, National Archives and Records Administration, www.nps.gov/museum/tmc/MANZ/handouts/Andrew_Jackson_Annual_Message.pdf

107 Wolfe, 'Settler Colonialism', 396.
108 Brockell, 'George Washington owned slaves'.
109 Andrew Jackson, 'Inaugural address, on being sworn into office, as President of the United States', United States Telegraph, 4 Mar 1829, www.loc.gov/item/rbpe.19301800/
110 Abraham Lincoln, 'First Annual Message, December 3, 1861', Gerhard Peters and John T. Woolley (eds), The American Presidency Project, www.presidency.ucsb.edu/node/202175
111 Abraham Lincoln, 'First Inaugural Address, March 4, 1861', The Avalon Project (Yale Law School: Lillian Goldman Law Library), avalon.law.yale.edu/19th_century/lincoln1.asp. Imagine a modern president, on his or her inauguration day, inviting the people to overthrow the government if they deem it necessary – the implausibility of that is surely one of the greatest refutations of the linear progress myth.
112 Abraham Lincoln, 'Response to a Pro-Slavery Friend, to Joshua. F. Speed. Springfield, August 24, 1855', *The Papers and Writings of Abraham Lincoln*, ed. Arthur Brooks Lapsley (2004), www.gutenberg.org/cache/epub/3253/pg3253-images.html
113 Of course this should be considered not an act of rare virtue, but the bare minimum expression of humanity. See 'Saved a Life', *The Papers and Writings of Abraham Lincoln*.
114 Lincoln, 'Annual Message to Congress, Washington, December 3, 1861', *The Papers and Writings of Abraham Lincoln*.
115 Lincoln, 'Call For 300,000 Volunteers, June 28, 1861', *The Papers and Writings of Abraham Lincoln*.
116 Text of the proclamation follows: 'Whereas reliable information has been received that hostile Indians, within the limits of the United States, have been furnished with arms and munitions of war by persons dwelling in conterminous foreign territory, and are thereby enabled to prosecute their savage warfare upon the exposed and sparse settlements of the frontier; . . . Now, therefore, be it known that I, Abraham Lincoln, President of the United States of America, do hereby proclaim and direct that all persons detected in that nefarious traffic shall be arrested and tried by court-martial at the nearest military post, and if convicted, shall receive the punishment due to their deserts . . . In witness whereof, I have hereunto set my hand, and caused the seal of the United States to be affixed . . .' Lincoln, 'Proclamation Concerning Indians, March 17, 1865', *The Papers and Writings of Abraham Lincoln*.
117 Lincoln, 'Annual Message to Congress, December 1, 1862', *The Papers and Writings of Abraham Lincoln*.
118 Dunbar-Ortiz, *An Indigenous Peoples' History*, 137.
119 Ibid.

120 Figures from Smil, *Energy and Civilisation*, 12.
121 E. A. Wrigley, *Energy and the English Industrial Revolution* (Cambridge, Cambridge University Press, 2014), 15.
122 Morris, *Foragers, Farmers, and Fossil Fuels*, 83.
123 Morris, *Foragers, Farmers, and Fossil Fuels*, 52.
124 Morris, *Foragers, Farmers, and Fossil Fuels*, 94.
125 Peter Brimblecombe, 'Attitudes and Responses Towards Air Pollution in Medieval England', *Journal of the Air Pollution Control Association*, 26:10, (1976), 941–5, doi.org/10.1080/00022470.1976.104 70341. In 1952, the Great Smog of London killed 12,000 people in just five days, when an anticyclone caused a cloud of cold air to settle over the city, trapping the pollutants produced by the city's cheap coal in the streets below. At the time, nobody panicked because they were used to the omnipresent smog in London; it was only later when the deaths were tallied in shocking numbers that people realised what had happened. See Alessandra Potenza, 'In 1952 London, 12,000 people died from smog – here's why that matters now', The Verge, 16 Dec 2017, www.theverge.com/2017/12/16/16778604/london-great-smog-1952-death-in-the-air-pollution-book-review-john-reginald-christie
126 Jordan Hanania *et al.*, 'Energy density', *Energy Education*, University of Calgary, 25 June 2018, energyeducation.ca/encyclopedia/Energy_density
127 Risks will only increase, and dramatically, as nation states' regulatory frameworks and the physical environment are destabilised by climate and ecological emergencies. See Samuel Miller McDonald, 'Is Nuclear Power Our Best Best Against Climate Change?', Boston Review, bostonreview.net/articles/is-nuclear-power-our-best-bet-against-climate-change/; Aaron Vansintjan, 'Where's the "eco" in ecomodernism?' *Red Pepper*, 4 Apr 2018, www.redpepper.org.uk/wheres-the-eco-in-ecomodernism/
128 'The Civil War', National Park Service, 6 May 2015, www.nps.gov/civilwar/facts.htm
129 Andreas Malm, *Fossil Capital: The Rise of Steam Power and the Roots of Global Warming* (London: Verso, 2016), 276, emphasis in original.
130 Malm, *Fossil Capital*, 67.
131 Karl Marx, *Capital: Volume I*, 262, Marxists.org archive, https://www.marxists.org/archive/marx/works/download/pdf/Capital-Volume-I.pdf
132 Colin Woodard, *American Character: A History of the Epic Struggle Between Individual Liberty and the Common Good* (London: Penguin, 2016), 115.
133 Alana Semuels, 'The Founding Fathers Weren't Concerned with Inequality', *The Atlantic*, 25 Apr 2016, www.theatlantic.com/

business/archive/2016/04/does-income-inequality-really-violate-us-principles/479577/
134 Leif Wenar, 'How to End the Oil Curse', *Foreign Affairs*, 3 June 2016, www.foreignaffairs.com/articles/2016-06-03/how-end-oil-curse.
135 I take this argument further at Sam Miller McDonald, 'Material Freedom', *Anthroposphere*, 31 May 2019, www.anthroposphere.co.uk/post/material-freedom
136 Clyde A. Haulman, 'The Panic of 1819: America's First Great Depression', *Financial History* (Winter 2010), 20–4, www.moaf.org/exhibits/checks_balances/andrew-jackson/materials/Panic_of_1819.pdf
137 'Global Crises Data by Country', Harvard Business School, www.hbs.edu/behavioral-finance-and-financial-stability/data/Pages/global.aspx; and John Taskinsoy, 'Financial Crises Continue to Strike amid Accelerated Evolution of Risk Management', *SSRN* (2022), doi.org/10.2139/ssrn.4038732.
138 Henry L. Tischler, *Introduction to Sociology*, 10th edition (Wadsworth, 2010), 12. For a fun profile of Spencer, see Steven Shapin, 'Man with a Plan', *New Yorker*, 6 Aug 2007, www.newyorker.com/magazine/2007/08/13/man-with-a-plan
139 Darwin also ate his own Galápagos tortoise specimens, contributing to the near total annihilation by sailors of a species still struggling for survival today.
140 Herbert Spencer, *The Man Versus the State* (1884; Indianapolis: Library of Economics and Liberty, 1992).
141 Herbert Spencer, *Social Statics: Or the Conditions Essential to Happiness Specified, and the First of Them Developed* (London: Williams and Norgate, 1868), 80.
142 Herbert Spencer, *Progress: Its Law and Cause* (New York: Humboldt Library, 1857).
143 It's not entirely clear if he means this in the most expansive way possible – that evolution created human beings who created the United States – or whether he was explicitly endorsing the view that the cut-throat nature of the American economy yielded 'progress'. Charles Darwin, *The Descent of Man, and Selection in Relation to Sex* (London: John Murray, 1871), 179.
144 Spencer's easy callousness and taste for cruel social policy may also have emerged from the emotional frostiness that haunted him all his life. His aloofness was so persistent that Eliot sympathetically referred to it as 'that tremendous glacier of yours'. George Eliot cited in Lyndall Gordon, 'Outlaw: George Eliot', *Hudson Review*, Summer 2017, hudsonreview.com/2017/07/outlaw-george-eliot/; Alain Jumeau, 'Progress in George Eliot's Novels', *Cahiers Victoriens & Édouardiens* (2002), www.researchgate.net/publication/297932697_Progress_in_George_Eliot%27s_novels

145 For an excellent updated look at the Luddites, see Matthew James Seidel, 'The Luddites Were Right', *Current Affairs*, 1 June 2021, www.currentaffairs.org/2021/06/the-luddites-were-right; Gavin Mueller, *Breaking Things at Work: The Luddites Are Right About Why You Hate Your Job* (London: Verso, 2021).
146 Rebecca Mead, 'How George Eliot's "Middlemarch" Resonates in the England of 2019', *New Yorker*, 21 Nov 2019, www.newyorker.com/culture/cultural-comment/how-george-eliots-middlemarch-resonates-in-the-england-of-2019; John Mullan, '*Middlemarch*: reform and change', British Library, 15 May 2014, www.bl.uk/romantics-and-victorians/articles/middlemarch-reform-and-change
147 Mary Wollstonecraft, 'A Vindication of the Rights of Woman', in *The Rights of Woman* (London: Scott, 1891).
148 Oscar Wilde, 'The Decay of Living', *Intentions* (New York: Brentano's, 1905).
149 Carnegie was so enamoured that he referred to Spencer as 'My Dear Master', and even declared, 'Few men have wished to know another more strongly than I to know Herbert Spencer.'
150 Jacob Davidson, 'The 10 Richest People of All Time', *Money*, 30 July 2015, money.com/the-10-richest-people-of-all-time-2/
151 'The Steel Business', PBS: Public Broadcasting Service, www.pbs.org/wgbh/americanexperience/features/carnegie-steel-business/
152 His use of 'Gospel' in the title is appropriate; this belief system, which persists today, more resembles religious faith than empirical science. See Eugene McCarraher, *The Enchantments of Mammon* (Cambridge, MA: Harvard University Press, 2019).
153 Andrew Carnegie, 'The Gospel of Wealth', two articles originally published in *The North American Review*, June 1889 and Dec 1889. See also Nathan J. Robinson, 'How Billionaires See Themselves', *Current Affairs*, 5 Jan 2021, www.currentaffairs.org/2021/01/how-billionaires-see-themselves
154 Not to be confused with the common anarchist practice of 'mutual aid' today, which involves communities organising to provide fundamental goods and services for those in need.
155 Peter Kropotkin, *Mutual Aid: A Factor of Evolution* (1902; Anarchy Archives 2021).
156 'Auguste Comte on the Natural Progress of Human Society', *Population and Development Review*, 37:2 (2011), 389–94, www.jstor.org/stable/pdf/23043288.pdf
157 This model superficially resembles the organisation of the present book, dividing progress myths between the 'mythic', 'secular' and 'economistic'. The big difference is, I am not arguing that this change occurs linearly or permanently, nor that it has necessarily yielded more desirable or positive outcomes.

158 Marx's theories, suffused in the beginning with a teleological history, inspired revolutions across the world, founded fields of study, and made him the most cited scholar in the world.
159 And he was prickly about it. See for instance Karl Marx, letter to editor of the *Otecestvenniye Zapisky*, 1877, *Marx and Engels Correspondence* (1968), www.marxists.org/archive/marx/works/1877/11/russia.htm
160 Ian Angus, 'Marx and Engels . . . and Darwin? The Essential Connection between Historical Materialism and Natural Selection', *International Socialist Review*, 65 (2012), isreview.org/issue/65/marx-and-engelsand-darwin
161 Karl Marx, 'Preface to the First German Edition', *Capital: Volume I* (1867), Marxists.org, www.marxists.org/archive/marx/works/1867-c1/p1.htm
162 Karl Marx, 'The British Rule in India', originally published in *New-York Daily Tribune*, 25 June 1853, www.marxists.org/archive/marx/works/1853/06/25.htm
163 Though that isn't to say India itself had no history of imperialism, colonisation, slavery and repression, but such a history shouldn't justify an increase in those things.
164 Fredrich Engels, *The Origin of the Family, Private Property and the State* (1884), quoted in Nancy Lindisfarne and Jonathan Neale, 'All Things Being Equal', *Climate and Capitalism*, 17 Dec 2021, climateandcapitalism.com/2021/12/17/the-dawn-of-everything-gets-human-history-wrong/
165 Marx, *Capital: Volume I*, 294–5.
166 Franklin Rosemont, 'Karl Marx and the Iroquois', in Nelson Algren *et al.*, *Arsenal: Surrealist Subversion* (Chicago: Black Swan, 1989), 205, libcom.org/files/Franklin_Rosemont_Karl_Marx_and_the_Iroquois.pdf
167 Karl Marx, letter to Vera Zasulich, 'the "First" Draft', Marx-Zasulich Correspondence February/March 1881, in Teodor Shanin (ed.), *Late Marx and the Russian Road: Marx and the 'Peripheries of Capitalism'*, (New York: Monthly Press, 1983), www.marxists.org/archive/marx/works/1881/zasulich/draft-1.htm
168 Rosemont, 'Karl Marx and the Iroquois', 207, emphasis in original.
169 Mary Bellis, 'The Most Important Inventions of the 19th Century: Innovations that Changed the World', ThoughtCo, 26 July 2019. www.thoughtco.com/inventions-nineteenth-century-4144740
170 Roosevelt loved killing things, especially 'for science'. He personally shot and donated a surprising number of specimens, big and small, to New York City's American Natural History Museum. He killed 296 animals of all sorts in Africa on a single safari. In his early life, he killed several now-extinct passenger pigeons. See: Hunter Lanier, 'Teddy Roosevelt once killed 296 animals in a single safari hunt',

Chron, 4 Feb 2016, www.chron.com/sports/outdoors/article/Teddy-Roosevelt-once-killed-296-animals-in-a-6807285.php. Roosevelt's racial notions should be noted for context here as well. He once said, for instance, 'Nineteenth-century democracy needs no more complete vindication for its existence than the fact that it has kept for the white race the best portion of the new world's surface.' See Crawford Kilian, 'Iconic Teddy, White Supremacist', *The Tyee*, 28 July 2010, thetyee.ca/Books/2010/07/28/TeddyWhiteSupremacist/

171 Theodore Roosevelt. 'Conservation as a National Duty,' speech delivered 13 May 1908, Voices of Democracy: The U.S. Oratory Project, voicesofdemocracy.umd.edu/theodore-roosevelt-conservation-as-a-national-duty-speech-text/.

172 When Foote went to present her findings before the prestigious American Association for the Advancement of Science, her study was read out by a male colleague, since women were not permitted to speak. See Clive Thompson. 'How 19th Century Scientists Predicted Global Warming', JSTOR Daily, 17 Dec 2019, daily.jstor.org/how-19th-century-scientists-predicted-global-warming/

173 Nina Strochlic, '100 Years Ago, Alexander Graham Bell Predicted Life in 2017', *National Geographic Magazine*, June 2017.

174 Deborah Coen, 'The 19th-century tumult over climate change – and why it matters today', The Conversation, 10 Sep 2018, theconversation.com/the-19th-century-tumult-over-climate-change-and-why-it-matters-today-101757

175 Karl Marx and Friedrich Engels, 'Manifesto of the Communist Party' (1848), in *Marx/Engels Selected Works, Vol. One* (Moscow: Progress Publishers, 1969), 98–137, www.marxists.org/archive/marx/works/1848/communist-manifesto/ch01.htm

176 See 'Records of the strike in Egypt under Ramses III, c1157 BCE', libcom.org, 26 Feb 2007, libcom.org/history/records-of-the-strike-in-egypt-under-ramses-iii; 'The so-called "Strike Papyrus" written by Amunnakht', Museo Egizio, collezioni.museoegizio.it/en-GB/material/Cat_1880

177 Joshua J. Mark, 'Beer in Ancient Egypt', World History Encyclopedia, 16 Mar 2017, www.worldhistory.org/article/1033/beer-in-ancient-egypt/

178 Morris, *Foragers, Farmers, and Fossil Fuels*, 76.

179 Colin G. Calloway, 'Gray Lock', *American National Biography* (Oxford: Oxford University Press, 2000), doi.org/10.1093/anb/9780198606697.article.2000407

180 Heather Zeppel. 'Who were Bennelong and Pemulwuy? Museums in Sydney and Interpretation of Eora Aboriginal Culture', *International Journal of Heritage Studies*, 5:3–4 (1999), 182–7, doi.org/10.1080/13527259908722265

181 K. Vlassopoulos, 'Slavery in Ancient Greece', in D. A. Pargas and J. Schiel (eds), *The Palgrave Handbook of Global Slavery throughout History* (New York: Palgrave Macmillan, 2023), doi.org/10.1007/978-3-031-13260-5

182 Anna Tsing, Heather Swanson, Elaine Gan and Nils Bubandt (eds), *Arts of Living on a Damaged Planet: Ghosts and Monsters of the Anthropocene* (Minneapolis: University of Minnesota Press, 2017), 65.

183 Historian Emma Griffin notes that in the United Kingdom, although poor children were sometimes employed before industrialisation, the boom in mines and factories created huge demand for child labour – workers who could be easily manipulated, paid little, and who could not unionise for better conditions. See Emma Griffin, 'Child Labour', British Library, 15 May 2014, www.bl.uk/romantics-andvictorians/articles/child-labour; In Britain, ninety new prisons were built in the mid-nineteenth century, often employing hard labour, isolation and flogging, and resulted in high suicide rates among prisoners.

184 Hickel, *Less is More*, 171.

185 Hickel, *Less is More*, 172.

186 Ibid.

187 Oscar Wilde, *The Soul of Man under Socialism* (1891), www.marxists.org/reference/archive/wilde-oscar/soul-man/

188 See more on this in my essay 'Oil Age', *Current Affairs*, 17 Mar 2020, www.currentaffairs.org/2020/03/oil-age

189 Timothy Mitchell, *Carbon Democracy: Political Power in the Age of Oil* (London: Verso, 2011), Introduction.

190 Mitchell, *Carbon Democracy*, 53.

191 Mitchell, *Carbon Democracy*.

192 'Girl Power in 1824: The First Factory Strike in America', New England Historical Society, www.newenglandhistoricalsociety.com/1824-factory-strike-1824/

193 That is to say, they almost certainly did, though accidents do happen. See Joey La Neve DeFrancesco, 'Pawtucket, America's First Factory Strike', *Jacobin*, June 2018, www.jacobinmag.com/2018/06/factory-workers-strike-textile-mill-women

194 Christopher Clark. 'Why should we think about the Revolutions of 1848 now?' *London Review of Books* 41:5, 7 Mar 2019, www.lrb.co.uk/the-paper/v41/n05/christopher-clark/why-should-we-think-about-the-revolutions-of-1848-now

195 Theodore Parker, 'Justice and the Conscience', *Ten Sermons on Religion* (Boston: Crosby, Nichols, and Co., 1853).

196 With some negative results as well, in the displacement of Indigenous peoples.

ENDNOTES

III SYSTEM

1 Giorgos Kallis, *Degrowth* (Newcastle upon Tyne: Agenda Publishing, 2018), 68.
2 Terry Eagleton, *Ideology: An Introduction* (London: Verso, 1991), 1.
3 George Orwell, 'Charles Dickens' (1939), *The Complete Works of George Orwell*, XXII, 597.
4 Vladimir Tismaneanu and Bogdan C. Iacob, 'Ideological Origins of World War II', in G. Kurt Piehler and Jonathan A. Grant (eds), *The Oxford Handbook of World War II* (Oxford Academic, 2023), doi.org/10.1093/oxfordhb/9780199341795.013.22.
5 For a nice mainstream profile of the futurists, see Jay Griffiths, 'Fire, hatred and speed!' *Aeon*, 8 Feb 2017, aeon.co/essays/the-macho-violent-culture-of-italian-fascism-was-prophetic
6 F. T. Marinetti, 'The Founding and Manifesto of Futurism 1909', *The Documents of 20th-Century Art: Futurist Manifestos*, ed. Umbro Apollonio (New York: Viking Press, 1973), 19–24, Article 4. It was also published in French daily newspaper *Le Figaro* on 20 Feb 1909.
7 Marinetti, 'Manifesto of Futurism', Article 8.
8 Marinetti, 'Manifesto of Futurism', Articles 9 and 10.
9 Lots of genuinely funny people – Hitler, Stalin, Trump – have made poor leaders. Some of Marinetti's lines even read like a Trump speech. But it's the dour responsibility of citizens to shrug off humorous tyrants, see past the chuckles, and meet their jokey nihilism with silent malice.
10 The rage of drivers towards cyclists and pedestrians that persists today has always been tinged with fascism.
11 Adam J. Sacks, 'The Fascist Sympathies of Britain's Aristocracy', *Tribune*, 25 Oct 2020, tribunemag.co.uk/2020/10/the-fascist-sympathies-of-britains-aristocracy
12 Benito Mussolini, *The Doctrine of Fascism* (1932), sjsu.edu/faculty/wooda/2B-HUM/Readings/The-Doctrine-of-Fascism.pdf, 7; parenthesis in original.
13 Ibid.
14 Jens Steffek and Francesca Antonini, 'Toward Eurafrica! Fascism, Corporativism, and Italy's Colonial Expansion', in Ian Hall (ed.), *Radicals and Reactionaries in Twentieth-Century International Thought* (New York: Palgrave Macmillan, 2015), 158, doi.org/10.1057/9781137520623_7
15 Steffek and Antonini, 'Toward Eurafrica!', 152.
16 Steffek and Antonini, 'Toward Eurafrica!', 157.
17 13 See: R. J. B. Bosworth, 'The Addis Ababa Massacre', *History Today* 68:2, 2 Feb 2018, www.historytoday.com/reviews/addis-ababa-massacre; Rory Carroll, 'Italy's bloody secret', *Guardian*, 25 Jun

2001, www.theguardian.com/education/2001/jun/25/ artsandhumanities.highereducation; Emanuele Ertola, 'Italians Committed Terrible Crimes, Then Forgot Them', *Jacobin*, 20 Feb 2024, jacobin.com/2024/02/addis-abada-ethiopia-italian-massacre

18 Giulia Barrera, 'Mussolini's Colonial Race Laws and State-Settler Relations in Africa Orientale Italiana (1935–41)', *Journal of Modern Italian Studies*, 8:3 (2003), 425–43, doi.org/10.1080/0958517032000011370

19 Mussolini, *The Doctrine of Fascism*, 9.

20 Steffek and Antonini, 'Toward Eurafrica!', 147.

21 17 Richard Brunies, 'The Fascist King: Victor Emmanuel III of Italy', The National WWII Museum, New Orleans, 14 July 2021, www.nationalww2museum.org/war/articles/fascist-king-victor-emmanuel-iii-italy

22 This brush with death-by-Soviet-execution may help to justify, or at least explain, what many see as Orwell's betrayal of the left, even his uncharacteristic cravenness, when he collaborated with a British government propaganda department to inform on people he thought might be sympathetic to Stalinism, or might even be Stalin's agents (as some did indeed turn out to be). The following article gives greater context: 'Orwell's idea was, if you are going to get somebody to write these kinds of pamphlets, they have to be genuine democrats. He was a democratic socialist who wanted democratic socialists, not right wingers, to be writing this propaganda. But he wanted them to be people who had seen through the Soviet illusion, not covert stooges for Stalin's Russia.' From Adam Lusher, 'Was George Orwell secretly a reactionary snitch? How the author became an internet meme and target of the hard left', *Independent*, 24 June 2018, www.independent.co.uk/news/uk/home-news/george-orwell-snitch-list-reactionary-grass-blacklist-communists-information-research-department-ird-government-celia-kirwan-a8414066.html

23 John Newsinger, 'Orwell and the Spanish Revolution', *International Socialism Journal*, 62 (1994), http://pubs.socialistreviewindex.org.uk/isj62/newsinger.htm

24 George Orwell, *Homage to Catalonia* (1938; London: Penguin Modern Classics, 2000), 217.

25 George Orwell, *Nineteen Eighty-Four* (1949; London: Harvill Secker, 2009), 242.

26 Gabriel Jackson, *The Spanish Republic and the Civil War, 1931–1939* (Princeton, NJ: Princeton University Press, 1967); Antony Beevor, *The Battle for Spain: The Spanish Civil War 1936–1939* (London: Penguin Books, 2006).

27 Eoghan Gilmartin, 'Franco's Stooges Still Dominate the Spanish Economy', *Jacobin*, 28 June 2020, www.jacobinmag.com/2020/06/francisco-franco-spain-monuments-fascism

28 Antonio Maestre, 'No Beauty in Defeat', *Jacobin*, 1 Apr 2019, jacobinmag.com/2019/04/spain-civil-war-franco-nationalist-fascism
29 Maestre, 'No Beauty in Defeat'.
30 For more, see Lucian M. Ashworth, 'Democratic Socialism and International Thought in Interwar Britain', in Ian Hall (ed.), *Radicals and Reactionaries in Twentieth-Century International Thought* (New York: Palgrave Macmillan, 2015), 79.
31 Evan Kindley, 'The Insanity Defense: Coming to Terms with Ezra Pound's Politics', *The Nation*, 28 Mar 2018, www.thenation.com/article/archive/coming-to-terms-with-ezra-pounds-politics/
32 Ernest Hemingway quoted in Josh Jones, 'Ernest Hemingway Writes of His Fascist Friend Ezra Pound: 'He Deserves Punishment and Disgrace' (1943), via *Letters of Note*, 22 Aug 2013, www.openculture.com/2013/08/hemingway-writes-of-his-friend-the-fascist-ezra-pound-he-deserves-punishment-and-disgrace-1943.html
33 Alan Munton, 'Modernist Politics: Socialism, Anarchism, Fascism', in Peter Brooker *et al.* (eds), *The Oxford Handbook of Modernisms* (Oxford: Oxford University Press, 2012), www.oxfordhandbooks.com/view/10.1093/oxfordhb/9780199545445.001.0001/oxfordhb-9780199545445-e-28
34 Wilde, *The Soul of Man under Socialism*.
35 Patrick McGuinness, 'Their Mad Gallopade', *London Review of Books*, 25 Jan 2018, www.lrb.co.uk/the-paper/v40/n02/patrick-mcguinness/their-mad-gallopade
36 Cunard was, furthermore, a high fashion icon and muse to Ezra Pound, Wyndham Lewis, Aldous Huxley, T. S. Eliot, Tristan Tzara, Louis Aragon and Ernest Hemingway.
37 Deborah Bird Rose, 'Anthropocene Noir', *ARENA Journal*, 41/42 (2013–14), 211–12.
38 Gerald Graff, 'Literary Modernism: The Ambiguous Legacy of Progress', *Social Research*, 41:1 (1974), 108.
39 Thomas Putnam, 'Hemingway on War and Its Aftermath', *Prologue Magazine* 38:1 (2006), National Archives, www.archives.gov/publications/prologue/2006/spring/hemingway.html
40 F. Scott Fitzgerald, *This Side of Paradise* (1920; London: Sirius Publishing, 2020), 268. The novel's final soliloquy, where this line appears, has, like a few of Fitzgerald's quotable turns of phrase, since been attributed to his wife Zelda. Fitzgerald sometimes published lines from Zelda's diaries and letters in his work, prompting her to playfully remark in a 1922 review of his novel *The Beautiful and Damned* for the *New York Tribune*, 'Mr. Fitzgerald . . . seems to believe that plagiarism begins at home.' Scholarly opinion remains divided about whether his use of her work was truly collaborative or plagiaristic.
41 Fitzgerald, *This Side of Paradise*, 253.

42 Ronald Berman, *Modernity and Progress: Fitzgerald, Hemingway, Orwell* (Tuscaloosa, AL: University of Alabama Press, 2005), 14.
43 Berman, *Modernity and Progress*, 14–15.
44 Growing up in northern Michigan had the same effect on me.
45 It had the opposite effect on me, however. 'He sulked like a child when, on his first safari, his wife Pauline shot a lion before he did,' reports the *Independent* of Hemingway. John Walsh, 'Being Ernest: John Walsh unravels the mystery behind Hemingway's suicide', *Independent*, 23 Oct 2011, www.independent.co.uk/news/people/profiles/being-ernest-john-walsh-unravels-mystery-behind-hemingway-s-suicide-2294619.html
46 John Banville, 'Frequent Gunfire', *The Nation*, 26 Oct 2017, www.thenation.com/article/archive/what-was-it-like-to-be-ernest-hemingway/
47 Robert McCrum, '1984: The masterpiece that killed George Orwell', *Guardian*, 10 May 2009, www.theguardian.com/books/2009/may/10/1984-george-orwell
48 Alex Ross, 'How American Racism Influenced Hitler', *New Yorker*, 30 Apr 2018, www.newyorker.com/magazine/2018/04/30/how-american-racism-influenced-hitler
49 Lia Mandelbaum, 'Hitler's Inspiration and Guide: The Native American Holocaust', Jewish Journal, 18 June 2013, jewishjournal.com/mobile_20111212/117824/hitlers-inspiration-and-guide-the-native-american-holocaust/
50 Quoted in James Whitman, *Hitler's American Model: The United States and the Making of Nazi Race Law* (Princeton, NJ: Princeton University Press, 2017), 10. For more on the comparison between American and German frontiers, see Carroll P. Kakel, *The American West and the Nazi East: A Comparative and Interpretive Perspective* (Basingstoke: Palgrave Macmillan, 2011).
51 Robert J. Miller, 'Nazi Germany's Race Laws, The United States, and American Indians', *St John's Law Review*, 3:94 (2021), 756.
52 Cited in Miller, 'Nazi Germany's Race Laws', 771–2.
53 More on May and Germany's fascination with Westerns at Rivka Galchen, 'Wild West Germany', *New Yorker*, 2 Apr 2012, www.newyorker.com/magazine/2012/04/09/wild-west-germany
54 Prized among Hitler's personal library was a complete collection of Karl May's books bound in vellum, a gift from Hitler's second-in-command Hermann Göring. They were said to be much thumbed and read, 'and usually one or two may be found in the small bedside bookcase with its green curtain in Hitler's bedroom'. W. Raymond Wood, 'The Role of the Romantic West in Shaping the Third Reich', *Plains Anthropologist*, 35:132 (1990), 316.
55 See: Anthony T. Grafton, 'Mein Buch', *New Republic*, 24 Dec 2008, newrepublic.com/article/64703/mein-buch
56 The antagonism may have been mutual: far from the upright,

uptight fascist ideal, May was a messy chaos agent who grew up poor and, as a young man, robbed frequently, landing him regularly in jail. He got into all sorts of mischief like having an affair with a married woman and dressing up in the frontier costumes of places he never visited. Plus, his closest friend was Jewish.
57 Karl Ove Knausgård, *The End – My Struggle: Book 6*, trans. Martin Aitken and Don Bartlett (2011; London: Vintage, 2019), 741.
58 Knausgård, *My Struggle*, Book 6, 487.
59 Adolf Hitler, *Mein Kampf* (Boston, MA: Houghton Mifflin, 1941), 405.
60 Hitler, *Mein Kampf*, 84.
61 Ibid.
62 Adolf Hitler, *Collection of Speeches: 1922–1945*, 911, ia801903.us.archive. org/17/items/AdolfHitlerCollectionOfSpeeches19221945/Adolf%20 Hitler%20-%20Collection%20of%20Speeches%201922-1945.pdf
63 Hitler, *Collection of Speeches*, 12.
64 A vast majority of European Jews killed in the Holocaust were not German. See 'Jewish Losses During the Holocaust: By Country', Holocaust Encyclopedia, United States Holocaust Memorial Museum, Washington, DC, 27 Mar 2018, encyclopedia.ushmm.org/content/en/article/jewish-losses-during-the-holocaust-by-country
65 Alexandra Minna Stern, 'Forced sterilisation policies in the US targeted minorities and those with disabilities – and lasted into the 21st century', *The Conversation*, 26 Aug 2020, theconversation.com/forced-sterilisation-policies-in-the-us-targeted-minorities-and-those-with-disabilities-and-lasted-into-the-21st-century-143144
66 Jane Lawrence, 'The Little-Known History of the Forced Sterilisation of Native American Women', JSTOR Daily, 25 Aug 2016, daily.jstor.org/the-little-known-history-of-the-forced-sterilisation-of-native-american-women/
67 IQ tests were and are culturally specific, often containing cultural references that would be unintelligible to people from other cultures; in examinations of delinquency, they would even be administered to people who did not speak the language the test was given in, making them notoriously unreliable for measuring intelligence; Lisa Ko, 'Unwanted Sterilisation and Eugenics Programs in the United States', PBS, 29 Jan 2016, www.pbs.org/independentlens/blog/unwanted-sterilisation-and-eugenics-programs-in-the-united-states/
68 Whitman, *Hitler's American Model*, xii.
69 Lorraine Boissoneault, 'A 1938 Nazi Law Forced Jews to Register Their Wealth – Making It Easier to Steal', Smithsonian, 26 Apr 2018, www.smithsonianmag.com/history/1938-nazi-law-forced-jews-register-their-wealthmaking-it-easier-steal-180968894/. See also Richard J. Overy, 'Making a Killing: The Economics of the

Holocaust', Fifth University of Glasgow Holocaust Memorial Lecture, 26 Jan 2005, www.gla.ac.uk/media/Media_767756_smxx.pdf

70 Hitler, *Collection of Speeches*, 15.
71 Whitman, *Hitler's American Model*, 2.
72 Whitman, *Hitler's American Model*, 5.
73 Whitman, *Hitler's American Model*, 140.
74 Sacks, 'The Fascist Sympathies of Britain's Aristocracy'.
75 Robert Philpot, 'The men and women who plotted to stab Britain in the back during WWII', *Times of Israel*, 17 Sep 2018, www.timesofisrael.com/the-mn-and-women-who-plotted-to-stab-britain-in-the-back-during-wwii/.
76 Perhaps the most aptly named man in history.
77 Here's a modern example: Rob Picheta, 'Meghan reveals "concerns" within royal family about her baby's skin color', CNN, 14 Mar 2021, edition.cnn.com/2021/03/08/uk/meghan-oprah-interview-racism-scli-gbr-intl/index.html
78 Julian Borger, 'Nazi documents reveal that Ford had links to Auschwitz', *Guardian*, 20 Aug 1999, www.theguardian.com/world/1999/aug/20/julianborger1
79 Michael Dobbs, 'Ford and GM Scrutinized for Alleged Nazi Collaboration', *Washington Post*, 30 Nov 1998, www.washingtonpost.com/wp-srv/national/daily/nov98/nazicars30.htm
80 Jack Beatty, 'Hitler's Willing Business Partners', *The Atlantic*, Apr 2001, www.theatlantic.com/magazine/archive/2001/04/hitlers-willing-business-partners/303146/
81 Ben Aris and Duncan Campbell, 'How Bush's grandfather helped Hitler's rise to power', *Guardian*, 25 Sep 2004, www.theguardian.com/world/2004/sep/25/usa.secondworldwar
82 For 'internal colonisation' see Lino Camprubí, *Engineers and the Making of the Francoist Regime* (Cambridge, MA: MIT Press, 2014). Spain resolved the tension of liberalism by ejecting political liberalism in favour of economic liberalism.
83 Charles Feinstein, 'The Equalizing of Wealth in Britain Since the Second World War', *Oxford Review of Economic Policy*, 12:1 (1996), 102–3, www.jstor.org/stable/pdf/23606413
84 Walter Scheidel, 'Inequality: Total war as a great leveller,' VoxEU and CEPR, 2 Sep 2019, voxeu.org/article/inequality-total-war-great-leveller
85 For a discussion of lessons on the New Deal and pitfalls we must avoid today, see Samuel Miller McDonald, 'The Green New Deal Can't Be Anything Like the New Deal', *New Republic*, 31 May 2019, newrepublic.com/article/153996/green-new-deal-cant-anything-like-new-deal
86 Nick Estes, *Our History is the Future: Standing Rock Versus the Dakota Access Pipeline, and the Long Tradition of Indigenous Resistance* (London: Verso, 2019), 9–11.

87 Adolf Reed Jr., 'The New Deal Wasn't Intrinsically Racist', *New Republic*, 26 Nov 2019, newrepublic.com/article/155704/new-deal-wasnt-intrinsically-racist
88 Will Steffen *et al.*, 'The Trajectory of the Anthropocene: The Great Acceleration', *Anthropocene Review*, 16 Jan 2015, doi.org/10.1177/2053019614564785
89 See Naomi Klein, *The Shock Doctrine: The Rise of Disaster Capitalism* (London: Penguin, 2008); and on the difference between ordo- and neoliberalism, Bob Jessop, 'Ordoliberalism and Neoliberalisation: Governing through Order or Disorder', *Critical Sociology*, 45:7–8 (2019), doi.org/10.1177/0896920519834068
90 John Maynard Keynes, 'Economic Possibilities for our Grandchildren', *Essays in Persuasion* (1930; New York: W. W. Norton, 1963), 358–73.
91 '1957: Britons "have never had it so good"', BBC: On This Day, 20 July 1957, http://news.bbc.co.uk/onthisday/hi/dates/stories/july/20/newsid_3728000/3728225.stm
92 British Pathé, 'Macmillan's Speech From 10 Downing Street (1957)', YouTube, 13 Apr 2014, www.youtube.com/watch?v=suyoCrj1XbA
93 John W. Young, 'Macmillan, the Free Trade Area and the First Application, 1957–63', in *Britain and European Unity, 1945–1992* (London: Palgrave Macmillan, 1993), 57–85.
94 Vaclav Smil, *Growth: From Microorganisms to Megacities* (London: MIT Press, 2019), 399.
95 Kallis, *Degrowth*, 68.
96 Smil, *Growth*, 411.
97 Christopher F. Jones, 'The Delusion and Danger of Infinite Economic Growth', *New Republic*, 1 Oct 2019, newrepublic.com/article/155214/delusion-danger-infinite-economic-growth
98 It is important to note that Summers has consistently been wrong about matters economic and otherwise, writing memos that include comments like 'the economic logic behind dumping a load of toxic waste in the lowest wage country is impeccable and we should face up to that. . . . I've always thought that under-populated countries in Africa are vastly underpolluted'; from 'Furor on Memo at World Bank', *New York Times*, 7 Feb 1992, www.nytimes.com/1992/02/07/business/furor-on-memo-at-world-bank.html. Summers was also a major voice in the Clinton administration arguing against taking action to curtail greenhouse gas emissions; see Robert Wampler (ed.), 'Kyoto Redux? Obama's Challenges at Copenhagen Echo Clinton's at Kyoto', National Security Archive Electronic Briefing Book No. 303, 11 Dec 2009, nsarchive2.gwu.edu/NSAEBB/NSAEBB303/index.htm; Lawrence Summers, cited by Professor William E. Rees, 'Footprints to Sustainability', University of British Columbia News, 6 Apr 2006, news.ubc.ca/2006/04/06/archive-ubcreports-2006-06apr06-footprints/

99 Hickel, *Less is More*, 91–2.
100 Quoted in John M. Hartwick, 'What Would Solow Say?' *Journal of Natural Resources Policy Research*, 1:1 (2008), 91–6, doi.org/10.1080/19390450802504451
101 As cited in Christopher F. Jones, 'The Delusion and Danger of Infinite Economic Growth', *New Republic*, 1 Oct 2019, newrepublic.com/article/155214/delusion-danger-infinite-economic-growth
102 'The March of Socialism 1928-1932', *Journal of the Seventeenth National Convention Socialist Party*, Milwaukee, Wisconsin, 1932 (Chicago: Italian Labor Publishing Co., 1932), ia800502.us.archive.org/33/items/ProceedingsOfThe1932NationalConventionOfTheSocialistParty/1932sp2_text.pdf
103 Dennis Kavanagh and Iain Dale, *Labour Party General Election Manifestos 1900–1997: Volume Two* (London: Routledge, 2015), 31.
104 For any bad-faith commentators still hoping to connect Nazism with socialism, Nathan J. Robinson definitively rebuts that argument in 'Putting The "Nazis Were Socialist" Nonsense To Rest', *Current Affairs*, 24 Jan 2020, www.currentaffairs.org/2020/01/putting-the-nazis-were-socialist-nonsense-to-rest
105 Ashworth, 'Democratic Socialism and International Thought', 75.
106 Robert C. Allen, 'The Rise and Decline of the Soviet Economy', *Canadian Journal of Economics*, 34:4 (2001), 859–81, www.jstor.org/stable/3131928
107 'United States GDP Annual Growth Rate', Trading Economics, U.S. Bureau of Economic Analysis, tradingeconomics.com/united-states/gdp-growth-annual
108 Massimo Livi-Bacci, 'On the Human Costs of Collectivisation in the Soviet Union', *Population and Development Review*, 19:4 (1993), 743–66, www.jstor.org/stable/2938412
109 Felix Wemheuer, *Famine Politics in Maoist China and the Soviet Union* (New Haven: Yale University Press, 2014; Yale Scholarship Online, 2015), 1, www.jstor.org/stable/j.ctt1bhknwh. Forced economic growth and the introduction and mechanisation of industrial agriculture in the Americas *also* yielded millions of deaths, as we have covered, but over a longer time span and in a more decentralised manner.
110 Alexander Berkman, 'Russian Revolution and the Communist Party' (1922), The Anarchist Library, 18 Nov 2020, theanarchistlibrary.org/library/alexander-berkman-russian-revolution-and-the-communist-party. See also Gabriel Kuhn, 'Ecological Leninism: Friend or Foe?' *LeftTwoThree Blog*, 2 Jan 2021, lefttwothree.org/ecological_leninism/.
111 J. Stalin, 'Speech Delivered by J. V. Stalin at a Meeting of Voters of the Stalin Electoral District, Moscow', 9 Feb 1946, Gospolitizdat, Moscow, from *Speeches Delivered at Meetings of Voters of the*

Stalin Electoral District, Moscow (Foreign Languages Publishing House, Moscow, 1950), 19–44. History and Public Policy Program Digital Archive, digitalarchive.wilsoncenter.org/document/116179

112 Martin McCauley, *Stalin and Stalinism* (London: Routledge, 2013), 43.
113 Cited in Bathsheba Demuth, 'When the Soviet Union Freed the Arctic from Capitalist Slavery', *New Yorker*, 15 Aug 2019, www.newyorker.com/news/dispatch/when-the-soviet-union-freed-the-arctic-from-capitalist-slavery
114 Igor Krupnik and Michael Chlenov, *Yupik Transitions: Change and Survival at Bering Strait, 1900–1960* (Fairbanks, AK: University of Alaska Press, 2013).
115 See a fascinating story in Demuth, 'When the Soviet Union Freed the Arctic from Capitalist Slavery'.
116 Dennis Bartels and Alice L. Bartels, 'Indigenous Peoples of the Russian North and Cold War Ideology', *Anthropologica*, 48:2 (2006), 265–79 (quoted from Uvachan, 1960), www.jstor.org/stable/pdf/25605315
117 Bartels and Bartels, 'Indigenous Peoples of the Russian North'.
118 Krupnik and Chlenov, *Yupik Transitions*, 289. Canadians likely treated circumpolar Indigenous peoples even worse, with thousands of children being kidnapped from their parents, forced into residential schools, beaten and starved, and dying in the hundreds or thousands without their parents being notified. Their shallow mass graves have only recently been uncovered; see 'Canada: 751 unmarked graves found at residential school', BBC News, 24 June 2021, www.bbc.co.uk/news/world-us-canada-57592243
119 Krupnik and Chlenov, *Yupik Transitions*.
120 Catriona Kelly, 'The New Soviet Man and Woman', in Simon Dixon (ed.), *The Oxford Handbook of Modern Russian History* (Oxford University Press, 2016), doi.org/10.1093/oxfordhb/9780199236701.013.204
121 See Lynne Attwood, *Creating the New Soviet Woman: Women's Magazines as Engineers of Female Identity, 1922–53* (New York: Palgrave Macmillan, 1999).
122 See '*Hujum*', in Bonnie G. Smith (ed.), *The Oxford Encyclopedia: Women in World History* (Oxford: Oxford University Press, 2008), www.oxfordreference.com/view/10.1093/acref/9780195148909.001.0001/acref-9780195148909-e-467
123 Russia's polar bears, Amur tigers and Caucasian leopards are nearing extinction. For the myriad environmental harms of the USSR, see Renfrey Clarke, 'The ecological disaster that was the USSR', *Socialist Alliance*, Dec 2011, socialist-alliance.org/alliance-voices/ecological-disaster-was-ussr-0; Ankit Panda, 'How the Soviet Union Created Central Asia's Worst Environmental Disaster', *The Diplomat*, 3 Oct 2014, thediplomat.com/2014/10/

how-the-soviet-union-created-central-asias-worst-environmental-disaster/; Paul Quinn-Judge, 'Soviet Union's Environmental Disasters Mount', *Seattle Times*, 30 Jan 1990, archive.seattletimes.com/archive/?date=19900130&slug=1053482; 'Environmental collapse before Soviet's fall', Down to Earth, 26 Dec 2016, www.downtoearth.org.in/coverage/environment/environmental-collapse-before-the-soviet-s-fall-56642; Armine Sahakyan, 'The Grim Pollution Picture in the Former Soviet Union', HuffPost, 6 Dec 2017, www.huffpost.com/entry/the-grim-pollution-pictur_b_9266764; Quirin Schiermeier, 'Soviet Union's collapse led to massive drop in carbon emissions', *Nature*, 1 July 2019, www.nature.com/articles/d41586-019-02024-6; Douglas Weiner, 'Ecology in the former Soviet Union', *Models of Nature: Ecology, Conservation, and Cultural Revolution in Soviet Union* (Indiana University Press, 1988); 'Changing Forest Cover Since the Soviet Era', NASA Earth Observatory, 16 July 2015, earthobservatory.nasa.gov/images/86221/changing-forest-cover-since-the-soviet-era

124 Maria Shahgedanova and Timothy Burt, 'New Data on Air Pollution in the Former Soviet Union', *Global Environmental Change*, 4:3 (1994), 201–27, doi.org/10.1016/0959-3780(94)90003-5; Arran Gare, 'The Environmental Record of the Soviet Union', *Capitalism Nature Socialism*, 13:3 (2002), 52–72.

125 Vaclav Smil, 'China's Great Famine: 40 Years Later', *British Medical Journal*, 319:7225 (1999), 1619–21, www.ncbi.nlm.nih.gov/pmc/articles/PMC1127087/

126 'Who, What, Why: What is the Little Red Book?' BBC News, 26 Nov 2015, www.bbc.co.uk/news/magazine-34932800

127 All quotations from The Little Red Book are from David Quentin and Brian Baggins (eds), *Quotations from Chairman Mao Tse-Tung ('The Little Red Book')*, transcribed from Peking Foreign Languages Press (Marxists Internet Archive, 2019), www.marxists.org/ebooks/mao/Quotations_from_Chairman_Mao_Tse-tung.pdf

128 Unknown artist, 'Collection of Chinese propaganda material from the mid-late 20th century', Art Fund, 2013, www.artfund.org/supporting-museums/art-weve-helped-buy/artwork/12202/collection-of-chinese-propaganda-material-from-the-mid-late-20th-century

129 David Shambaugh, 'China's Propaganda System: Institutions, Processes and Efficacy', *China Journal*, 57 (2007), 25–58, www.jstor.org/stable/20066240

130 George Dvorsky, 'China's Worst Self-Inflicted Environmental Disaster: The Campaign to Wipe Out the Common Sparrow', Gizmodo, 18 July 2012, gizmodo.com/china-s-worst-self-inflicted-environmental-disaster-th-5927112

131 Vladimir Lenin, 'The Historical Destiny of the Doctrine of Karl

Marx', *Lenin Collected Works*, Volume 18 (Progress Publishers, 1975), 582–5, Marxists Internet Archive, www.marxists.org/archive/lenin/works/1913/mar/01.htm

132 Thomas Piketty and Emmanuel Saez, 'Inequality in the Long Run', *Science*, 344:6186 (2014), 838–43.
133 Piketty and Saez, 'Inequality in the Long Run'.
134 For a definitive history of neoliberalism, see Daniel Stedman Jones, *Masters of the Universe: Hayek, Friedman, and the Birth of Neoliberal Politics* (Princeton: Princeton University Press, 2012). For its role in international institutions, see David Harvey, *A Brief History of Neoliberalism* (Oxford: Oxford University Press, 2005).
135 Patrick Iber, 'Worlds Apart', *New Republic*, 23 Apr 2018, newrepublic.com/article/147810/worlds-apart-neoliberalism-shapes-global-economy
136 Manfred B. Steger and Ravi K. Roy, 'Three Waves of Neoliberalism', *Neoliberalism: A Very Short Introduction*, 2nd edition (Oxford: Oxford University Press, 2021).
137 Friedrich Hayek, 'Extracts from an Interview with Friedrich von Hayek (El Mercurio, Chile, 1981)', *Punto de Vista Economico*, 21 Dec 2016, puntodevistaeconomico.com/2016/12/21/extracts-from-an-interview-with-friedrich-von-hayek-el-mercurio-chile-1981/. For more on Hayek's comfort with dictatorship, see Benjamin Selwyn, 'Friedrich Hayek: in defence of dictatorship', openDemocracy, 9 June 2015, www.opendemocracy.net/en/friedrich-hayek-dictatorship/.
138 Stedman Jones, *Masters of the Universe*, 9.
139 Stedman Jones, *Masters of the Universe*, 63.
140 Friedrich Hayek, 'The Common Sense of Progress', Foundation for Economic Education, 1 Nov 1960, fee.org/articles/the-common-sense-of-progress/
141 Eugene F. Miller, *Hayek's* The Constitution of Liberty: *An Account of Its Argument* (London: Institute of Economic Affairs, 2010), 83.
142 Friedrich Hayek, *The Road to Serfdom* (London: Routledge Classics, 2001), 61.
143 For discussions of this, see Marshall Sahlins' 1966 'Man the Hunter' symposium, for 'original affluent society', or more here: Marshall Sahlins, *Stone Age Economics* (Chicago: Aldine Atherton, 1972); Harris, *Cannibals and Kings*; Graeber and Wengrow, *The Dawn of Everything*; and Morris, *Foragers, Farmers, and Fossil Fuels*.
144 See Scott, *Against the Grain*, Chapter 5.
145 Wengrow, 'Cities before the State in Early Eurasia'.
146 See Scott, *Against the Grain*, Chapter 6.
147 'Chile's Caravan of Death: Ex-army chief Cheyre convicted for Pinochet-era crimes', BBC News, 10 Nov 2018, www.bbc.com/news/world-latin-america-46160437

148 See Marella Oppenheim, 'Excavations at Chile torture site offer new hope for relatives of disappeared', *Guardian*, 2 May 2018, www.theguardian.com/world/2018/may/02/chile-disappeared-excavations-colonia-dignidad; Kyle Swensen, 'Chilean Victims of ex-Naxi's Cult of Horrors May Finally Get Some Answers', *Washington Post*, 14 July 2017, www.washingtonpost.com/news/morning-mix/wp/2017/07/14/chilean-victims-of-ex-nazis-cult-of-horrors-may-finally-get-some-answers/

149 Hayek quoted in John M. Geddes, 'New Vogue for Critic of Keynes', *New York Times*, 7 May 1979, www.nytimes.com/1979/05/07/archives/new-vogue-for-critic-of-keynes-von-hayek-still-abhors-big.html

150 Richard Davies, 'Why is inequality booming in Chile? Blame the Chicago Boys', *Guardian*, 13 Nov 2019, www.theguardian.com/commentisfree/2019/nov/13/why-is-inequality-booming-in-chile-blame-the-chicago-boys

151 Stedman Jones, *Masters of the Universe*.

152 Stedman Jones, *Masters of the Universe*, 212.

153 Ibid.

154 It was during the Keynesian period, for instance, that many of the United States' most rigorous environmental regulations, like the Clean Air and Water Acts and Endangered Species Act, were passed. This 'invisible hand' is only rendered invisible, it should be noted, by the deliberately dense and obfuscating bureaucracies that rule most corporations and disperse accountability.

155 George Monbiot, 'Neoliberalism – the ideology at the root of all our problems', *Guardian*, 15 Apr 2016, www.theguardian.com/books/2016/apr/15/neoliberalism-ideology-problem-george-monbiot

156 Albert Einstein, letter to Heinrich Zangger, Dec 1917, *Collected Papers Vol. 8*, 412, as cited in Jürgen Neffe, *Einstein: A Biography* (Cambridge: Polity Press, 2007), 256.

157 Albert Einstein, 'Why Socialism?', *Monthly Review* (May 1949), online 1 May 2009, monthlyreview.org/2009/05/01/why-socialism/

158 Ibid.

159 James Vincent, 'SpaceX rocket debris creates a fantastic light show in the Pacific Northwest sky', The Verge, 26 Mar 2021, www.theverge.com/2021/3/26/22351956/oregon-washington-meteor-shower-explanation-spacex-falcon-9-rocket-debris

160 Associated Press, 'SpaceX Rocket Breaks Apart Before Touchdown at South Texas "Starbase"', *Dallas Morning News*, 30 Mar 2021, www.dallasnews.com/business/technology/2021/03/30/spacex-rocket-breaks-apart-before-touchdown-at-south-texas-starbase/

161 A major fan of bitcoin mining, mining is simply in his blood: Phillip de Wet, 'Elon Musk's family once owned an emerald mine in Zambia – here's the fascinating story of how they came to own it', *Business Insider*, 28 Feb 2018, www.businessinsider.co.za/

how-elon-musks-family-came-to-own-an-emerald-mine-2018-2; Neil Strauss, 'Elon Musk: The Architect of Tomorrow', *Rolling Stone*, 15 Nov 2017, www.rollingstone.com/culture/culture-features/elon-musk-the-architect-of-tomorrow-120850/; Adam Smith, 'Elon Musk's 50th: Taking a Look into the Billionaire's Wealth – From Emeralds to SpaceX and Tesla', *Independent*, 28 June 2021, www.independent.co.uk/life-style/gadgets-and-tech/elon-musk-birthday-ceo-tesla-b1874017.html

162 Clare O'Connor, 'Jeff Bezos' Spacecraft Blows Up in Secret Test Flight; Locals Describe "Challenger-Like" Explosion', *Forbes*, 2 Sep 2011, www.forbes.com/sites/clareoconnor/2011/09/02/jeff-bezos-spacecraft-blows-up-in-secret-test-flight-locals-describe-challenger-like-explosion/

163 A. Tarantola, 'Blue Origin has been trying to get the hell off this planet for 20 years now', Engadget, 8 Sep 2020, www.engadget.com/blue-origin-celebrates-its-20th-anniversary-163055108.html

164 Michael Sheetz, 'What's behind SpaceX's $74 billion valuation: Elon Musk's two "Manhattan Projects"', CNBC News, 19 Feb 2021, www.cnbc.com/2021/02/19/spacex-valuation-driven-by-elon-musks-starship-and-starlink-projects.html

165 Reuters, 'An unleashed Jeff Bezos looks to shift space venture Blue Origin into hyperdrive', NBC News, 8 Feb 2021, www.nbcnews.com/science/space/unleashed-jeff-bezos-looks-shift-space-venture-blue-origin-hyperdrive-rcna266

166 Zachary Shahan, 'Attacks On Elon Musk For His Wealth Are Truly Ridiculous', *CleanTechnica*, 20 Mar 2021, cleantechnica.com/2021/03/20/attacks-on-elon-musk-for-his-wealth-are-ridiculous/

167 Elon Musk [@elonmusk], 'I am accumulating resources to help make life multiplanetary & extend the light of consciousness to the stars,' Twitter, 21 Mar 2021, twitter.com/elonmusk/status/1373507545315172357

168 'Home', Blue Origin, www.blueorigin.com/

169 The wealthy emit vastly more carbon and consume significantly more energy and resources than the non-rich. See Sara Schonhardt, 'Rich Americans Have Higher Carbon Footprints Than Other Wealthy People', *Scientific American*, 8 Dec 2021, www.scientificamerican.com/article/rich-americans-have-higher-carbon-footprints-than-other-wealthy-people/; Laura Paddison, 'How the rich are driving climate change', BBC News, 28 Oct 2021, www.bbc.com/future/article/20211025-climate-how-to-make-the-rich-pay-for-their-carbon-emissions; Anna Ratcliff, 'Carbon emissions of richest 1 percent more than double the emissions of the poorest half of humanity', Oxfam, 21 Sep 2020, www.oxfam.org/en/press-releases/carbon-emissions-richest-1-percent-more-double-emissions-poorest-half-humanity; Helena Horton, '"Carbon footprint gap" between rich and poor expanding, study finds', *Guardian*, 4 Feb

2022, www.theguardian.com/environment/2022/feb/04/carbon-footprint-gap-between-rich-poor-expanding-study; Roger Harrabin, 'Climate change: The rich are to blame, international study finds', BBC News, 16 Mar 2020, www.bbc.co.uk/news/business-51906530; Yannick Oswald, Anne Owen and Julia K. Steinberger, 'Large Inequality in International and Intranational Energy Footprints between Income Groups and Across Consumption Categories', *Nature Energy*, 5 (2020), 231–9, www.nature.com/articles/s41560-020-0579-8

170 Cade Metz, Karen Weise, Nico Grant and Mike Isaac, 'Ego, Fear and Money: How the A.I. Fuse Was Lit', *New York Times*, 3 Dec 2023, www.nytimes.com/2023/12/03/technology/ai-openai-musk-page-altman.html

171 This adolescent mindset does not plague exclusively male technological fetishists. Vox blogger Kelsey Piper posted on Twitter on 6 Nov 2022, in response to discourses concerning the negative impacts of very large luxury cruise ships: 'We should put [nuclear] reactors on our cruise ships so we can make them even bigger. On-board downhill skiing. Skyscrapers connected with ziplines so you can play Spiderman. I want billions of people entering the global middle class and I want them to all get whatever fun shit they want', twitter.com/KelseyTuoc/status/1589094830365630464. This offers a further demonstration of the infantile psychological mode of those who put their faith in fantastical technological development.

172 Stuart Clark, 'Elon Musk is building Starbase, a city with a spaceport to the Moon, Mars and beyond. Here's what's inside', BBC Science Focus, November 4 2021, https://www.sciencefocus.com/news/starbase-texas-inside-elon-musks-plans-to-build-the-city-of-the-future/; Rachel Thomas, 'SpaceX's rockets and spacecraft have really cool names. But what do they mean?' Florida Today, https://eu.floridatoday.com/story/tech/science/space/2019/04/29/spacex-names-of-course-i-still-love-you-millennium-falcon-dragon-meaning/3621453002/ ; 'X Æ A-12: Elon Musk and Grimes confirm baby name,' BBC News, 6 May 2020; Akin Olla, 'Elon Musk declared himself "technoking". He's just a hyper-capitalist clown', *Guardian*, 20 Mar 2021, www.theguardian.com/commentisfree/2021/mar/20/elon-musk-declared-himself-technoking-hes-just-a-hyper-capitalist-clown; 'Tesla CEO Elon Musk Smokes Weed During Joe Rogan Podcast Interview | Velshi & Ruhle | MSNBC', uploaded by MSNBC, YouTube, 7 Sep 2018, www.youtube.com/watch?v=1rS8fFbW57o; Taylor Lorenz, 'Elon Musk: Memelord or Meme Lifter?' *New York Times*, 7 May 2021, www.nytimes.com/2021/05/07/style/elon-musk-memes.html; Luke Winkie, 'Memes, grudges and moving to Mars: the week in Elon Musk', *Guardian*, 20 Mar 2022, www.theguardian.com/technology/2022/mar/20/the-week-in-elon-musk-memes-grudges-mars

173 On 20 Nov 2022, Musk claimed on Twitter that his 'firstborn child died in [his] arms'. Three days later, on the same platform, his ex-wife clarified that, in fact, it was she who held their son as he died: twitter.com/esjesjesj/status/1733533720559182198?s=20. See Justine Musk, '"I Was a Starter Wife": Inside America's Messiest Divorce', *Marie Claire*, 10 Sep 2010, www.marieclaire.com/sex-love/a5380/millionaire-starter-wife/; Nathan J. Robinson, 'Surely We Can Do Better Than Elon Musk', *Current Affairs*, 7 Apr 2021, www.currentaffairs.org/2021/04/surely-we-can-do-better-than-elon-musk

174 Peter Whoriskey, 'For Bezos, The Post represents new frontier', *Washington Post*, 10 Aug 2013, www.washingtonpost.com/business/economy/for-bezos-the-post-represents-new-frontier/2013/08/10/ba7cfeb6-013c-11e3-9a3e-916de805f65d_story.html

175 Paul Rincon, 'Jeff Bezos launches to space aboard New Shepard rocket ship', BBC News, 20 July 2021, www.bbc.co.uk/news/science-environment-57849364. Bezos isn't wrong to say that we should shoot industry into space.

176 'Elon Musk's SpaceX vs Jeff Bezos's Blue Origin: Which Space Project Will Dominate the Cosmos?' Robb Report, 29 Aug 2020, robbreport.com/lifestyle/news/spacex-blue-origin-head-to-head-2944342/

177 Andrew Wong, 'Space mining could become a real thing – and it could be worth trillions', CNBC News, 15 May 2018, www.cnbc.com/2018/05/15/mining-asteroids-could-be-worth-trillions-of-dollars.html

178 See Anthony Cuthbertson, 'Elon Musk Says First Mars Colony Settlers Will Live in "Glass Domes" Before Terraforming Planet', *Independent*, 21 Nov 2020, www.independent.co.uk/life-style/gadgets-and-tech/elon-musk-mars-colony-spacex-starship-b1759074.html; Adam Bensaid, 'Elon Musk's astonishing mission to colonise Mars: here's how he'll do it', TRT World, 10 Dec 2020, www.trtworld.com/magazine/elon-musk-s-astonishing-mission-to-colonise-mars-here-s-how-he-ll-do-it-42246; Tom McKay, 'Elon Musk: A New Life Awaits You in the Off-World Colonies – for a Price', Gizmodo, 17 Jan 2020, gizmodo.com/elon-musk-a-new-life-awaits-you-on-the-off-world-colon-1841071257

179 'One of the central allegations in the lawsuit is that Apple, Google, Dell, Microsoft and Tesla were aware and had "specific knowledge" that the cobalt they use in their products is linked to child labour performed in hazardous conditions, and were complicit in the forced labour of the children.' Annie Kelly, 'Apple and Google named in US lawsuit over Congolese child cobalt mining deaths', *Guardian*, 16 Dec 2019, www.theguardian.com/global-development/2019/dec/16/apple-and-google-named-in-us-lawsuit-over-congolese-child-cobalt-mining-deaths

180 Elon Musk [@elonmusk], 'Cybervikings of Mars', Twitter, 5 Mar 2021, twitter.com/elonmusk/status/1367784992961531905. Musk would like to be associated with Serbian-American inventor Nikola Tesla. In an episode of *Star Trek*, Musk is compared with the fictional inventor of the 'warp drive', Zefram Cochrane. But Musk has not invented anything. His closer analogue is P. T. Barnum; he is a talented self-promoter and circus master. With parasitism, there is no 'final' frontier.

181 Alexander Koch *et al.*, 'European colonisation of the Americas killed 10 percent of world population and caused global cooling', The Conversation, 31 Jan 2019, www.pri.org/stories/2019-01-31/european-colonisation-americas-killed-10-percent-world-population-and-cause

182 See Olla, 'Elon Musk declared himself "technoking"'; Van Badham, 'Here's why Elon Musk's robot is electrified neoliberalism', *Guardian*, 31 Aug 2021, www.theguardian.com/commentisfree/2021/aug/31/the-tesla-bot-reveals-the-alarming-truth-billionaires-dont-know-what-its-like-for-the-rest-of-us; Tom McKay, 'Elon Musk: A New Life Awaits You in the Off-World Colonies – for a Price', Gizmodo, 17 Jan 2020, gizmodo.com/elon-musk-a-new-life-awaits-you-on-the-off-world-colon-1841071257

183 Adam Mann, 'Is Mars Ours?' *New Yorker*, 11 May 2021, www.newyorker.com/science/elements/is-mars-ours.

184 A sentiment echoed by 2016 and 2020 presidential candidate Bernie Sanders: @BernieSanders, 'We are in a moment in American history where two guys – Elon Musk and Jeff Bezos – own more wealth than the bottom 40% of people in this country. That level of greed and inequality is not only immoral. It is unsustainable.' Twitter, 18 Mar 2021, twitter.com/BernieSanders/status/1372641707829829632

185 See A. J. Dellinger, 'Amazon's carbon footprint goes beyond shipping millions of Prime packages', Mic, 13 Sep 2019, www.mic.com/p/amazons-carbon-footprint-goes-beyond-shipping-millions-of-prime-packages-18739965; Nicole Nguyen, 'The Hidden Environmental Cost of Amazon Prime's Free, Fast Shipping', BuzzFeed News, 21 July 2018, www.buzzfeednews.com/article/nicolenguyen/environmental-impact-of-amazon-prime

186 Ken Klippenstein, 'Documents Show Amazon is Aware Drivers Pee in Bottles and Defecate En Route, Despite Company Denial', The Intercept, 25 Mar 2021, theintercept.com/2021/03/25/amazon-drivers-pee-bottles-union/

187 See Emily Guendelsberger, 'I Worked at an Amazon Fulfillment Center; They Treat Workers Like Robots', *TIME*, 18 July 2019, time.com/5629233/amazon-warehouse-employee-treatment-robots/; Eric Spitznagel, 'Inside the hellish workday of an Amazon warehouse employee', *New York Post*, 13 July 2019, nypost.com/2019/07/13/inside-the-hellish-workday-of-an-amazon-warehouse-employee/; Will

Evans, 'Ruthless Quotas at Amazon are Maiming Employees', *The Atlantic*, 25 Nov 2019, www.theatlantic.com/technology/archive/2019/11/amazon-warehouse-reports-show-worker-injuries/602530/; Michael Sainato, '"I'm not a robot": Amazon workers condemn unsafe, grueling conditions at warehouse', *Guardian*, 5 Feb 2020, www.theguardian.com/technology/2020/feb/05/amazon-workers-protest-unsafe-grueling-conditions-warehouse

188 See Alexia Fernández Campbell, 'Elon Musk broke US labor laws on Twitter', Vox, 30 Sep 2019, www.vox.com/identities/2019/9/30/20891314/elon-musk-tesla-labor-violation-nlrb; Julia Carrie Wong, 'Tesla factory workers reveal pain, injury and stress: "Everything feels like the future but us"', *Guardian*, 18 May 2017, www.theguardian.com/technology/2017/may/18/tesla-workers-factory-conditions-elon-musk; Bradley Berman, 'Tesla employees fear unsafe conditions at factory, call it "modern-day sweatshop"', Electrek, 21 May 2020, electrek.co/2020/05/21/tesla-employees-fear-unsafe-conditions-at-factory-call-it-modern-day-sweatshop/

189 The pain behind the smile of this article is particularly dystopian: 'A phrase we threw around a lot was, *"You are your own slavedriver."*' Josh Boehm, 'I Worked At SpaceX, And This Is How Elon Musk Inspired A Culture Of Top Performers', Forbes, 8 Nov 2017, www.forbes.com/sites/quora/2017/11/08/i-worked-at-spacex-and-this-is-how-elon-musk-inspired-a-culture-of-top-performers/. See also Jonathan Chadwick, 'Elon Musk says SpaceX's Starship reusable rocket that could one day carry humans to Mars is the company's "top priority" from now on', *Daily Mail*, 8 June 2020, www.dailymail.co.uk/sciencetech/article-8398449/Elon-Musk-says-SpaceXs-Starship-rocket-priority-on.html; Richard Feloni, 'Why Elon Musk sets nearly impossible goals for SpaceX employees', Yahoo! Finance, 27 May 2015, uk.finance.yahoo.com/news/why-elon-musk-sets-nearly-150708269.html

190 This tweet reveals how Musk sees his project: as a form of entertainment spectacle to entrance his masses of followers and keep their attention away from his wealth hoarding, union busting, and many broken promises: Elon Musk [@elonmusk] 'cgi irl @spacex', Twitter, 4 Mar 2021, twitter.com/elonmusk/status/1367330165144059906

191 Elon Musk [@elonmusk], 'those who attack space maybe don't realize that space represents hope for so many people', Twitter, 13 July 2021, twitter.com/elonmusk/status/1414782972474048516?lang=en-GB

192 The Problem With Jon Stewart, 'Jon's Dinner With Obama And Bezos | The Problem With Jon Stewart Podcast | Apple TV+', YouTube, 17 Jan 2022, www.youtube.com/watch?v=Vewz1v4rQYg

193 Lloyd Alter, 'A SpaceX Launch Puts Out as Much CO_2 as Flying 341 People Across the Atlantic', Treehugger, 14 Nov 2019,

www.treehugger.com/spacex-launch-puts-out-much-co-flying-people-across-atlantic-4857958
194 Samuel Miller McDonald, 'Lifecycle of a Leaf', *Current Affairs*, 23 Aug 2019, www.currentaffairs.org/2019/08/lifecycle-of-a-leaf
195 See Katharine Sanderson, 'We're running out of lithium for batteries – can we use salt instead?' *New Scientist*, 20 Jan 2021, www.newscientist.com/article/mg24933180-600-were-running-out-of-lithium-for-batteries-can-we-use-salt-instead/; James Attwood, 'The World Will Need 10 Million Tons More Copper to Meet Demand', *Bloomberg*, 19 Mar 2021, www.bloomberg.com/news/articles/2021-03-19/the-world-will-need-10-million-tons-more-copper-to-meet-demand; Frank Swain, 'The wonder material we all need but is running out', BBC Future, 9 Mar 2021, www.bbc.com/future/article/20210308-rubber-the-wonder-material-we-are-running-out-of; Xiaozhi Lim, 'The World Is Running Out Of Elements, And Researchers Are Looking In Unlikely Places For Replacements', *Discover*, 16 May 2020, www.discovermagazine.com/planet-earth/the-world-is-running-out-of-elements-and-researchers-are-looking-in-unlikely
196 See Thomas Wiedmann *et al.*, 'Scientists' Warning on Affluence', *Nature Communications*, 11:3107 (2020), doi.org/10.1038/s41467-020-16941-y; Akshat Rathi, 'These Folks Think Eternal Economic Growth Will Lead to Unstoppable Climate Change', Bloomberg Green, 18 Aug 2020, www.bloomberg.com/news/articles/2020-08-18/why-degrowth-may-be-necessary-to-prevent-climate-catastrophe; Jason Hickel, 'What Does Degrowth Mean? A Few Points of Clarification', *Globalisations*, 18:7 (2020), doi.org/10.1080/14747731.2020.1812222; Viviana Asara *et al.*, 'Socially Sustainable Degrowth as a Social–ecological Transformation: Repoliticizing Sustainability', *Sustainability Science*, 10 (2015), 375–84, doi.org/10.1007/s11625-015-0321-9; Ben Robra and Pasi Heikkurinen, 'Degrowth and the Sustainable Development Goals', in Walter Leal Filho *et al.* (eds), *Decent Work and Economic Growth*, Encyclopedia of the UN Sustainable Development Goals (Cham: Springer, 2019), doi.org/10.1007/978-3-319-71058-7_37-1
197 See Patrick Donnelly [@BitterWaterBlue], faa.gov/about/office_org/headquarters_offices/ast/environmental/nepa_docs/review/launch/spacex_texas_launch_site_environmental_impact_statement/ . . . What a disaster', Twitter, 1 Apr 2021, twitter.com/BitterWaterBlue/status/1377681007516688384
198 Soledad Calvino, 'U.S. EPA settles with Tesla over hazardous waste violations at Fremont, Calif., facility', United States Environmental Protection Agency, 1 Apr 2019, archive.epa.gov/epa/newsreleases/us-epa-settles-tesla-over-hazardous-waste-violations-fremont-calif-facility.html
199 Carl Zimmer, 'The Lost History of One of the World's Strangest

Science Experiments', *New York Times*, 2019, www.nytimes.com/2019/03/29/sunday-review/biosphere-2-climate-change.html
200 Samantha Cole, 'The Strange History of Steve Bannon and the Biosphere 2 Experiment', Vice, 15 Nov 2016, www.vice.com/en/article/qkjn87/the-strange-history-of-steve-bannon-and-the-biosphere-2-experiment
201 For more on this, see my essay, Samuel Miller McDonald, 'Climate Kings', *New Republic*, 30 July 2018, newrepublic.com/article/148861/climate-change-authoritarian-leaders
202 Joseph Hincks, 'How the Giant Boat Blocking the Suez Canal Was Freed: Dredgers, Tugboats, and a Full Moon', *TIME*, 29 Mar 2021, time.com/5950888/suez-canal-boat-freed-explained/; James Stavridis, 'The Blocked Suez Canal Isn't the Only Waterway the World Should Be Worried About', *TIME*, 29 Mar 2021, time.com/5950791/suez-canal-ship-blockage-trade/
203 Or potentially human technological hubris. See Nathan J. Robinson, 'The Stakes of Finding COVID-19's Origins', *Current Affairs*, 14 May 2021, www.currentaffairs.org/2021/05/the-stakes-of-finding-covid-19s-origins
204 Arooba Ahmed, 'COVID-19 and Biodiversity Loss: How Destruction of the Environment Leads to Pandemics', State of the Planet, Columbia Climate School Blog, 24 Nov 2020, blogs.ei.columbia.edu/2020/11/24/covid-19-biodiversity-loss-pandemics/
205 Christopher Ingraham, 'World's richest men added billions to their fortunes last year as others struggled', *Washington Post*, 1 Jan 2021, www.washingtonpost.com/business/2021/01/01/bezos-musk-wealth-pandemic/
206 Vaclav Smil interviewed by Nathan Gardels, 'Want Not, Waste Not', *Noēma Magazine*, 25 Feb 2021, www.noemamag.com/want-not-waste-not/
207 Damian Carrington, 'Climate emergency: world "may have crossed tipping points"', *Guardian*, 27 Nov 2019, www.theguardian.com/environment/2019/nov/27/climate-emergency-world-may-have-crossed-tipping-points
208 Damian Carrington, 'Humanity has wiped out 60% of animal populations since 1970, report finds', *Guardian*, 30 Oct 2018, www.theguardian.com/environment/2018/oct/30/humanity-wiped-out-animals-since-1970-major-report-finds
209 My essay in The Baffler develops a similar argument in further detail: Samuel Miller-McDonald, 'Carbon Omissions', The Baffler, 18 Jan 2018, thebaffler.com/latest/climate-omissions-miller-mcdonald
210 See Will Steffen *et al.*, 'Trajectories of the Earth System in the Anthropocene', *PNAS*, 115:33 (2018), 8252–9, www.pnas.org/content/115/33/8252; Richard Betts, 'Hothouse Earth: here's what the science actually does – and doesn't – say', The Conversation, 9 Aug 2018, theconversation.com/hothouse-earth-heres-what-the-science-actually-

does-and-doesnt-say-101341; Jan Zalasiewicz, Mark Williams and Thomas Hearing, 'Hothouse Earth: our planet has been here before – here's what it looked like', The Conversation, 13 Aug 2018, theconversation.com/hothouse-earth-our-planet-has-been-here-before-heres-what-it-looked-like-101413

211 Ajit Niranjan, 'Earth on verge of five catastrophic climate tipping points, scientists warn', *Guardian*, 6 Dec 2023, www.theguardian.com/environment/2023/dec/06/earth-on-verge-of-five-catastrophic-tipping-points-scientists-warn

212 Grace Palmer, 'Greenland ice sheet reached tipping point 20 years ago, new study finds', Phys.org, 2 Sep 2020, phys.org/news/2020-09-greenland-ice-sheet-years.html

213 Miller McDonald, 'Climate Kings'.

214 Institute for Economics and Peace, 'Ecological Threat Register 2020: Understanding Ecological Threats', *Resilience and Peace*, Sep 2020, http://visionofhumanity.org/reports.

215 Joe Phelan, 'How Many Nuclear Weapons Exist, and Who Has Them?' *Scientific American*, 22 Mar 2022, www.scientificamerican.com/article/how-many-nuclear-weapons-exist-and-who-has-them/

216 S. Willcock, G. S. Cooper, J. Addy, et al., 'Earlier Collapse of Anthro-pocene Ecosystems Driven by Multiple Faster and Noisier Drivers', *Nature Sustainability*, 6 (2023), 1331–42, doi.org/10.1038/s41893-023-01157-x

217 Chris Arsenault, 'Only 60 Years of Farming Left If Soil Degradation Continues', *Scientific American*, 5 Dec 2014, www.scientificamerican.com/article/only-60-years-of-farming-left-if-soil-degradation-continues/

218 Jeremy Engle, 'A Mysterious Infection, Spanning the Globe in a Climate of Secrecy', *New York Times*, 6 Apr 2019, www.nytimes.com/2019/04/06/health/drug-resistant-candida-auris.html

219 The exact figures included 324 grey wolves, 64,131 coyotes, 433 black bears, 200 mountain lions, 605 bobcats, 3,014 foxes and 24,687 beavers. Oliver Milman, '"A barbaric federal program": US killed 1.75m animals last year – or 200 per hour', *Guardian*, 25 Mar 2022, www.theguardian.com/world/2022/mar/25/us-government-wildlife-services-animals-deaths

220 'The State of the World's Forests 2020', Food and Agriculture Organisation of the United Nations, http://www.fao.org/state-of-forests/en. For a visual of a hectare, see Jim Csek, 'How Big is a Hectare? A Better Way to Visualize The Size', *Kelowna Now*, 17 July 2014, www.kelownanow.com/galavanting/news/Tourist_Information/14/07/17/How_Big_Is_A_Hectare_A_Better_Way_to_Visualize_The_Size/

221 'Extinction of Plants and Animals', National Museum of Natural History, naturalhistory.si.edu/education/teaching-resources/paleontology/extinction-over-time

222 Christina Larson, 'As endangered birds lose their songs, they can't find mates', AP News, 17 Mar 2021, apnews.com/article/endangered-birds-lose-song-no-mates-f38f5c97cccc50cba1e884a907db13d8
223 'Male Singing To Female That Will Never Come | Racing Extinction', YouTube, uploaded by Discovery Channel Southeast Asia, 25 Nov 2015, www.youtube.com/watch?v=5THqAY3u5oY
224 Donnachadh McCarthy, 'BP's "green" promises are anything but – their latest investments are a slap in the face for humanity and nature', *Independent*, 13 Aug 2020, www.independent.co.uk/voices/bp-green-promises-renewables-fossil-fuels-climate-crisis-a9667371.html
225 'Lawyers take action against BP's climate "greenwashing" advertising campaign', press release, ClientEarth, 4 Dec 2019, www.clientearth.org/latest/press-office/press/lawyers-take-action-against-bp-s-climate-greenwashing-advertising-campaign/
226 'Shell accelerates drive for net-zero emissions with customer-first strategy', media releases, Shell, 11 Feb 2021, www.shell.com/media/news-and-media-releases/2021/shell-accelerates-drive-for-net-zero-emissions-with-customer-first-strategy.html
227 Molly Taft, 'Chevron Celebrates Pride While Funding Bigots in Congress', Gizmodo, 2 June 2021, https://gizmodo.com/chevron-s-pride-celebration-is-greenwashing-at-its-mo-1847018001
228 For examples, see Sharon Lerner, 'How the Media Launders Fossil Fuel Industry Propaganda through Branded Content', The Intercept, 3 Apr 2019, theintercept.com/2019/04/03/branded-content-fossil-fuel-companies/; Emily Atkin, 'Introducing: The Fossil Fuel Ad Anthology', Heated, 13 Dec 2019, heated.world/p/introducing-the-fossil-fuel-ad-anthology?s=r; Nathan J. Robinson, 'Turning Down the Money', *Current Affairs*, 16 May 2019, www.currentaffairs.org/2019/05/turning-down-the-money. For ads beyond native advertising that openly deny climate change, see Geoffrey Supran and Naomi Oreskes, 'The forgotten oil ads that told us climate change was nothing', *Guardian*, 18 Nov 2021, www.theguardian.com/environment/2021/nov/18/the-forgotten-oil-ads-that-told-us-climate-change-was-nothing
229 Jessica Murray, 'Half of emissions cuts will come from future tech, says John Kerry', *Guardian*, 16 May 2021, www.theguardian.com/environment/2021/may/16/half-of-emissions-cuts-will-come-from-future-tech-says-john-kerry
230 There are plenty of reasons for scientists to say geoengineering is potentially dangerous. See Alan Robock, '20 Reasons Why Geoengineering May Be a Bad Idea', *Bulletin of the Atomic Scientists*, 64:2 (2008), 14–8, climate.envsci.rutgers.edu/pdf/20Reasons.pdf
231 John Fialka, 'Scientists find analogue to growing smoke clouds:

Nuclear war', E&E News, 18 May 2021, www.eenews.net/climatewire/2021/05/18/stories/1063732805

232 Marton Dunai and Geert De Clercq, 'Nuclear energy too slow, too expensive to save climate: report', Reuters, 24 Sep 2019, www.reuters.com/article/us-energy-nuclearpower-idUSKBN1W909J

233 See Amory B. Lovins, 'Does Nuclear Power Slow Or Speed Climate Change?' Forbes, 18 Nov 2019, www.forbes.com/sites/amorylovins/2019/11/18/does-nuclear-power-slow-or-speed-climate-change/amp/; Jim Polson, 'More Than Half of America's Nuclear Reactors Are Losing Money', Bloomberg, 14 June 2017, http://www.bloomberg.com/news/articles/2017-06-14/half-of-america-s-nuclear-power-plants-seen-as-money-losers

234 Lisa Martine Jenkins, 'Nuclear Energy Among the Least Popular Sources of Power in the U.S., Polling Shows', Morning Consult, 9 Sep 2020, morningconsult.com/2020/09/09/nuclear-energy-polling/

235 Steve Wing, David B. Richardson and Wolfgang Hoffmann, 'Cancer Risks near Nuclear Facilities: The Importance of Research Design and Explicit Study Hypotheses', *Environmental Health Perspectives*, 119:4 (2011), doi.org/10.1289/ehp.1002853

236 David Thorpe, 'Extracting a disaster', *Guardian*, 5 Dec 2008, www.theguardian.com/commentisfree/2008/dec/05/nuclear-greenpolitics

237 Ibid.

238 'Europe's heatwave affects NPPs', *Nuclear Engineering International*, 30 July 2018, www.neimagazine.com/news/newseuropes-heatwave-affects-npps-6271432

239 Keith Barnham, 'False solution: Nuclear power is not "low carbon"', *Ecologist*, 5 Feb 2015, theecologist.org/2015/feb/05/false-solution-nuclear-power-not-low-carbon

240 Justine Calma, 'Microsoft is going nuclear to power its AI ambitions', The Verge, 26 Sep 2023, www.theverge.com/2023/9/26/23889956/microsoft-next-generation-nuclear-energy-smr-job-hiring

241 Nick Estes, 'A Red Deal', *Jacobin*, 6 Aug 2019, www.jacobinmag.com/2019/08/red-deal-green-new-deal-ecosocialism-decolonisation-indigenous-resistance-environment

242 Traci Brynne Voyles, *Wastelanding: Legacies of Uranium Mining in Navajo Country* (Minneapolis: University of Minnesota Press, 2015), 2–3.

243 I wrote a comprehensive cost-benefit analysis of nuclear energy here: Samuel Miller McDonald, 'Is Nuclear Power Our Best Bet Against Climate Change?' *Boston Review*, 12 Oct 2021, bostonreview.net/articles/is-nuclear-power-our-best-bet-against-climate-change/

244 Bob Ward, 'A closer examination of the fantastical numbers in Bjorn Lomborg's new book', *LSE Commentary*, 10 Aug 2020, www.lse.ac.uk/granthaminstitute/news/a-closer-examination-of-the-fantastical-numbers-in-bjorn-lomborgs-new-book/

245 A review dismantling Epstein's previous book, which is making essentially the same argument, can be found here: Rob Hopkins, 'Review: "The Moral Case for Fossil Fuels" – Really?' Our World, United Nations University, 5 Feb 2015, ourworld.unu.edu/en/review-the-moral-case-for-fossil-fuels-really

246 See Eric Niiler, 'Do Carbon Offsets Really Work? It Depends on the Details', *Wired*, 14 Jan 2020, www.wired.com/story/do-carbon-offsets-really-work-it-depends-on-the-details/; Lisa Song and Paula Moura, 'An Even More Inconvenient Truth', ProPublica, 22 May 2019, features.propublica.org/brazil-carbon-offsets/inconvenient-truth-carbon-credits-dont-work-deforestation-redd-acre-cambodia/; Tim McDonnell, 'Carbon offsets are going primetime and they're not ready', Quartz, 20 May 2021, qz.com/2009746/not-all-carbon-offsets-are-a-scam-but-many-still-are/; Ben Elgin, 'This Timber Company Sold Millions of Dollars of Useless Carbon Offsets', Bloomberg, 17 Mar 2022, www.bloomberg.com/news/articles/2022-03-17/timber-ceo-wants-to-reform-flawed-carbon-offset-market

247 Fiona Harvey, 'Carbon emissions to soar in 2021 by second highest rate in history', *Guardian*, 20 Apr 2021, www.theguardian.com/environment/2021/apr/20/carbon-emissions-to-soar-in-2021-by-second-highest-rate-in-history

248 Adnan Al-Daini, 'Global Warming: Delusions, Hubris and Wishful Thinking', HuffPost, 28 June 2013, www.huffingtonpost.co.uk/adnan-aldaini/global-warming-delusions-_b_3175118.html

249 This is developed further in my essay in *Boston Review*: Miller McDonald, 'Is Nuclear Power Our Best Bet Against Climate Change?'.

250 Michael Graetz, *The End of Energy: The Unmaking of America's Environment, Security, and Independence* (MIT Press, 2011), 3.

251 Oliver Milman, '"Invisible killer": fossil fuels caused 8.7m deaths globally in 2018, research finds', *Guardian*, 9 Feb 2021, www.theguardian.com/environment/2021/feb/09/fossil-fuels-pollution-deaths-research; Leah Burrows, 'Deaths from fossil fuel emissions higher than previously thought', *Harvard School of Engineering and Applied Sciences News*, 9 Feb 2021, www.seas.harvard.edu/news/2021/02/deaths-fossil-fuel-emissions-higher-previously-thought

252 'Road Traffic Injuries and Deaths – A Global Problem', Centers for Disease Control and Prevention, 14 Dec 2020, www.cdc.gov/injury/features/global-road-safety/index.html?CDC_AA_refVal=https%3A%2F%2F. See Ben Goldfarb, *Crossings: How Road Ecology Is Shaping the Future of Our Planet* (W. W. Norton, 2023).

253 It's the perfect public persona for such a man, with a name straight out of a Galaxy Far Far Away, a vaguely android-shaped face, and the uncanny valley accent one might expect to hear in the boardrooms of the twenty-second century. But it's all a lie.

254 Polls suggest a significant majority of the public prefer NASA to private companies for space exploration, and also believe that space missions should prioritise monitoring Earth (for asteroids and other dangers) rather than going to Mars. And given the sloppiness of Musk's creations – the many exploding Teslas, the breakable Cybertrucks, crash-landing rockets and murderous autonomous vehicles – it will be a while before many self-preservation-minded members of the public climb aboard his spaceships. See 'Majority of Americans Believe It Is Essential That the U.S. Remain a Global Leader in Space', Pew Research Center, 6 June 2018, www.pewresearch.org/science/2018/06/06/majority-of-americans-believe-it-is-essential-that-the-u-s-remain-a-global-leader-in-space/; Marcia Smith, 'Poll Shows Public's Space Priority is Monitoring Earth, Not Sending People to the Moon or Mars', Space Policy Online, 25 Feb 2021, spacepolicyonline.com/news/poll-shows-publics-space-priority-is-monitoring-earth-not-sending-people-to-the-moon-or-mars/. For examples of Musk's hazardous contraptions, see Katie Shepherd and Faiz Siddiqui, 'A driverless Tesla crashed and burned for four hours, police said, killing two passengers in Texas', *Washington Post*, 19 Apr 2021, www.washingtonpost.com/nation/2021/04/19/tesla-texas-driverless-crash/; Evelyn Cheng, 'Tesla Model 3 reportedly explodes in Shanghai parking garage', CNBC News, 20 Jan 2021, www.cnbc.com/2021/01/21/tesla-model-3-reportedly-explodes-in-shanghai-parking-garage.html; Darren Orf, 'Tesla Model 3 Blows Up . . . Twice . . . on Busy Highway', *Popular Mechanics*, 13 Aug 2019, www.popularmechanics.com/cars/hybrid-electric/a28687473/tesla-model-3-explosion/; 'Watch the Tesla Cybertruck's Windows Get Smashed', uploaded by Bloomberg Technology, YouTube, 22 Nov 2019, www.youtube.com/watch?v=BqGf8pqCZi0; Kenneth Chang, 'The SpaceX Test Rocket for Mars Goes Up Again, and Explodes Again', *New York Times*, 30 Mar 2021, www.nytimes.com/2021/03/30/science/space/spacex-starship-launch.html; Tom Krisher, '3 crashes, 3 deaths raise questions about Tesla's Autopilot', AP News, 3 Jan 2020, apnews.com/article/ca5e62255bb87bf1b151f9bf075aaadf

255 For more, see Samuel Miller McDonald, 'The Last Great Forests', *Current Affairs*, www.currentaffairs.org/2021/11/the-last-great-forests

256 See Marlowe Hood, 'Close to tipping point, Amazon could collapse in 50 years', Phys.org, 11 Mar 2020, phys.org/news/2020-03-amazon-collapse-years.html; Robert Toovey Walker, 'Collision Course: Development Pushes Amazonia Toward Its Tipping Point', *Environment: Science and Policy for Sustainable Development*, 63:1 (2021), 15–25, doi.org/10.1080/00139157.2021.1842711

257 Agence France-Presse, 'Brazilian Amazon released more carbon than it absorbed over past 10 years', *Guardian*, 30 Apr 2021,

www.theguardian.com/environment/2021/apr/30/brazilian-amazon-released-more-carbon-than-it-absorbed-over-past-10-years

258 Forestry Department of the Food and Agriculture Organisation of the United Nations (FAO) and the International Tropical Timber Organisation (ITTO), 'The State of Forests in the Amazon Basin, Congo Basin and Southeast Asia', Summit of the Three Rainforest Basins Brazzaville, Republic of Congo, 31 May–3 June 2011, http://www.fao.org/3/i2247e/i2247e00.pdf

259 Morgan Erickson-Davis, 'Congo Basin rainforest may be gone by 2100, study finds', Mongabay, 7 Nov 2018, news.mongabay.com/2018/11/congo-basin-rainforest-may-be-gone-by-2100-study-finds/

260 Chikezie Omeje, 'Do You Know Where Your Grilling Charcoal Comes From?' New Republic, 3 May 2021, newrepublic.com/article/161947/nigeria-where-does-grill-charcoal-come-from

261 'The Heart of Borneo Under Siege', WWF, wwf.panda.org/discover/knowledge_hub/where_we_work/borneo_forests/borneo_deforestation/?#deforestation

262 '8 Things to Know About Palm Oil', WWF, 17 Jan 2020, www.wwf.org.uk/updates/8-things-know-about-palm-oil

263 Kazimierz Becek and Aline B. Horwath, 'Is Vegetation Collapse on Borneo Already in Progress?' *Natural Hazards*, 85 (2017), 1279–90, doi.org/10.1007/s11069-016-2623-3

264 They killed enough emperors to have their own separate Wikipedia list: 'Roman emperors murdered by the Praetorian Guard', Wikipedia, 22 Dec 2017, en.wikipedia.org/wiki/Category:Roman_emperors_murdered_by_the_Praetorian_Guard

265 Roger Boesche, *The First Great Political Realist: Kautilya and His Arthashastra* (Lanham, MD: Lexington Books, 2002), 83.

266 Natasha Sheldon, 'Eight of the Deadliest Assassin Groups in History', History Collection, 27 Nov 2017, historycollection.com/assassination-bureau-8-groups-assassins-history/6/

267 'Secret Intelligent Service: the spies before James Bond', Royal Museums Greenwich, 21 Jan 2020, www.rmg.co.uk/stories/blog/library-archive/secret-intelligent-service-spies-james-bond

268 Dana Priest and William M. Arkin, 'Top Secret America: A Hidden World, Growing Beyond Control', *Washington Post*, 19 July 2010, archive.globalpolicy.org/pmscs/50501-top-secret-america-a-hidden-world-growing-beyond-control.html

269 'Executive Orders: Executive Order 12333–United States intelligence activities', Federal Register, National Archives, updated 15 Aug 2016, www.archives.gov/federal-register/codification/executive-order/12333.html

270 William M. Arkin, 'Exclusive: Inside the Military's Secret Undercover Army', *Newsweek*, 17 May 2021, www.newsweek.com/exclusive-inside-militarys-secret-undercover-army-1591881

271 William D. Hartung, 'The Pentagon's dark money: Billions of federal dollars are vanishing into thin air', Salon, 28 May 2016, www.salon.com/2016/05/28/the_pentagons_dark_money_billions_of_federal_dollars_are_vanishing_into_thin_air_partner/
272 Hartung, 'The Pentagon's dark money'.
273 See 'Digital National Security Archive (DNSA): CIA Covert Operations: From Carter to Obama, 1977-2010', *Digital National Security Archive*, ProQuest and The National Security Archive, proquest.libguides.com/dnsa/covert1977
274 Dan Kedmey, 'Report: NSA Authorized to Spy on 193 Countries', *TIME*, 1 July 2014, time.com/2945037/nsa-surveillance-193-countries/
275 See 'FAQ: What You Need to Know About the NSA's Surveillance Programs', ProPublica, 5 Aug 2013, www.propublica.org/article/nsa-data-collection-faq; Sarah Taitz, 'Five Things to Know About NSA Mass Surveillance and the Coming Fight in Congress', ACLU, 11 Apr 2023, www.aclu.org/news/national-security/five-things-to-know-about-nsa-mass-surveillance-and-the-coming-fight-in-congress
276 'Executive Orders: Executive Order 12333–United States intelligence activities, Section 1.1(c)', Federal Register, National Archives, updated 15 Aug 2016, www.archives.gov/federal-register/codification/executive-order/12333.html
277 Spencer Ackerman, Dominic Rushe and Julian Borger, 'Senate report on CIA torture claims spy agency lied about "ineffective" program', *Guardian*, 9 Dec 2014, www.theguardian.com/us-news/2014/dec/09/cia-torture-report-released
278 TeleSUR English, 'Mike Pompeo About CIA: We lied, We cheated, We stole', YouTube, 26 Apr 2019, www.youtube.com/watch?v=6RmEsPE7iq0; Lee Ferran, 'Ex-NSA Chief: "We Kill People Based on Metadata"', ABC News, 12 May 2014, abcnews.go.com/blogs/headlines/2014/05/ex-nsa-chief-we-kill-people-based-on-metadata
279 Some of their most famous attempts at grand geopolitical meddling have been embarrassing failures, from the Bay of Pigs to Vietnam to Iran, Iraq and Afghanistan, to the recent coup attempts.
280 'GHOSTS IN THE MACHINE', video uploaded by 4th PSYOP Group, YouTube, 2 May 2022, www.youtube.com/watch?v=VA4e0NqyYMw
281 Haley Britzky, 'This unsettling Army recruitment video is a master class in psychological warfare', Task and Purpose, 13 May 2022, taskandpurpose.com/news/what-is-psyops-army/
282 See Matt Taibbi, 'Reporters Once Challenged the Spy State. Now, They're Agents of It', Substack, 11 May 2021, taibbi.substack.com/p/reporters-once-challenged-the-spy; Glenn Greenwald, 'Journalists, Learning They Spread a CIA Fraud About Russia, Instantly Embrace

a New One', Substack, 16 Apr 2021, greenwald.substack.com/p/journalists-learning-they-spread
283 Jack Shafer, 'The Spies Who Came in to the TV Studio', Politico, 6 Feb 2018, www.politico.com/magazine/story/2018/02/06/john-brennan-james-claper-michael-hayden-former-cia-media-216943/
284 Max Blumenthal, 'Reuters, BBC, and Bellingcat participated in covert UK Foreign Office-funded programs to "weaken Russia," leaked docs reveal', The Gray Zone, 20 Feb 2021, thegrayzone.com/2021/02/20/reuters-bbc-uk-foreign-office-russian-media/
285 Gopnik, 'The Caging of America'; Steven Anthony Barnes, *Death and Redemption: the Gulag and the Shaping of Soviet Society* (Princeton, NJ: Princeton University Press, 2011), 1.
286 Laura M. Maruschak and Todd D. Minton, 'Correctional Populations in the United States, 2017–2018', The Bureau of Justice Statistics, U.S. Department of Justice, Aug 2020, bjs.ojp.gov/content/pub/pdf/cpus1718.pdf
287 Gopnik, 'The Caging of America'. As a brief aside, in this very article, the writer claims, 'The normalization of prison rape – like eighteenth-century japery about watching men struggle as they die on the gallows – will surely strike our descendants as chillingly sadistic, incomprehensible on the part of people who thought themselves civilized.' The piece is very good, but here is an example of how presumptions of inevitable progress sneak their way into the worldviews of ostensibly impartial journalists and are dashed off as simple matters of common sense. First, it's not necessarily more barbaric to hold public executions than it is to hold secret ones. One could argue the reverse is true. Extreme violence is treated with irreverence, humour and indignity in virtually all media today, from books, to video games, to music, movies and television. And while presumptions of progress are comforting, unfortunately, there is no good reason to believe our descendants will view prison rape as 'incomprehensible'. They will surely have their own daily savagery that may well be even more 'chillingly sadistic' than what we can imagine today.
288 James Cullen, 'The History of Mass Incarceration', Brennan Center for Justice, 20 July 2018, www.brennancenter.org/our-work/analysis-opinion/history-mass-incarceration
289 Cameron Smith, '"Authoritarian Neoliberalism" and the Australian Border-Industrial Complex', *Competition and Change*, 23:2 (2019), 192–217, doi.org/10.1177/1024529418807074; Denis Frank, 'Changes in Migration Control During the Neoliberal Era: Surveillance and Border Control in Swedish Labour Immigration Policy', *Journal of Political Power*, 7:3 (2014), 413–32, doi.org/10.1080/2158379X.2014.963381
290 Harvey, *A Brief History of Neoliberalism*, 195.

291 Vikram Dodd and Jamie Grierson, 'Terrorism police list Extinction Rebellion as extremist ideology', *Guardian*, 10 Jan 2020, www.theguardian.com/uk-news/2020/jan/10/xr-extinction-rebellion-listed-extremist-ideology-police-prevent-scheme-guidance

292 The police have since recalled the guide and stated they do not consider Extinction Rebellion extremist, but there's little reason to believe this position will be sustained and far more reason to believe it will be reversed; such a move was likely intended simply to test the waters of public approval and move the Overton window. How Just Stop Oil protestors have been treated more recently is evidence of that. Dodd and Grierson, 'Terrorism police list Extinction Rebellion as extremist ideology'.

293 'Policy Paper: Police, Crime, Sentencing and Courts Bill 2021: protest powers factsheet', Home Office, GOV.UK, 7 July 2021, www.gov.uk/government/publications/police-crime-sentencing-and-courts-bill-2021-factsheets/police-crime-sentencing-and-courts-bill-2021-protest-powers-factsheet

294 Damien Gayle, 'Just Stop Oil activist jailed for six months for taking part in slow march', *Guardian*, 15 Dec 2023, www.theguardian.com/environment/2023/dec/15/just-stop-oil-activist-is-first-to-be-jailed-under-new-uk-protest-law

295 This may also give credence to the idea that the CIA puts on a malevolent face as a form of social control. But the article itself is a brazen display of pro-CIA propaganda (see these choice quotes: 'there is no question that today's Agency is still necessary . . . It is easy to focus on failures and forget the dangers and drudgery of intelligence gathering, or to take time to celebrate the varied, but mostly secret victories of a service working hard to defend America. . . . America needs more credible, analytic, insightful, accessible, high level and grassroots deep intelligence. . . . Steinem chose to do an honorable duty.'). From Markos Kounalakis, 'The feminist was a spook', *Chicago Tribune*, 25 Oct 2015, www.chicagotribune.com/opinion/commentary/ct-gloria-steinem-cia-20151025-story.html

296 'Humans of CIA', Central Intelligence Agency, YouTube, 25 Mar 2021, www.youtube.com/watch?app=desktop&v=X55JPbAMc9g. See commentary at CIA [@CIA], '#WednesdayWisdom "I am unapologetically me. I want you to be unapologetically you, whoever you are. Whether you work at #CIA, or anywhere else in the world. Command your space. Mija, you are worth it."' Twitter, 28 Apr 2021, twitter.com/CIA/status/1387431636732633089

297 Dean Spade and Sarah Lazare, 'Women Now Run the Military-Industrial Complex. That's Nothing To Celebrate', *In These Times*, 12 Jan 2019, inthesetimes.com/article/women-military-industrial-complex-gina-haspel-trump-feminism-lockheed-marti.

298 Jefferson Morley, 'Raytheon's profits boom alongside civilian deaths in Yemen', Salon, 27 June 2018, www.salon.com/2018/06/27/raytheons-profits-boom-alongside-civilian-deaths-in-yemen_partner/
299 Raytheon Company, 'Raytheon Named "Best Place to Work" for LGBT Equality for Tenth Straight Year', news release, RTX News Release Archive, 19 Nov 2014, raytheon.mediaroom.com/2014-11-19-Raytheon-named-Best-Place-to-Work-for-LGBT-equality-for-tenth-straight-year
300 Medea Benjamin and Marcy Winograd, 'Biden's pick for intelligence chief, Avril Haines, is tainted by drones and torture', Salon, 30 Dec 2020, www.salon.com/2020/12/30/bidens-pick-for-intelligence-chief-avril-haines-is-tainted-by-drones-and-torture/
301 Adam Federman, 'Revealed: how the FBI targeted environmental activists in domestic terror investigations', *Guardian*, 24 Sep 2019, www.theguardian.com/us-news/2019/sep/23/revealed-how-the-fbi-targeted-environmental-activists-in-domestic-terror-investigations
302 Astra Taylor, 'In Defense of Liberal Conspirators', *New Republic*, 6 May 2021, newrepublic.com/article/162318/astra-taylor-defense-liberal-conspirators
303 Branko Marcetic, 'The FBI's Secret War', *Jacobin*, 31 Aug 2016, www.jacobinmag.com/2016/08/fbi-cointelpro-new-left-panthers-muslim-surveillance
304 'COINTELPRO', FBI Records: The Vault, vault.fbi.gov/cointel-pro
305 Protests took place in 2014 in Ferguson, Missouri after the high-profile killing of Michael Brown. There have since been an eerie number of deaths and suicides of Ferguson activists; it's unclear how many of them have been targeted by the FBI for such treatment. See Robert Cohen, 'Deaths of six men tied to Ferguson protests alarm activists', NBC News, 18 Mar 2019, www.nbcnews.com/news/us-news/puzzling-number-men-tied-ferguson-protests-have-died-n984261
306 Alec Karakatsanis, *Usual Cruelty: The Complicity of Lawyers in the Criminal Injustice System* (New York: The New Press, 2019).
307 Olivia Laughland, 'Malcolm X family says letter shows NYPD and FBI conspired in his murder', *Guardian*, 21 Feb 2021, www.theguardian.com/us-news/2021/feb/21/malcolm-x-death-family-letter-nypd-fbi
308 See for example Jim Salter, 'A puzzling number of men tied to the Ferguson protests have since died', *Chicago Tribune*, 18 Mar 2019, www.chicagotribune.com/nation-world/ct-ferguson-activist-deaths-black-lives-matter-20190317-story.html
309 Ken Bensinger and Jessica Garrison, 'Watching the Watchmen', Buzzfeed News, 20 July 2021, www.buzzfeednews.com/article/kenbensinger/michigan-kidnapping-gretchen-whitmer-fbi-informant
310 Branko Marcetic, 'The FBI's Domestic "War on Terror" is an

Authoritarian Power Grab', *Jacobin*, 28 July 2021, jacobinmag.com/2021/07/fbi-domestic-war-on-terror-authoritarianism-gretchen-whitmer-civil-liberties

AFTER PROGRESS

1. Randall Schaetzl, 'Native Americans in the Great Lakes Region', Michigan State University, project.geo.msu.edu/geogmich/paleo-indian.html
2. Kyle Whyte, Jared L. Talley and Julia D. Gibson, 'Indigenous Mobility Traditions, Colonialism, and the Anthropocene', *Mobilities*, 14:3 (2019), 321, doi.org/10.1080/17450101.2019.1611015
3. Larissa Swedell, 'Primate Sociality and Social Systems', *Nature Education Knowledge*, 3:10 (2012), www.nature.com/scitable/knowledge/library/primate-sociality-and-social-systems-58068905/
4. See Morris, *Foragers, Farmers, and Fossil Fuels*.
5. Pedro Carrera-Bastos et al., 'The Western Diet and Lifestyle and Diseases of Civilisation', *Research Reports in Clinical Cardiology*, 2:2 (2011), 2–15, doi.org/10.2147/RRCC.S16919.
6. Sahlins, *Stone Age Economics*.
7. Nathan Cohen in *Health and the Rise of Civilisation* (New Haven: Yale University Press, 2012) notes 'the major trend in the quality and quantity of human diets has been downward' from highs in generally what I am terming commensalistic societies (which overlap with but are not the same as the foraging societies represented in Cohen's study), to lows in parasitic and intensive agricultural societies.
8. See: Douglas P. Fry, 'Peace in Our Time', *Book Forum*, Dec/Jan 2012, www.bookforum.com/print/1804/steven-pinker-offers-a-curiously-foreshortened-account-of-humanity-s-irenic-urges-8575; Pasquale Cirillo and Nassim Nicholas Taleb, 'The Decline of Violent Conflicts: What Do The Data Really Say?' Nobel Foundation Symposium 161: The Causes of Peace, 27 Nov 2016, www.fooledbyrandomness.com/violencenobelsymposium.pdf; R. Brian Ferguson, 'Pinker's List: Exaggerating Prehistoric War Mortality', in Douglas P. Fry (ed.), *War, Peace, and Human Nature: The Convergence of Evolutionary and Cultural Views* (Oxford: Oxford University Press, 2013), oxford.universitypressscholarship.com/view/10.1093/acprof:oso/9780199858996.001.0001/acprof-9780199858996-chapter-7
9. C40 Cities, 'Why Cities are the Solution to Global Climate Change', C40.org, 2012, www.c40.org/ending-climate-change-begins-in-the-city
10. David Wallace-Wells, 'How to Live in a Climate "Permanent Emergency"', *New York Intelligencer*, 1 July 2021, nymag.com/intelligencer/2021/07/how-to-live-in-a-climate-permanent-emergency.html

11 See Ashley Dawson, *Extreme Cities: The Peril and Promise of Urban Life in the Age of Climate Change* (London: Verso, 2019).
12 Institute for Economics and Peace, 'Ecological Threat Register 2020: Understanding Ecological Threats', *Resilience and Peace*, Sep 2020, http://visionofhumanity.org/reports.
13 'Cities are at the frontline of the energy transition', International Energy Agency, 7 Sep 2016, www.iea.org/news/cities-are-at-the-frontline-of-the-energy-transition; 'Energy', UN Habitat, unhabitat.org/topic/energy
14 For more on the fossil city, see my essay 'The City of Tomorrow', *Current Affairs*, 30 May 2019, www.currentaffairs.org/2019/05/the-city-of-tomorrow-is-the-city-of-yesterday
15 Farshid Aram et al., 'Urban Green Space Cooling Effect in Cities', *Heliyon*, 5:4 (2019), doi.org/ 10.1016/j.heliyon.2019.e01339
16 Jake M. Robinson, Martin Breed and Ross Cameron, 'How the trees in your local park help protect you from disease', The Conversation, 5 May 2021, theconversation.com/how-the-trees-in-your-local-park-help-protect-you-from-disease-160312
17 See more at The Michigan Urban Farming Initiative, 2013, www.miufi.org/
18 Nathan J. Robinson, 'The Need for a New Garden City Movement', *Current Affairs*, 15 July 2021, www.currentaffairs.org/2021/07/the-need-for-a-new-garden-city-movement
19 'Cities are at the frontline of the energy transition', International Energy Agency, 7 Sep 2016, www.iea.org/news/cities-are-at-the-frontline-of-the-energy-transition. See also David McDermott Hughes, 'Use Sunlight Locally (or Lose It)', *Boston Review*, 5 Mar 2021, bostonreview.net/articles/david-mcdermott-hughes-solar-homesteading-tk/
20 See more of this political argument in Sam Miller McDonald, 'Material Freedom', *Anthroposphere*, 24 Apr 2018, www.anthroposphere.co.uk/post/material-freedom
21 Wengrow, 'Cities before the State in Early Eurasia'.
22 'Make cities and human settlements inclusive, safe, resilient and sustainable', Sustainable Development Goals, United Nations, 2021, unstats.un.org/sdgs/report/2019/goal-11
23 Nick Turse, 'Pentagon Video Warns of "Unavoidable" Dystopian Future for World's Biggest Cities', The Intercept, 13 Oct 2016, theintercept.com/2016/10/13/pentagon-video-warns-of-unavoidable-dystopian-future-for-worlds-biggest-cities/
24 Mike Davis, *Planet of Slums* (London: Verso, 2007), 19.
25 Christien Pheby, 'Third of Brits would reintroduce wolves and lynxes to the UK, and a quarter want to bring back bears', YouGov, 28 Jan 2020, yougov.co.uk/topics/science/articles-reports/2020/01/28/third-brits-would-reintroduce-wolves-and-lynxes-uk

26 Christine Peterson, '25 years after returning to Yellowstone, wolves have helped stabilize the ecosystem', *National Geographic*, 10 July 2020, www.nationalgeographic.com/animals/article/yellowstone-wolves-reintroduction-helped-stabilize-ecosystem
27 Alex Woolf, *From Pictland to Alba, 789–1070* (Edinburgh: Edinburgh University Press, 2007), 13.
28 No, a lynx did not write this. See Urs Breitenmoser *et al.*, 'Action Plan for the Conservation of the Eurasian Lynx (*Lynx lynx*) in Europe', *Nature and Environment*, 112 (2000), 1–70.
29 Center for Human-Carnivore Coexistence, 'Wolves and Human Safety', Colorado State University, May 2020, extension.colostate.edu/topic-areas/people-predators/wolves-and-human-safety-8-003/
30 'What are the world's deadliest animals?' BBC News, 15 June 2016, www.bbc.co.uk/news/world-36320744
31 Stephen Herrero *et al.*, 'Fatal Attacks by American Black Bear on People: 1900–2009', *Journal of Wildlife Management*, 75:3 (2011), 596–603.
32 Ed Yong, 'The Disturbing Sound of a Human Voice', *The Atlantic*, 17 July 2019, www.theatlantic.com/science/archive/2019/07/humans-predators-mountain-lions-landscape-of-fear/594187
33 For public opinion, see Pheby, 'Third of Brits would reintroduce wolves and lynxes to the UK'; Jody Harrison, 'Rewilding Scotland: Public hugely in favour', *Herald Scotland*, 17 Feb 2021, www.heraldscotland.com/news/19095394.rewilding-scotland-public-hugely-favour/. On the growing threat of ticks in the United Kingdom, see Michelle Roberts, 'Brain illness spread by ticks has reached UK', BBC News, 29 Oct 2019, www.bbc.co.uk/news/health-50206382
34 FAOSTAT Analytical Brief 18, 'Emissions due to agriculture: Global, regional and country trends 2000–2018', Food and Agricultural Organisation of the United Nations, 2021, www.fao.org/3/cb3808en/cb3808en.pdf
35 Amanda Garris, 'Industrial methane emissions are underreported, study finds', *Cornell Chronicle*, 6 June 2019, news.cornell.edu/stories/2019/06/industrial-methane-emissions-are-underreported-study-finds
36 Adrian Muller and Lin Bautze, 'Agriculture and deforestation: The EU Common Agricultural Policy, soy, and forest destruction: Proposals for Reform', Fern, May 2017, www.fern.org/publications-insight/agriculture-and-deforestation-264/
37 'Burden of Foodborne Illness: Overview', Centers for Disease Control and Prevention, 5 Nov 2018, www.cdc.gov/foodborneburden/estimates-overview.html
38 See 'Leading Causes of Death', Centers for Disease Control and Prevention, 13 Jan 2022, www.cdc.gov/nchs/fastats/leading-causes-of-death.htm; 'Heart Disease Facts', Centers for Disease Control and Prevention, 7 Feb 2022, www.cdc.gov/heartdisease/facts.htm

39 Eric Holthaus, 'Deadly air pollution has a surprising culprit: Growing corn', Grist, 4 Apr 2019, grist.org/article/deadly-air-pollution-has-a-surprising-culprit-growing-corn/
40 Debbie Weingarten, 'Why are America's farmers killing themselves?' Guardian, 11 Dec 2018, www.theguardian.com/us-news/2017/dec/06/why-are-americas-farmers-killing-themselves-in-record-numbers
41 Salimah Shivji, 'Burdened by debt and unable to eke out a living, many farmers in India turn to suicide', CBC News, 30 Mar 2021, www.cbc.ca/news/world/india-farmers-suicide-1.5968086
42 Sarah Yang, 'Human security at risk as depletion of soil accelerates, scientists warn', *Berkeley News*, 7 May 2015, news.berkeley.edu/2015/05/07/soil-depletion-human-security/
43 Industrial agriculture is, as we have seen throughout this book, simply another central aspect of modern parasitism. See Georgina Gustin, 'Industrial Agriculture, an Extraction Industry Like Fossil Fuels, a Growing Driver of Climate Change', *Inside Climate News*, 25 Jan 2019, insideclimatenews.org/news/25012019/climate-change-agriculture-farming-consolidation-corn-soybeans-meat-crop-subsidies/
44 Davis, *Planet of Slums*, 9.
45 For a good overview of land policy reform, see Susan Sunhee Volz, 'It Begins With the Land', *The Trouble*, 26 Apr 2019, www.the-trouble.com/content/2019/4/26/in0ipnptnhqqbot1zvqv6ubfnrculh.
46 Statistics from the World Wildlife Fund, 2024.
47 The long-form article, 'Lifecycle of a Leaf', I wrote for *Current Affairs* in 2019 argues for a major shift in the global food system, and includes illustrative personal anecdotes from my time in food production.
48 For good introductions, see Fredrik Moberg, 'How are Agroecological Farmers Challenging the Industrial Way of Farming?' Organic Without Boundaries, 8 Aug 2018, www.organicwithoutboundaries.bio/2018/08/08/agroecological-farmers-rethink/; Henrietta Moore, 'Can agroecology feed the world and save the planet?' *Guardian*, 9 Oct 2016, www.theguardian.com/global-development-professionals-network/2016/oct/09/agroecological-farming-feed-world-africa; Chris Williams, 'The Future of Farming is Agroecological', New Economics Foundation, 1 Feb 2021, neweconomics.org/2021/02/the-future-of-farming-is-agroecological
49 Miguel A. Altieri et al., 'Agroecology and the Design of Climate Change-Resilient Farming Systems', *Agronomy for Sustainable Development*, 35 (2015), 869–90, doi.org/10.1007/s13593-015-0285-2
50 See 'The Amazing World of Apple Varieties', The Orchard Project, 2022, www.theorchardproject.org.uk/blog/the-amazing-world-of-apple-varieties/
51 Frank Newport, 'Americans Big on Idea of Living in the Country',

Gallup News, 7 Dec 2018, news.gallup.com/poll/245249/americans-big-idea-living-country.aspx

52 There are many good analyses of the UBI debate. Here is my contribution to the discussion, which I believe makes a novel argument: Samuel Miller McDonald, 'A Jobless Economy', Resilience, 22 May 2018, www.resilience.org/stories/2018-05-22/a-jobless-economy/

53 See Elizabeth Anderson, *Private Government: How Employers Rule Our Lives (and Why We Don't Talk about It)* (Princeton, NJ: Princeton University Press, 2017).

54 Alexandra Kelley, 'DOI returns more than 18k acres of land to Native American tribes', *Changing America*, 23 June 2021, thehill.com/changing-america/respect/equality/559902-doi-returns-over-18000-acres-of-land-to-native-american

55 Reported by Jim Robbins, 'How Returning Lands to Native Tribes Is Helping Protect Nature', Yale Environment 360, 3 June 2021, e360.yale.edu/features/how-returning-lands-to-native-tribes-is-helping-protect-nature

56 Nick Estes, 'A Red Deal', *Jacobin*, 6 Aug 2019, www.jacobinmag.com/2019/08/red-deal-green-new-deal-ecosocialism-decolonisation-indigenous-resistance-environment

57 Here I place the agent as the whole species. Some reasonably balk at this. Many scholars have criticised terms like 'Anthropocene' for holding the human species as an undifferentiated mass, collectively and equivalently responsible for climate and environmental crises, rather than identifying those particular classes, corporations or nation-states that are most culpable. We should blame the rich, specifically, yes, but we should also recognise that we non-rich accept their existence and profligacy, however much we grumble. And while the rich consume disproportionately much more, the fact remains that every human body needs at least 2,000 k/cal a day to live, and those calories must come out of the mouths of other living things. As more human and human-use biomass has been grown, non-human biomass has proportionately shrunk.

58 In fact, one argument suggests that divestment could *increase* carbon emissions by giving more market shares to state-owned companies, which tend to be less transparent and less susceptible to pressure from activists. See Stefan Andreasson, 'Fossil fuel divestment will increase carbon emissions, not lower them – here's why', The Conversation, 25 Nov 2019, theconversation.com/fossil-fuel-divestment-will-increase-carbon-emissions-not-lower-them-heres-why-126392

59 'About National Oil Companies', Stanford Program on Energy and Sustainable Development, pesd.fsi.stanford.edu/theme/national-oil-companies-

60 Marisa Guerrero, 'New York State Codifies Fracking Ban in Budget', NRDC, 3 Apr 2020, www.nrdc.org/experts/marisa-guerrero/new-york-state-codifies-fracking-ban-budget

61 'Scottish government confirms "no fracking" policy', BBC News, 3 Oct 2019, www.bbc.com/news/uk-scotland-scotland-politics-49924749

62 For why 'natural' gas is so bad, see Adam Morton, 'Booming LNG industry could be as bad for climate as coal, experts warn', *Guardian*, 2 July 2019, www.theguardian.com/environment/2019/jul/03/booming-lng-industry-could-be-as-bad-for-climate-as-coal-experts-warn; David Roberts, 'More natural gas isn't a "middle ground" – it's a climate disaster', Vox, 30 May 2019, www.vox.com/energy-and-environment/2019/5/30/18643819/climate-change-natural-gas-middle-ground; 'Gas', Oil Change International, http://priceofoil.org/program-areas/stopping-carbon-lock-in/gas/

63 Thomas W. Murphy, 'Limits to economic growth', *Nature Physics* 18 (2022), 844–7, doi.org/10.1038/s41567-022-01652-6

64 There is abundant media covering the topic of how people can affect these issues. I have published several pieces focused on 'what you can do' and several more on how the movements themselves should change. These sorts of articles are fine starting points, but strategic and tactical writing for a very broad audience can only go so far. All good strategy and tactics are local and hyper-specific. Nevertheless, if you're interested, here are some articles of my own and others with ideas about how to get involved in movements and how those movements can improve:

1) Samuel Miller McDonald, 'What We Should Really Do for the Climate', The Trouble, 16 Mar 2019, www.the-trouble.com/content/2019/3/16/what-we-should-really-do-for-the-climate

2) Samuel Miller McDonald, 'What Must We Do to Live?' The Trouble, 14 Oct 2018, www.the-trouble.com/content/2018/10/14/what-must-we-do-to-live

3) Samuel Miller McDonald, 'Rage Gap', The Trouble, 11 June 2019, www.the-trouble.com/content/2019/6/11/rage-gap

4) Samuel Miller McDonald, 'The Climate Movement Needs More Radicals', The Trouble, 4 May 2019, www.the-trouble.com/content/2019/5/4/the-climate-movement-needs-more-radicals

5) Samuel Miller McDonald, 'The Green New Deal Can't Be Anything Like the New Deal', *New Republic*, 31 May 2019, newrepublic.com/article/153996/green-new-deal-cant-anything-like-new-deal

6) Samuel Miller McDonald, 'The Climate Movement Needs More Creative Tactics', *New Republic*, 18 Nov 2019, newrepublic.com/article/155764/climate-movement-needs-creative-tactics

7) Samuel Miller McDonald, 'Beyond Fluorescent Bulbs: 4 Things Millennials Can Do To Fight Climate Change', *In These Times*, 19

Apr 2018, inthesetimes.com/article/4-things-millennials-can-do-to-fight-climate-change-divest-solar-energy

8) Samuel Miller McDonald, 'Carbon Omissions', *The Baffler*, 18 Jan 2018, thebaffler.com/latest/climate-omissions-miller-mcdonald

9) Samuel Miller McDonald, 'Strong and Stable Socialism for a Climate-Changed World', *Socialist Forum*, Winter 2019, socialistforum.dsausa.org/issues/winter-2019/strong-and-stable-socialism-for-a-climate-changed-world/

10) Samuel Miller McDonald, 'Collapse Despair', *Epilogue*, 11 Jan 2020, http://epiloguemag.com/2020/01/collapse-despair/

11) Jonathan Guy and Sam Zacher, 'What the Sunrise Movement Can Do Better', *Jacobin*, 23 Aug 2021, www.jacobinmag.com/2021/08/sunrise-movement-green-new-deal-left-politics-local-organizing

12) Ashwin Ravikumar, 'To Save the Rainforests, Provide Healthcare, Education and Services for Those Who Protect Them,' The Trouble, 18 Jan 2022, www.the-trouble.com/content/2022/1/17/to-save-the-rainforests-provide-healthcare-education-and-services-for-those-who-protect-them.

13) Jason Hickel, 'What Would It Look Like If We Treated Climate Change as an Actual Emergency?' *Current Affairs*, 15 Nov 2021, www.currentaffairs.org/2021/11/what-would-it-look-like-if-we-treated-climate-change-as-an-actual-emergency

14) Nick Engelfried, 'It's Time to Challenge Animal Agriculture – Without the "Go Vegan" Campaigns', The Trouble, 5 Jan 2022, www.the-trouble.com/content/2022/1/5/its-time-to-challenge-animal-agriculturewithout-the-go-vegan-campaigns

15) George Monbiot, 'After the failure of Cop26, there's only one last hope for our survival', *Guardian*, 14 Nov 2021, www.theguardian.com/commentisfree/2021/nov/14/cop26-last-hope-survival-climate-civil-disobedience

16) George Monbiot, 'Capitalism is killing the planet – it's time to stop buying into our own destruction', *Guardian*, 30 Oct 2021, www.theguardian.com/environment/2021/oct/30/capitalism-is-killing-the-planet-its-time-to-stop-buying-into-our-own-destruction

65 Bill McKibben, 'The New Climate Math: The Numbers Keep Getting More Frightening', Yale Environment 360, 25 Nov 2019, e360.yale.edu/features/the-new-climate-math-the-numbers-keep-getting-more-frightening

66 'Cuba opens up its economy to private businesses', BBC News, 7 Feb 2021, www.bbc.com/news/world-latin-america-55967709

67 We can point to punitive intervention by the US and France in the wake of Haiti's revolution as a major source of its failures, but the facts remain: it cannot be said to be a successful revolution given its

failure to deliver liberty, equality or its stated aims to a majority of the population, even if external forces had an outsize role in that failure.
68 A vision that is often encapsulated in the growing degrowth literature and movement. See degrowth.org/ and www.degrowth.info/
69 See groups like the Sunrise Movement, the Democratic Socialists of America (Ecosocialists), the Climate Justice Alliance, Earth First! and Extinction Rebellion.
70 Leah Temper *et al.*, 'Movements Shaping Climate Futures: A Systematic Mapping of Protests Against Fossil Fuel and Low-Carbon Energy Projects', *Environmental Research Letters*, 15:12 (2020), doi.org/10.1088/1748-9326/abc197
71 Canada's record is very bad: Jaskiran Dhillon and Will Parrish, 'Exclusive: Canada police prepared to shoot Indigenous activists, documents show', *Guardian*, 20 Dec 2019, www.theguardian.com/world/2019/dec/20/canada-indigenous-land-defenders-police-documents; Jorge Barrera, 'Military's counter-intelligence unit monitored Elsipogtog anti-fracking protests: documents', APTN National News, 30 May 2014, www.aptnnews.ca/national-news/militarys-counter-intelligence-unit-monitored-elsipogtog-anti-fracking-protests-documents/. This is not surprising from a country that recently dug up mass graves of children at the schools built to imprison and assimilate Indigenous peoples.
72 Barrera, 'Military's counter-intelligence unit monitored Elsipogtog anti-fracking protests'.
73 For 2019 figures, see Patrick Greenfield and Jonathan Watts, 'Record 212 land and environment activists killed last year', *Guardian*, 29 July 2020, www.theguardian.com/environment/2020/jul/29/record-212-land-and-environment-activists-killed-last-year; for 2020, see Kate Hodal, 'At least 331 human rights defenders were murdered in 2020, report finds', *Guardian*, 11 Feb 2021, www.theguardian.com/global-development/2021/feb/11/human-rights-defenders-murder-2020-report
74 In Madagascar, defenders like Clovis Razafimalala, Armand Marozafy, Augustin Sarovy and many others have fought on the front lines of attempting to prevent environmental destruction and have been harassed, jailed and exiled.
75 See also Daisy Dunne, 'Indigenous leadership key to curbing deforestation, says UN report', *Independent*, 25 Mar 2021, www.independent.co.uk/independentpremium/indigenous-leadership-deforestation-un-b1822279.html
76 'Indigenous Resistance Against Carbon', Oil Change International, Indigenous Environmental Network and Indigenous Climate Action, Aug 2021, www.indigenousclimateaction.com/entries/indigenous-land-defenders-are-the-best-defence-against-the-climate-crisis. Emphasis in original.

77 Arnim Scheidel *et al.*, 'Global Impacts of Extractive and Industrial Development Projects on Indigenous Peoples' Lifeways, Lands, and Rights', *Science Advances*, 9:23 (2023), oi.org/10.1126/sciadv.ade9557
78 Kimmerer, *Braiding Sweetgrass*, 307.
79 Morris, *Foragers, Farmers, and Fossil Fuels*, 43.
80 Matthew J. Hornsey and Kelly S. Fielding, 'A Cautionary Note about Messages of Hope: Focusing on Progress in Reducing Carbon Emissions Weakens Mitigation Motivation', *Global Environmental Change*, 39 (2016), 26–34, doi.org/10.1016/j.gloenvcha.2016.04.003; Thea Gregersen, Gisle Andersen and Endre Tvinnereim, 'The Strength and Content of Climate Anger', *Global Environmental Change*, 82 (2023), doi.org/10.1016/j.gloenvcha.2023.102738. Further, my own doctoral research contributes to scholarship suggesting that hope and optimism are either neutral or negative as motivators to participate in movements and other actions. See S. Miller-McDonald, 'Degrowth and Progress: a Critical Genealogy and Contemporary Analysis of Growthism', PhD thesis, University of Oxford, 2023.

MEANING BEYOND PROGRESS

1 'Declaration of Independence: A Transcription', America's Founding Documents, National Archives, www.archives.gov/founding-docs/declaration-transcript
2 Martin Hägglund, *This Life: Why Mortality Makes Us Free* (London: Profile Books, 2020), 95.
3 Hägglund, *This Life*, 275. Emphasis in original.
4 Quoted in Hägglund, *This Life*, 330.
5 Hägglund, *This Life*, 331.
6 Karl Ove Knausgård, *A Death in the Family – My Struggle: Book 1*, trans. Don Bartlett (New York: Farrar, Straus and Giroux, 2013), 441. Throughout his series, Knausgård frequently sprinkles environmental sublime, despite his stated lack of interest in nature, and he opens the final volume with a character who laments the sudden disappearance of a rare frog species and nightingales. His more recent novel, *The Morning Star*, joins an emerging group of writers engaging in ecoGothic fiction to better understand and contend with the crisis of human ecology.
7 For more on this argument, see Samuel Miller McDonald, 'The Left Must Adapt or Die', *The Nation*, 20 July 2023, www.thenation.com/article/environment/the-left-warming-planet/

Index

Abbott, Derek, 269
Abenaki people, 44, 45–6, 182, 183
Achuar people, 44–5
Africa, 14–16, 63, 64, 82, 97, 99, 120, 277; as birthplace of humanity, 40, 42, 82; European colonialism in, 15, 142, 198, 204
Agricola (Roman general), 83, 85
agriculture/food production: 'agricultural revolution', 43; agroecological farming, 306–9, 311; in ancient civilisations, 59, 63, 64–5, 68, 70, 82; in communist states, 237, 238, 239, 241–3; consolidation in agribusiness, 304, 306, 307; deforestation for, 24, 36, 68–9, 82, 277–8, 304; draining of the Everglades for, 154; and energy density, 163–4, 165, 305–6; food safety legislation, 191, 314; food sources vulnerable to climatic instability, 12; global land coverage of farms, 26; impact of colonialism on European 'host classes', 121–2; livestock, 27, 51, 64, 76, 90, 100, 161, 162, 163, 165, 295, 302, 310; and mass, mechanistic slaughter, 29, 64, 265–6; Native American, 2, 32, 33–5, 36, 43, 59, 147–8, 150, 155, 159, 160, 293; need to eat seasonally/locally, 299, 300, 307; parasitic/intensive over-production, 62–5, 76, 82, 102, 110, 160–1, 237, 238, 242–3, 265, 272, 304–6; and spread of Christianity/Islam, 76, 101, 110; transition to capitalism/industrial methods, 114–15, 166, 188; unhealthy/toxic food, 165, 304

Alaric (Visigoth King), 87
Allen, Paula Gunn, 54
Aly, Götz, 222
Amazon (company), 259
Amazon rainforest, 36, 152, 277
American Enterprise Institute, 247–8
American Revolutionary War, 150
anarchism, 173, 190–1, 206, 207, 211, 237
Angell, Norman, 236
Anglo-Saxons, 92–3, 133–4
animal life, 27, 40, 41–4, 301–4, 305, 323; big game hunting, 215; domesticated, 12, 26, 27, 49, 51, 52, 59, 64, 100, 130, 164, 195, 294, 302–3; eradicated/harmed by human 'progress', 23–4, 26, 28–9, 38, 47, 64–5, 82, 143–5, 192, 261, 265–6, 272, 277, 312; habitat destruction, 22, 24, 29, 50, 234, 243, 272, 277, 291, 304; meat corporations' treatment of livestock, 51, 265, 302; thoughts/emotions/quality of life, 13, 43, 49–52, 64, 134–6

animism, 43–8, 55, 61, 332–3
Anishinaabe people, 46–7, 293
ants, 42, 51
Arabian Peninsula, 93–5
Arawak peoples, 115–16, 120
artificial intelligence, 255–61, 278
Ashoka, 16
Asia, European colonialism in, 142, 174–5, 198
Asimov, Isaac, 258
Assyria, 60, 70
Attlee, Clement, 227, 245
Audubon, John James, 143–4
Augustine of Hippo, 88, 242
Augustus, Roman emperor, 82, 84
Australia, 45, 47, 182–3, 308–9
Axial Age, 104, 125
Aztec empire, 118, 152, 279

Backman, Clifford, 78
bacteria, 26, 40, 41, 42, 334
Baghdad, 95, 97
Baillie, John, 128
Bambery, Chris, 122
Bancroft, George, 145
Banerjee, Sikata, 133
Bannon, Steve, 262
Barnham, Keith, 270
Barrera, Giulia, 205
Bartels, Dennis and Alice, 239
bears, 44, 265, 302, 303
Beddoe, John, 133
Bedouin, 94, 99
bees, 42, 51
Bell, Alexander Graham, 179
Beothuk people, 89
Beringia, land bridge at, 293
Berman, Ronald, 214
Bezos, Jeff, 254–5, 257–8, 259–60, 261, 262, 266, 273–4
biosphere: biocentricity, 75–7, 320–1, 333, 334–5; biodiversity, 26, 33, 41, 43, 45, 60, 76, 178–9, 295–7, 302, 306–9, 310, 319, 326, 333; biomes/biomass, 26, 27, 38, 41–4, 63, 65–6, 143, 161–3, 233, 277, 305, 310, 316; destruction/shrinking of, 25, 26, 28–9, 178–80, 192, 262–6, 272, 277, 304, 310, 320–1, 333–5; human/animal instinct towards biophilia, 43–4, 64–7, 333, 334–5; need for transition to commensalism, 293–7, 301–2, 305–9, 322–6; sustainability paradox, 261–2, 273–4, 317 *see also* ecological crisis; ecological systems
'Biosphere 2' project, 262
birds, 42, 50, 135, 143–4
Al-Biruni, Abu Rayhan, 95–6
bison, 135, 144–5
Black Death, 114
Black Lives Matter movement, 19, 284
Blue Origin, 254–5, 258, 259–60
BP, 266–7, 311
Britain: apple varieties in, 307; Brexit debate, 230–1; Celtic regions of, 114, 121–2, 132–3; elite support for Nazism/fascism, 203, 205–6, 211, 223–4; industrialisation in, 115, 161, 189; maritime empire of, 119, 121–2, 132–3, 142–3, 161, 174–5; nineteenth-century debate on progress, 168–72, 177; nineteenth-century progress, 187, 189; and rewilding of predators, 302, 303; Roman rule in, 83, 84–5, 102; Saxon rule after fall of Rome, 92–3; under social democracy, 7, 227–8, 230–1, 244–5, 251–2; state targeting of environmentalists in, 284–5; Thatcher government, 21
Brockell, Gillian, 150
Brooks, Lisa, 45–6
Buddhism, 72, 104, 125
Burj Khalifa, Dubai, 5
Burns, Robert, 125–6

INDEX

Bush, George W., 287
Bush, Prescott, 225

Caledonians, 83, 85
Calhoun, John C., 222
Calloway, Colin G., 182
Canaan, 77
Canada, 89, 124, 311, 319
capitalism, 111–15, 120, 122, 143, 148, 161, 165–7, 171–3, 178, 191–2, 197; Einstein's view of, 253–4; financial, 167–8, 195, 199, 216, 220–1, 245–6, 250; and Hitler's anti-Semitism, 220–1, 222–3; and liberal ideology, 138, 139, 140, 142, 148, 197, 208, 229, 235, 245–50; and Karl Marx, 173–8, 243–4; under social democracy, 227, 229, 235, 245–6, 247, 282
carbon capture and sequestration, 266, 272–3, 275
Carnegie, Andrew, 171–2
Çatal Hüyük, Anatolia, 301
Chevron, 267
Chile, 250, 319
China, 102, 104, 199, 237, 241–4, 276, 279, 315
Chlenov, Michael, 240
Christianity, 72, 73, 74, 75, 85–6, 88–9, 91–2, 101, 109, 134, 137–8; missionaries in New World, 45, 111–12, 123, 124; progress narrative of, 67, 85–8, 91, 93, 96, 103, 110, 111–12, 115–19, 128, 145, 274; Roman conversion to, 86–8, 92, 103, 130
Churchill, Winston, 189
Civil War, American, 158, 160, 165, 313
Clark, Christopher, 189–90
class issues, 7, 11, 12, 19, 54, 173–8, 188–9, 222–3, 244–50, 259–60
ClientEarth, 266–7

climate emergency, 16–17, 24–8, 265–78, 288–9, 297–8, 306–13, 314–21, 327–34; 'hothouse earth' scenario, 263–4, 276 see also ecological crisis
climatology, 179–80
clownfish, 41
Coen, Deborah, 180
colonialism, 110–13, 115–26, 131–5, 141, 174–5, 191–2, 197–8, 239, 250; in ancient civilisations, 59–60, 63–4; European in Africa and Asia, 15, 142, 174–5, 198, 203–5; European in the Americas, 30, 31–5, 37–9, 54, 111–12, 123–5, 141, 142–3, 192; of Fascist Italy, 199, 203–5, 217; Indigenous rebellions/revolts, 182–3; Viking, 89–92, 258
Columbus, Christopher, 115–16, 120
commensalistic/mutualistic relationships, 41–2, 43–8, 53–5, 60–1, 114, 119, 125, 318–20, 325–6; '*calculated*-commensalist' societies, 294–7; foraging communities, 32, 33, 42–3, 52, 163–4, 239–40, 294, 305–6; Mesopotamian move away from, 59–60, 67–72, 82, 93, 101–2, 103, 249, 275; mutual extraction principle, 321, 323–4; need for wide-scale shift to, 293–7, 301–2, 305–9, 322–6
communications technologies, 26, 110–11, 143, 178, 179, 191–2, 195, 201, 254–61
communism, 157, 173–8, 180–1, 197, 201, 206–7, 209, 226–7, 236–43, 276
Comte, Auguste, 173
Condorcet, Marquis of, 131
Confucianism, 104, 125
Conservative Party, 230–1, 232, 285
Constantine I, Emperor, 86–7, 88, 92, 93, 103

corporate world: anti-trust laws in USA, 191, 209, 314; consolidation in agribusiness, 304, 306, 307; cynical propaganda on climate, 266–7; elite opposition to social democracy, 244–50, 251–3; and fascism, 208, 225; habitat destruction by, 50, 234, 252, 277; and market utility of fossil fuels, 311; meat corporations' treatment of livestock, 51, 265, 302; multinational corporations, 195; and 'social Darwinism', 169–70, 171–2, 177, 185; and US intelligence community, 280, 281–3, 284; use of violence by, 167, 172, 319, 320
cosmology, 3, 55, 67–78, 79, 102–5
Coudenhove-Kalergi, Richard von, 84, 204
Covid-19 pandemic, 262
Crévecoeur, Hector de, 37
Crossley-Holland, Kevin, 90–1
Crowder, Henry, 212
Cuba, 215, 315
Cunard, Nancy, 211–12, 216
Curtis, Charles, 20
Cyrus II, King of Persia, 16, 74

Darwin, Charles, 125, 168–70, 171, 174, 179, 185
Davis, Mike, 301, 305
Dawson, Ashley, 82
Dearborn, Henry, 147
DeepMind, 256–7
Demuth, Bathsheba, 239
Denmark, 11, 91, 189
Descartes, René, 134–6
'development' terminology, 31–2, 36, 55
Diamond, Jared, 90
Dimmock, Spencer, 115
diversity, cultural, 94, 123, 153, 192, 250, 294, 296, 297, 309–10, 325–6

dogs, 64, 302–3
Douglass, Frederick, 188
Du Bois, W. E. B., 212
Dubar-Ortiz, Roxanne, 132–3
Dùthchas concept, 114

Eagleton, Terry, 196
ecological crisis: achieving future non-parasitic ecology mode, 292, 293–7, 299–304, 305–13, 316–21, 322–6; deforestation, 24, 36, 68–9, 82, 90, 266, 277–8, 304; fundamental change as already inevitable, 264–5, 274–6, 277–8, 297–8, 304–5; greenhouse gases/carbon emissions, 14, 20, 24, 179–80, 228–9, 252, 263, 269–70, 272, 304, 312; and idea of infinite growth, 233, 234, 253–4, 287; increasing number of emergencies, 262, 272; mass extinctions driven by parasitism, 24, 26, 179, 192, 266, 272, 304; overfishing, 24, 26, 144, 265; proposed actions of individual engagement, 292, 327–34; rising sea levels, 24, 264, 297, 298; severity of/risk of global collapse, 16–17, 24–8, 264–5, 273–6, 277–8, 287–9, 291–2, 298, 304–5, 314–21; wilderness decline, 210, 212, 303–4 *see also* climate emergency
ecological systems: adverse effects of industrialisation, 165–7, 175, 178–80, 192, 198, 212–13; communist destruction of, 241, 243; corporate destruction of, 50, 234, 252, 277; destruction during conquest of Native American land, 23–4, 36, 38–9, 121, 143–6, 147–56, 160–1, 192; dominion over in Abrahamic creation myths, 75–8, 100, 103, 130, 134–5; fire ecologies, 24, 47, 298; and

INDEX

Indigenous cultures, 32–4, 36, 38–9, 40, 43–8, 49–50, 53, 54–5, 60–1, 65, 102, 160; last great forests, 277–8; 'naturalistic fallacy', 170; 'nature' or 'wilderness' concept, 48–9, 52, 303–4; and notion of 'profit', 66, 93; progress narratives justifying destruction, 135–6, 143–6, 147–56, 158–61; recalibration in post-Black Death period, 114; space as non-viable frontier, 260–3 *see also* biosphere

ecology, discipline of, 41–3, 50

economic growth, 7; classical economists see as finite, 232; GDP measure, 13, 231–5

economic systems: in Bolshevik Russia, 237–41; debt, 65, 66, 85, 112, 141, 147–8, 159, 185, 252; expansion in early USA, 34–5, 36, 38–9; Friedman's monetarist analysis, 251; inflation in, 230, 251; Italian fascist corporatism, 203, 204–5; Keynesianism, 229–35, 236, 244, 245, 248, 251–2, 277; Ibn Khaldun's ideas, 98; maritime economies from fifteenth-century, 63–4, 110–12, 115–22, 191, 195; monopolies/trusts, 172, 209, 314; Native American, 54; notion of 'profit', 60, 66, 93, 113, 122; oil shock (1973), 251; rent-seeking, 65, 249; under social democracy, 227–35, 236, 246, 251–2; tenuous distribution networks, 262 *see also* commensalistic/mutualistic relationships; market economics; neoliberalism; parasitic economies/societies; trade and commerce

economics, scholarly discipline of, 196, 231, 232–5

education, 95, 146, 169, 171, 187, 227

Edward VIII, King, 224
Egypt, ancient, 5, 181
Einstein, Albert, 218, 253–4
electrification, 6, 164, 178, 189, 195, 196
Eliot, George, 170, 177, 178
Eliot, T.S., 210, 211
Ellis, Caitlin, 91
energy: 'abstract energy capture', 60–5, 90, 94, 110, 122, 148, 196, 203, 216–17, 222, 278, 294–5, 332; achieving future non-parasitic ecology mode, 292, 293–7, 299–304, 305–13, 316–21, 322–5; capture relationships *within* societies, 53–5, 61–4, 65–7, 221–3, 225, 239; 'concrete energy capture', 40–3, 52–3, 63–4, 89–90, 110, 122, 148, 156, 195–6, 203, 216–17, 294–5, 305, 310; density, 161–5, 201, 298–9, 305–6; nuclear, 268–71, 273; release of by modern technology, 6–7, 24–5, 161, 163–4, 165–8, 176, 195–6, 231, 234–5; 'secondary' energy, 53–4; use by human brains/bodies, 162–3, 262, 305–6 *see also* fossil fuel exploitation

Engels, Friedrich, 174, 175, 180–1, 232, 243

Enlightenment thinking, 111–15, 122–9, 140–3, 145–9, 154, 168, 171–2, 225, 247, 250; and colonial conquest, 29–30, 101, 122–5, 130, 131–4, 143, 145, 148–9, 154, 225; idea of civilisational progress, 79, 99, 113–15, 122, 125–34, 136–42, 148–9, 154, 168, 171–2, 247, 250; Islamic influence on, 95, 97–9, 104; 'state of nature' concept, 31, 39, 49, 99, 127, 136–7 *see also* liberalism, ideology of

Epstein, Alex, 271
Eriksson, Leif, 89

413

Eusebius, 86
evolutionary science, 168–70, 171, 173, 174, 175, 177, 185–6, 192
Extinction Rebellion, 284
ExxonMobil, 311

famine, 121, 237, 238, 288
fascism, 199, 201–9, 211, 217, 223–6, 236, 237 *see also* Nazi Germany
Fatima al-Fihri, 95
feudalism, 90, 113–14
Fialka, John, 268
financial crisis (2008), 19
First World War, 168, 197–8, 210, 227, 253, 288
Fitzgerald, F. Scott, 213–15, 216
Foote, Eunice Newton, 179
Ford, Henry, 225
fossil fuel exploitation, 20, 164, 260–1, 266–7, 273, 278, 306, 311–12, 316–17, 319; explosion in during 'networked' parasitism, 24–5, 195–6, 226, 228–9, 231, 234–5, 287; and fascism, 203–5, 226; internal combustion engine, 178, 202; nineteenth-century growth in, 143, 161, 165–8, 176, 178–80, 188–9, 191–2; 'resource curse'/authoritarian outcomes, 167, 310–11; steam engine invented, 164, 166, 178
Fourier, Joseph, 179
Fowden, Garth, 86, 94–5
fracking, 312, 319
France, 119, 124–5, 142, 189, 315
Franco, Francisco, 208–9, 211, 226
Franklin, Benjamin, 37, 125
Freamon, Bernard K., 97
Freud, Sigmund, 210
Friedman, Milton, 247, 250, 251
frogs, 51
Fukuyama, Francis, 72–3
futurism, Italian, 202–3, 205, 208, 209, 240, 255–6

Gaddafi, Muammar, 16, 20
Gagarin, Yuri, 4
Gates, Bill, 8
Gazzaniga, Valentina, 11
gender, 146, 189, 240–1, 279, 285; in ancient Greece, 79–80, 146, 183; futurist scorn for women, 202, 240; in Indigenous cultures, 32, 33, 34, 35, 54, 124, 176–7, 249; in late-medieval period, 114, 115; women's rights, 7, 14, 21, 171, 191, 209
General Motors, 225
Gentile, Giovanni, 204
geography, 10–11, 12, 22–3
Germany, 142, 189
Gibbon, Edward, 129–31, 149
Gibson, Julia D., 293
Gini coefficient, 11
Gizmodo, 267
Global Tipping Points report (2023), 264
Gopnik, Adam, 17
Graeber, David, 54, 123, 124, 153
Graetz, Michael J., 273
Graff, Gerald, 212
Grandi, Count Dino, 206
Gray Lock (Abenaki war chief), 182, 183
Great Acceleration, 7, 24, 84, 228–9, 268, 277, 287
Great Depression, 191, 216, 227, 229, 245
Greece, ancient, 67, 72, 79–81, 103, 132, 146, 183, 200
Greenland, 89, 90, 264
Greenwald, Glenn, 282–3
gunpowder, 6, 131, 316

Hadza people of Tanzania, 41–2
Hägglund, Martin, 328–9, 331, 333
Haines, Avril, 285
Haiti, 115, 315
Harald Bluetooth, 91

Harari, Yuval Noah, 104
Harris, Kamala, 20
Harris, Marvin, 102
Hartung, William D., 280
Harvey, David, 284
Hassabis, Demis, 256–7
Hayden, Michael, 281, 283
Hayek, Friedrich von, 244–50, 283
health and safety legislation, 191
health/medicine, 7, 9, 10, 11, 12, 25, 298–9, 302–3; colonial introduction of diseases, 33, 38, 120; decreases in infant mortality, 10, 13–14, 187; impact of fracking, 312; and industrialisation, 165, 192; Islamic scholarship, 95; overuse of antibiotics in meat industry, 265; in 'Progressive Era' (1890-1920), 191, 209; sanitation, 6, 10, 14, 187, 209; toxic/unhealthy products, 165, 172, 304; twentieth century triumphs in, 6–7, 227, 245; zoonotic diseases, 12, 33, 295
Heemstra, Baroness Ella van, 224
Heinlein, Robert, 136
Hemingway, Ernest, 210–11, 212, 215, 286
Hepburn, Audrey, 224
Hepburn-Ruston, Joseph, 224
Hesiod, 79
Hickel, Jason, 8, 44–5, 89, 114, 187, 234
Hinduism, 48
history/historians, professional, 23
Hitler, Adolf, 216, 217–20, 221, 222–3, 224, 225, 236, 240, 277
Hobbes, Thomas, 31, 39, 49, 51, 98–9, 127, 136–7
Hodal, Kate, 16
Hoover, J. Edgar, 215
housing, 19, 24, 165, 187
Hubbard, L. Ron, 256
Hughes, Langston, 212

Hume, David, 126, 129, 132
Hurston, Zora Neale, 212

IBM, 225
Iceland, 89, 90
ideology, 8–9, 197–209, 216–30, 234–44, 276–7, 288; futurist modernity, 202–3, 205, 208, 209, 240, 255–6; modernist scepticism, 209–16; neoliberalism disguised as absence of, 252–3; working definition of, 196–7 *see also* communism; fascism; liberalism, ideology of; social democracy
Inca empire, 152
India, 16, 102, 104, 174–5, 279, 304
Indigenous cultures: Aboriginal Australians, 47, 182–3, 308–9; animist beliefs, 43–8, 55, 61, 332–3; conquest of as inversion of genuine progress, 39, 250; Enlightenment ideas taken from, 123, 146; false narratives of historical violence, 295; Hobbesian view of, 31, 39; and Karl Marx, 176–8, 239, 244; punishing of overconsumption/hoarding, 321, 323; relationship with ecological systems, 32–4, 36, 38–9, 40, 44–8, 49–50, 53, 54–5, 65, 102, 160; relationships within societies, 53–4; resistance movements, 19, 182–3, 268, 284, 309, 318–20; return of stolen land to, 308–9, 310; in Soviet Union/Russia, 239–40, 308; Spanish treatment of in Americas, 115–19, 120; tradition of oral histories, 44, 47, 68–9, 78 *see also* Native Americans
industry/manufacturing, 60, 160, 165–8, 171–2, 186–7, 228–9, 231; adverse effects of industrialisation, 165–7, 172, 175, 178–9, 192, 198,

212–13; in communist states, 237, 238, 241–3; textile industries, 60, 154, 189
inequality, 8, 11–12, 18, 19, 21, 25, 230, 296, 300–1; as falling under social democracy, 227–8, 229, 244–5, 251; and fossil fuel exploitation, 165–7, 176; as increasing in 1920s, 191, 209, 214–15; neoliberal desire for, 129, 247–8, 252–3, 276; and wealth taxes/UBI idea, 228, 307–8
Institute for Economics and Peace, 264
Institute of Economic Affairs, 247–8
intelligence/espionage services, 278–83, 284, 285–7
International Monetary Fund (IMF), 147–8, 246, 306
Inuit Thule people, 89, 90
Ireland, 121, 132–3
Iroquois nations, 54, 150, 152, 176–8, 244
Isaac, Benjamin, 80
Islam, 72, 75, 91, 93–5; Islamic Golden Age, 67, 95–101, 103–4, 109, 111, 128, 132
Italy, 142, 199, 201–6, 207, 217, 223, 224–5, 236

Jackson, Andrew, 145, 146, 153–6, 158, 225
James, Henry, 213
James I, King, 33
Japan, 199, 225, 236, 279
Jefferies, Richard, 180
Jefferson, Thomas, 1–2, 3, 23–4, 31, 35–6, 39, 126, 145, 146–9, 156, 225, 327
Jesup, Thomas S., 154, 156
Jesus of Nazareth, 85, 86
Jones, Christopher F., 232
Joyce, James, 210, 211
Judaism, 48, 72, 75–8, 274

Judea, 67, 75, 85
Julius Caesar, 82
Jung, Carl, 210
Just Stop Oil, 284–5

Kallis, Giorgos, 196, 231
Kandiaronk, 123–4
Kaw Nation, 20
Kerry, John, 267–8
Keynes, John Maynard, 229–30, 234, 236, 244, 245, 251, 253, 277
Ibn Khaldun, 97–100, 142
Kimmerer, Robin Wall, 46–7
King, Martin Luther, 190, 286
Klein, Naomi, 229
Knausgård, Karl Ove, 219, 328, 332–3
Knox, Henry, 35
Knox, Robert, 133
Korea, 199
Kropotkin, Peter, 173, 177, 184
Krupnik, Igor, 240
Kurlansky, Mark, 144

labour, 60, 65, 112–15, 121, 141, 165–7, 172, 195, 203, 307–8; division of labour/specialisation, 98; exploitation of children, 187, 189; forced, 16–17, 65, 71, 80, 90, 120, 166, 188, 217; in Indigenous cultures, 32, 53–4, 65; labour/union movement, 166, 167, 172, 181, 182, 188–9, 190, 191, 209, 227, 228, 259, 314; Marx on mechanisation, 176; Paine on workers' rights, 127; strikes/unrest, 167, 172, 181, 182, 188–9, 192
Labour Party, 230, 235, 236, 245
Lafeyette, Marquis de, 151
Lao Tse, 125
Las Casas, Bartolomé de, 118–19
Leibiger, Stuart, 150
Lenin, Vladimir, 237, 238, 244

INDEX

Levin, Samuel M., 138
LGBTQ community, 267, 285
Libanius, 87
liberalism, ideology of, 138–42, 151, 157, 168, 175, 239; economic liberalism, 129, 140–2, 148, 192, 197, 200–1, 204, 208–9, 226, 229, 244–53, 315; Fukuyama's 'End of History' notion, 72–3; in interwar period, 201, 203, 204, 206, 208–9, 212, 216, 226–7, 229; irreconcilable contradictions within, 139–40, 141, 208, 229, 245–53, 315; political liberalism, 140–2, 146, 148, 204, 206, 208, 229, 245–6, 315; and social democratic ideology, 201, 226, 227–33, 236, 245, 250–1; technocratic liberals today, 318, 320
libertarians, 129
life expectancy, 6, 10–13, 25, 102, 187, 245, 276, 300
Lincoln, Abraham, 145, 146, 157–60, 225
Lincoln Cathedral, 5, 6
Lindisfarne, Nancy, 175
Linnaeus, Carl, 41
literacy, 26, 93
literary forms, 67–72, 180, 210–16; dystopian fictions, 180, 207–8, 218
'Little Ice Age', 121
Locke, John, 141–2
Lomborg, Bjørn, 271
Long Depression (1873-96), 167
Lovins, Amory, 269
Lucretius, 81–2
Luddite movement, 170
lynx, 44, 302

Macmillan, Harold, 230–1, 232, 238
macrohistorical narratives: cyclical histories/theories, 78, 79, 98, 177; epochal breaks or stages of development, 43, 59, 101–2, 110–12, 191; and Ibn Khaldun, 97–8; nineteenth-century debate on progress, 168–78; origins of progressive historiography, 79–81; Whig historiography, 127–34, 136–42, 318 *see also* progress, grand narratives of
Maestre, Antonio, 208–9
Malcolm X, 14, 15, 286–7
Malm, Andreas, 166
Malthus, Thomas, 138–9, 247
Mamluks, 100
Mao Zedong, 241–3, 244, 276
Marcetic, Branko, 286, 287
Marinetti, F. T., 202–3, 255–6
market economics, 17, 131, 137–8, 140, 177, 197, 213–14, 324; first market economies, 59–61, 67, 93, 113, 114–15; and self-interested impulses, 129, 137, 138–9, 171–2, 247–8
Marx, Karl, 99, 104, 127, 157, 166, 173–8, 180–1, 912, 230, 239, 243–4, 277, 331
May, Karl, 218
McKinnon, Charlie, 126
Mecca, 94
media networks, 246, 267, 282–3
Meek, Ronald L., 128, 129
mercantilism, 113
Mesoamerican empires, 117–19
Mesopotamia, 3, 29, 59, 93, 96, 101–2, 249, 275; Babylonian Enuma Elish/cult of Marduk, 69–72, 74, 75, 78, 79, 103, 202, 208; *The Epic of Gilgamesh*, 67–9, 70, 71, 72, 75, 78, 79, 82, 103; Uruk/Sumerian culture, 59–60, 67–9, 70, 82, 93, 103
metallurgy, 59, 137
Michelis, Giuseppe di, 204
Microsoft, 270
migration, human, 20, 42, 82, 147, 221, 224, 293, 305; climate refugees, 264, 298

Mi'kmaq people, 124, 319
military technologies, 6, 38, 165, 168, 189, 285, 311, 316
Mill, John Stuart, 125, 232
Miller, Robert J., 217
mind, human, 278
mining/miners, 16, 94, 110, 166, 175, 187, 188–9, 198, 278
Mises; Ludwig von, 247
Mitchell, Timothy, 188
modernism, 209–16, 226
Mommsen, Theodor E., 85–6
Monbiot, George, 252
Mongolian empire, 97
Mons Graupius, Battle of (83 CE), 83
Mont Pèlerin meeting (1947), 244–5, 246
Moore, G. E., 170
Morris, Ian, 11, 71, 84–5, 92, 93, 163–4, 321
Mosley, Oswald, 205–6, 211, 224, 236
Muhammad, 94–5, 96, 100
Murphy, Thomas, 312
Musk, Elon, 254, 255, 256–7, 258–61, 262–3, 266, 273–4
Mussolini, Benito, 204, 205, 206, 217, 224, 236, 277
mutualistic relationships *see* commensalistic/mutualistic relationships

Nari Nari people, 309
National Health Service, 227, 245
nationalism, 185, 206
Native Americans: colonial introduction of diseases, 33, 38, 120; colonists' impulse to join communities of, 37–8, 249; Dakota War (1862), 159; 'Five Civilized Tribes', 20, 147, 155–6; gender relationships, 32, 33, 34, 35, 54; genocide and land theft against, 20, 23–4, 33–9, 115–21, 141, 142–4, 147–56, 158–61, 192, 225, 228, 250, 258; Hitler influenced by US treatment of, 217–18, 221, 225; impact of nuclear industry on, 270–1; 'indigenous critique', 123–5; 'Long Walk of the Navajo' (1864), 159; mythic narrative of Euroamerican superiority, 34, 35–9, 155–7, 158–60, 250; reservations, 159, 217, 239, 270, 308; Sand Creek massacre (1864), 159; Seminole Wars in Florida, 154–5; and Spanish Requirement (1513), 116–17; state violence against today, 19, 149, 284; total deaths caused by European colonisation, 120–1; Trail of Tears, 155; variety/diversity of peoples, 32–3, 152, 153, 250; Whitestone Hill massacre (1863), 159; Wounded Knee massacre (1890), 158
'nature deficit disorder', 43
Naumkeag people, 144
Navajo people, 270
navigation techniques, 47
Nazi Germany, 199, 201, 205, 206–7, 216–21, 222–5, 236; the Holocaust, 221, 231, 319; USA as inspiration for, 217–18, 221–2, 223
Neale, Jonathan, 175
neoliberalism, 129, 201, 244–53, 269, 276–86, 318
Netherlands, 119, 142
Nigeria, 277
Nisbet, Robert, 3, 79, 87, 88, 125, 147
Nixon, Richard, 314
Norway, 11, 89, 311
nuclear weapons, 231, 263, 264
N~~u~~m~~u~~n~~uu~~ (Comanche) peoples, 152

Obama, Barack, 18–20
October Sky (film, 1999), 258
octopuses, 50

oil palm production, 277–8
OilChange International, 319–20
Olaf Haraldsson, 92
Olaf Tryggvason, 92
Old Testament, 75–8, 103
OpenAI, 257
orangutans, 50, 277
Orwell, George, 201, 206, 207–8, 211, 215, 218, 236
O'Shaughnessy, Eileen, 207
Ostler, Jeffrey, 148
O'Sullivan, John, 146, 158
Overbye, Dennis, 95

Page, Larry, 255–6, 257, 266
Paine, Thomas, 126–7
Paiute people, 270
parasitic economies/societies, 61–7; advent of with first empires, 59–61, 63–5, 67–75, 76–8, 82–7, 96, 101–2, 103, 109, 249, 275; agricultural over-production, 62–5, 76, 82, 102, 110, 160–1, 237, 238, 242–3, 265, 272, 304–6; boom-and-bust of economies, 167–8, 199, 245–6, 250; and 'capitalist' principles, 113, 122, 143, 148, 178; collapse of, 61, 90, 181, 271, 287–8, 320–1, 334; 'contiguous parasitism', 63–75, 102–3, 109–10, 143–56, 158–61, 195, 217, 221, 225–6; as contrary to instinctual human tendencies, 43–4, 64–7, 333, 334–5; disparate mode from 1400s, 110–12, 115–23, 127–39, 140–2, 179, 191–2, 195, 197–8, 246; European settling of the Americas, 30, 31–5, 37–9, 54, 111–12, 115–21, 123–5, 141, 142–3, 192; expanding American state, 35–6, 143–56, 158–61, 168, 192; expansion of Christianity and Islam, 87, 88, 91–2, 93, 96–101, 103–4, 115–19; hierarchies intrinsic to, 61–7, 71–2, 79–85, 90–4, 98–105, 124–5, 132–4, 141–2, 185, 203–5, 224–5, 246, 249; host class, 71, 80, 84, 104–5, 112, 121–2, 149, 156, 174–5, 183–92, 206, 222–3, 238–41, 334; host class terminology, 65–7, 228; Indigenous forms in the Americas, 119, 152; and Marxian progress myths, 236; mass extinctions driven by, 24, 26, 179, 192, 266, 272, 304; need for permanent dismantling of, 292, 309–21, 323–6, 335; 'networked' parasitism, 195–6, 198–200, 216, 220–1, 226, 230–6, 274–5, 277, 288, 334; and new technologies in nineteenth century, 143, 160–1, 163–4, 165–8, 171–2, 176, 195; psychological engineering in, 29, 66–7, 104–5, 184, 185, 186, 281–2, 326; quantity as more important than quality in human life, 102, 154–5; resilience of parasitic human-ecology, 181–4; sedentarisation in, 12, 67–8, 94, 99, 101; turn to anthropocentrism, 75–7, 102
parasitism (ecological relationship), 41, 42, 53, 61, 65
Parchami, Ali, 83, 84
Parker, Theodore, 190
Patel, Priti, 285
Pemulwuy (Eora warrior), 182–3
Persia, 16, 67, 72, 73–4, 103, 104
personhood principle, 323
petroleum, 162, 164, 167, 178, 188, 202, 261, 266–7, 306, 310–11
pharmaceutical/drug industry, 165, 191, 314
Pinochet, Augusto, 250
plant life, 26, 29, 40, 42, 59, 69–70, 97; trees, 24, 34, 36, 42, 47, 59, 61, 68–9, 82, 90, 266, 277–8, 304

plastic waste, 24–5, 26
police forces/policing, 17–18, 19, 22, 165, 167, 184, 278
political/social structures: agriculture and hierarchy, 32, 43, 63; ancient Greek proto-democracy, 79–81, 103, 146, 183; in ancient Mesopotamia, 59–60, 67–72, 101–2, 103; in ancient Rome, 63–5, 81–8, 102, 146, 182; centralised bureaucracies, 32, 61, 89, 98–9, 110, 111, 112, 136, 182, 195, 203, 204; city states, 59–60, 63, 66, 67–72, 79–82, 101, 110; dictatorships in twentieth-century, 203–9, 216–26, 236–44; of early/medieval Christianity, 86–8, 89, 91–3, 103, 109–10; Hayek's 'primitive' societies, 248–50; Hobbes' centralised state, 49, 98–9, 136–7; of Islamic Golden Age, 95–100, 103–4, 111, 128, 132; nation-states, 131–2, 142–3, 146, 192, 197–8, 199, 249; Native American, 32–3, 34, 36–7, 43, 54, 124–5, 149, 153, 155, 176–8; oligarchies, 60, 64, 81–2, 94, 142, 146, 161, 183, 185, 188, 233, 250, 276, 315; social trust in institutions as decreasing, 25; structures of city governance, 300, 301; Viking, 89–92, 111; welfare states, 227–8, 229, 245, 251, 314; world's first empires, 59–61, 67–74, 101–3, 275 *see also* capitalism; commensalistic/mutualistic relationships; parasitic economies/societies
politics, 25, 30, 139–40, 157–60, 328–32, 333; achieving future non-parasitic ecology mode, 292, 293–7, 299–304, 305–13, 316–21, 322–6; climate targets/deadlines, 271–2; egalitarianism, 190, 328–9, 333; idea of infinite/perpetual growth, 7, 84, 231, 232–4, 244, 247, 253–4, 287; narrative of growth-as-progress, 6–7, 8, 24, 217, 230–5, 236–44, 247, 252–4, 277, 281–3, 287; positive change as occurring rapidly, 313–15; 'Progressive Era' (1890-1920), 190–1, 209, 313–14; suffrage movements, 190, 191 *see also* revolutionary politics
pollution, 24, 180, 234, 257, 273, 276–7, 284, 291, 304, 311; in communist countries, 241, 243; Indigenous resistance against, 19, 284, 319–20; New Deal creation of, 228; Theodore Roosevelt on, 179; unregulated in nineteenth-century, 172, 178; in urban areas, 164, 165, 172, 175, 178, 192, 300–1
Polynesia, 47
Pompeo, Mike, 281
population data, global, 24, 25–6
Portugal, 119, 142–3
positivism, 173
Potawatomi Nation, 46, 321
Pound, Ezra, 210–11
poverty, 8, 13, 21, 25, 85, 278, 300–1
Powhatan people, 32–3, 34
prison systems, 17–18, 283
progress, grand narratives of: ancient Roman, 81–4, 87, 103; banishing of wilderness, 48–9, 52, 303–4; book of Genesis, 75–8, 274; civilised-savage binary in, 2, 7, 31–2, 35–9, 67–8, 79–81, 99, 103, 127–34, 154–6, 239; as collective mass delusion, 156; communist, 236–44; of early Christian thinkers, 85–6, 87–8, 91; erasure of past instances to create newness, 20; and eschatology, 72–3, 74, 91, 103; European secular tradition, 79–84, 95, 97,

99, 111–12, 125–39, 143, 145, 169–70, 200; faith in inevitable technological advancement, 242, 266, 267–8, 271, 275, 315–17; formula of, 2–3, 29–30, 55, 65–6, 67, 128, 130–1, 331; frontier spaces, 2, 32–6, 77–8, 160, 260–3; future severed from history, 70–1, 75, 202, 208; futurist modernity, 202–3, 205, 208, 209, 240, 255–6; growth-as-progress mythos, 6–7, 8, 24, 217, 230–5, 236–44, 247, 252–4, 277, 281–3, 287; Islamic, 95, 96, 97–101; of Thomas Jefferson, 1–2, 3, 23–4, 31, 35–6, 39, 147–9; modernist scepticism, 209–16; as mythic in classical and medieval periods, 85–8, 89–91, 95, 96, 100, 103–4, 111; myths of Mesopotamia, 67–72, 75, 78, 79, 82, 103; Nazi/fascist, 204–5, 218–20, 222–3; need for total break from lineage of, 318, 322–6; progressive historiography, 78–80, 127–34, 136–42, 248–50, 318; 'promised land' notion, 18–19, 77–8, 103; science fiction fantasies of Silicon Valley moguls, 254–61, 273–4, 275; scientist in industrial era, 143, 168–78, 185, 192; selfishness/self-interest promoted by, 128–9, 137, 138–9, 171–2, 247–8; and 'social Darwinism', 168–72, 173, 177, 185, 202; social justice issues today, 7, 14–22; and technologies, 3–6, 14, 160–1, 176, 178, 192, 205, 228–30, 234, 242, 251–63, 266–75, 287–8; teleological aspects, 31–2, 148, 165, 177–8, 200, 212, 229, 238–9, 243–4, 248–50; and US expansionism, 143, 145–6, 147–9, 150–2, 154–6, 157–60, 192; and use of quantitative data, 3–7, 8–14, 21–2; vague/ever-future paradise of, 2, 158, 242, 256, 274, 288, 297, 326, 331–2, 335

psychology, human: the individual in daily life, 327–34; 'nature deficit disorder', 43; prosocial/cooperative nature, 29, 47, 64–5, 66–7, 105, 173, 177, 184–5; psychological manipulation, 29, 66–7, 104–5, 184, 185, 186, 281–2, 326; and religious faith, 104–5
psychology, science of, 210
Putnam, Thomas, 212

al-Qarawiyyin, University of, 95
Quraysh tribe, 93–5

race/ethnicity: and ancient Athens/Rome, 80–1, 132; Anglo-Saxon supremacism, 133–4; elite trickle-down progress idea, 18–20; Enlightenment racism, 132, 133; Harlem Renaissance, 212; and Hitler, 219–21, 222–3; material aspect of racism, 221–3; 'race science', 133, 175, 222; racial hierarchies in progress narratives, 80–1, 99, 103, 132–4, 141–2, 175, 204–5, 217–18, 220, 222, 225; racial injustice in USA, 17–20, 146, 221–2, 223, 228; and US intelligence community, 285
Ramses III, Pharaoh, 181, 182
Raytheon, 254, 285
Reagan, Ronald, 279–80, 281, 282
Red Deal manifesto, 309
religion: Abrahamic, 72, 74–8, 84, 87, 91, 94, 101, 103–4, 130; end times/process of judgment, 72–3, 74, 91; and human psychology, 104–5; ideas of the afterlife, 74, 87, 91, 92, 96, 100, 102, 103, 128, 274; Marx on, 104, 331; monotheism, 67, 72, 86–7, 94–6, 101, 102, 103;

myths of Mesopotamia, 67–72, 75, 78, 79, 82; Native creation stories, 45–6; 'pagan', 87, 88–9, 91, 103; polytheism of Rome, 67, 84, 86, 87, 89 *see also* Christianity; Islam; Judaism; Zoroastrianism
Renaissance period, 113
renewable energy, 266, 268–9, 273, 299, 316
Republican Party, US (GOP), 267
republicanism, 81–2, 192
resistance/activism/protest today, 19, 22, 284, 291, 309, 317–18, 324, 327, 331–2; environmentalists targeted by police/intelligence, 282, 284–5, 319; Indigenous, 19, 182–3, 268, 284, 309, 318–20; state crackdowns on, 282, 284–6, 319; willingness to use force, 311, 320, 325
revolutionary politics, 125, 127, 175–7, 180–2, 315, 318; 1848 Revolutions, 187–8, 189–90
rewilding concept, 299, 301–4
Ricardo, David, 232
Rich, Norman, 217
Richeson, Jennifer A., 18
robotics/bioengineering, 255–61
Rollo of Normandy, 91
Rome, ancient, 63–5, 67, 80, 81–9, 103, 109, 150, 182, 278–9; conversion to Christianity, 86–8, 92, 103; Italian fascist fantasies of, 203, 205; Pax Romana, 82–3, 84–6, 88, 103, 130, 146, 217; Republican, 11, 63, 81–2, 146; rule in Britain, 83, 84–5, 102
Roosevelt, Franklin Delano, 227, 314
Roosevelt, Theodore, 179, 190–1, 314
Rose, Deborah Bird, 212
Rosemont, Franklin, 177
Rothermere, 1st Viscount, 224
Rountree, Helen, 32
Rousseau, Jean-Jacques, 99, 136–7, 141

Roxanne Dunbar-Ortiz, 114–15
Russian Federation, 276, 282, 283, 308, 315
Russo-Ukraine war (from 2022), 282

Sacks, Adam J., 224
Sackville-West, Vita, 211
Saladin, 101
Salish and Kootenai Tribes, 308
Sámi people, 90, 92, 268, 308
Saxe-Coburg and Gotha, Charles Edward, Duke of, 224
Scandinavian peoples/nations, 11, 89–92, 103, 189, 268, 308
Scheidel, Walter, 228
science fiction, 180, 254–61, 273–4, 275
Scotland, 121–2, 125–6, 132, 133, 312
Scott, Charles C., 35
sea anemone, 41
seals, 50
Second World War, 168, 198–9, 201, 206, 211–12, 226–8, 229, 231, 238, 288
Shambaugh, David, 243
Shell (oil company), 267
Shelley, Mary, 180
shipbuilding, 89, 109, 191
Shoshone people, 270
Siberia, 239, 308
Silicon Valley moguls, 254–61, 273–4, 275
Sioux Natives, 159, 319
slavery: Amerindian, 153, 155; in ancient Greece, 79–80, 183; in ancient Mesopotamia, 71; Atlantic slave trade, 14–16, 120, 192, 258; and fossil fuel exploitation, 166, 188; in Islamic world, 96–7, 100; in medieval Europe, 90, 93; modern day, 16, 20; Thomas Paine on, 127; and 'parasitic' societies, 65; progress narratives justifying, 14–16, 141, 192; return of, 16–17,

20; Spartacus revolt, 182; in USA, 14–16, 146, 154, 155, 157, 166, 172, 188, 192, 313
Sleeper-Smith, Susan, 33–4
smallpox, 12, 33, 38
Smil, Vaclav, 231, 262–3, 272
Smith, Adam, 98, 126, 130, 137–8, 232
social Darwinism, 168–72, 173, 177, 185, 199, 202
social democracy, 7, 201, 226, 227–35, 236, 244–52, 277
socialism, 201, 206, 209, 211, 213, 215, 235, 236, 253–4, 328–9, 331
Sojourner Truth, 188
Solow, Robert, 232, 233, 234, 244
South America, 44–5, 142–3, 152, 280
Soviet Union, 206–7, 226–7, 236–41, 244, 263, 276, 283
space travel, 4, 254–63, 273–4, 275, 278, 316
Spain: Civil War (1936-9), 206–7, 208–9, 211, 215, 226; fascism in, 201, 206–7, 208–9, 211, 226; maritime empire of, 115–19, 120, 142–3
Spencer, Herbert, 168–71, 175, 185, 202, 209, 215, 318; and laissez-faire economics, 169, 170, 171–2, 173, 177, 178, 185, 191, 245–7, 248; 'perfect man' idea, 177, 213, 240
Stalin, Joseph, 226–7, 237, 238
Stangle, Col. Chris, 282
Stedman Jones, Daniel, 247–8, 251
Stein, Gertrude, 213
Steinem, Gloria, 285
Stern, Alexandra Minna, 221
Stewart, Jon, 260
Still, William, 188
Summers, Lawrence, 233
Sweden, 268

Tacitus, 83–4, 85
Taibbi, Matt, 282–3
Talley, Jared L., 293
Taylor, Astra, 286
Tertullian, 86
Tesla, 259, 261, 273
Thatcher, Margaret, 21
Theodosius I, Emperor, 87
Thorpe, David, 269
Thorvaldsson, Erik 'the Red', 90
Thucydides, 79, 200
Thutmose III, Pharaoh, 181
timber, 68, 89, 90
trade and commerce, 31, 34, 35, 59–60, 109–10, 141, 195
transportation systems/technologies, 13, 14, 25–6, 229, 306, 307; aviation, 3–4, 162, 228, 229, 278; growth of maritime economies from 1400s, 110–12, 115–22, 191, 195; internal combustion engine, 178, 202; Norse use of sea routes, 89, 111
Trotsky, Leon, 240
Trump, Donald, 20, 259, 283
tsar bomba, 6, 263
Tubman, Harriet, 188
Turchin, Peter, 104
Turgot, Anne Robert Jacques, 127–9, 138, 247

United Nations Sustainable Development Group, 7
United States: American Revolution (1776), 35, 126–7, 143, 315; and classical Rome/Greece, 81, 84, 146; Declaration of Independence, 146, 147, 327; elite support for Nazi Germany, 225; Gilded Age, 245–6; Great Lakes region, 292–3; high inequality in, 11, 18, 19; Homestead Act (1862), 159; incarceration rates in, 17–18, 283; industrialisation, 160, 165–8, 171–2, 192, 228–9; as inspiration for Nazi Germany, 217–18, 221–2, 223;

intelligence community/network, 279–83, 284, 285–7; manifest destiny idea, 38–9, 145–6, 157–60, 192, 217–18, 221, 225; National Park system in, 191, 302, 314; neoconservative foreign policy, 284; Ohio River Valley, 33–4, 35, 40, 43, 59, 150, 152, 156; 'Progressive Era' (1890-1920), 190–1, 209, 313–14; racial injustice in, 17–20, 146, 221–2, 223, 228; rewilding in, 302–3; Roaring Twenties, 209, 213–15; Roosevelt's New Deal, 227, 228, 245, 314–15; slavery in, 14–16, 146, 154, 155, 157, 166, 172, 188, 192, 313; under social democracy, 7, 227–9, 231, 234–5, 244–5, 251–2; Wagner Act, 314; westward expansion, 35–6, 143–56, 158–61, 186, 192, 217–18, 221

universal basic income policy, 307–8
uranium, 164, 269, 270
urban areas, 11–12, 25–6, 165–8, 178, 196, 198, 209, 241, 297–304, 305–6, 310; in ancient Mesopotamia, 59–60, 69; and insulation/'escape' from the wild, 48–9, 52; Native American towns, 32, 36; pollution in, 164, 165, 172, 175, 178, 192, 300–1; poverty and slums, 25, 278, 300–1; sanitation in, 6, 10, 14, 187, 209

Victor Emmanuel III, King of Italy, 205, 206
Vikings, 67, 89–92, 103, 111, 258
Virgil, 81, 84, 146
Voltaire, 132, 141
Voyager 1 spacecraft, 4
Voyles, Traci Brynne, 38, 270

Wales, 132, 133
Wall Kimmerer, Robin, 321
walrus populations, 90
Walsh, John, 215
warfare, 12, 63, 65, 84–5, 116–19, 155–6, 158–60, 206–9, 211, 215, 264; Augustine's just war theory, 88, 242; 'Thucydides trap', 200; twentieth century world wars, 168, 197–203, 206, 210–12, 226–8, 231, 238, 288
Washington, George, 35–6, 145, 146, 150–1, 153–4, 155–6, 225
Watson, Thomas, 225
Waugh, Evelyn, 211
Weiss, Dieter, 98
Wells, H.G., 211
Wendat people, 123–4
Wengrow, David, 43, 59–60, 123, 124, 153, 300
whales, 42, 50
Whitman, James Q., 222, 223
Whitmer, Gretchen, 287
Whyte, Kyle, 293
Wilde, Oscar, 171, 177, 187, 211
Wolfe, Patrick, 147, 155
Wollstonecraft, Mary, 170–1, 180
wolves, 64, 265, 302
Woodard, Colin, 145
Woolf, Leonard, 236
Woolf, Virginia, 211, 213, 215–16
World Bank, 147–8, 246
World Trade Organisation, 246
Wright, Orville and Wilbur, 3–4
Wrigley, E. A., 162–3

Zinn, Howard, 148
Zoroastrianism, 72, 73–4, 78, 91, 103, 104, 125